TRACERS IN THE OCEAN

TRACERS IN THE OCEAN

PROCEEDINGS OF
A ROYAL SOCIETY DISCUSSION MEETING
HELD ON 21 AND 22 MAY 1987

ORGANIZED AND EDITED BY
H. CHARNOCK, F.R.S., J. E. LOVELOCK, F.R.S.,
P. S. LISS AND M. WHITFIELD

Princeton University Press
Princeton, New Jersey

Published by Princeton University Press, 41 William Street,
Princeton, New Jersey 08540

 First published in *Philosophical Transactions of the Royal Society of London*, series A, volume 325 (No. 1583), pages 1–236

First Princeton University Press printing, 1990

Library of Congress Cataloging-in-Publication Data

Royal Society Discussion Meeting (1987: London, England)
Tracers in the ocean : proceedings of a Royal Society Discussion Meeting held on 21 and 22 May 1987/ organized and edited by H. Charnock . . . [et al.]
p. cm.
Reprint. Originally published: London: Royal Society, 1988.
ISBN 0-691-09969-3 (alk. paper)
ISBN 0-691-08571-4 (pbk.: alk. paper)
1. Radioactive tracers in oceanography—Congresses. I. Charnock, Henry. II. Title.
GC10.4.R3R68 1987 89-49496
551.46—dc20 CIP

Princeton University Press books are printed on acid-free paper, and meet the guidelines for permanence and durability of the Committee on Production Guidelines for Book Longevity of the Council on Library Resources

10 9 8 7 6 5 4 3 2 1
10 9 8 7 6 5 4 3 2 1 (pbk.)

Printed in the United States of America by Princeton University Press, Princeton, New Jersey

PREFACE

The distribution of tracers in the ocean can provide information about its general circulation that cannot be obtained in any other way. Classical oceanographers used maps of temperature, salinity and oxygen to make qualitative inferences that are still generally accepted: later improvements in chemical analytical techniques have allowed an extension of their work to other conservative tracers. A major development in geochemical tracing came when anthropogenically derived transient tracers (chlorofluoromethanes, tritium and ^3He) became available. Maps of the distribution of these tracers are suggestive not only of water movement but of the timescales involved.

Tracer chemistry is telling us more about the source and sinks, about scavenging by uptake at the solid–solution interface and about the difficult problem posed by the effect of biological processes. Special attention has been paid to their air–sea exchange of particles and gases, particularly of CO_2 because of its importance to climatic change.

To deduce quantitative information from observations of tracer distributions is proving difficult, partly because the measurements are irregular in space and time, but there are improving methods for using tracer data as constraints when calculating ocean currents and to deduce something of the mixing processes involved. The detailed mixing processes remain poorly understood but there is the exciting possibility of using deliberate releases of chemicals such as sulphur hexafluoride and perfluorodecalin to study them directly.

The organizers of the meeting at which these contributions were presented are most grateful to the authors for providing an up-to-date account of ocean tracer chemistry, its problems and its potential. The meeting was capably arranged by Miss C. A. Johnson and the proceedings edited by Mrs B. L. Lawrence; we are grateful for their help and patience.

February 1988

H. Charnock
J. E. Lovelock
P. S. Liss
M. Whitfield

CONTENTS

CONTENTS

Ocean carbon-cycle dynamics and atmospheric p_{CO_2}

By J. L. Sarmiento, J. R. Toggweiler and R. Najjar

Program in Atmospheric and Oceanic Sciences, James Forrestal Campus, Princeton University, Princeton, New Jersey 08542, *U.S.A.*

Mechanisms are identified whereby processes internal to the oceans can give rise to rapid changes in atmospheric p_{CO_2}. One such mechanism involves exchange between the atmosphere and deep ocean through the high-latitude outcrop regions of the deep waters. The effectiveness of communication between the atmosphere and deep ocean is determined by the rate of exchange between the surface and deep ocean against the rate of biological uptake of the excess carbon brought up from the abyss by this exchange. Changes in the relative magnitude of these two processes can lead to atmospheric p_{CO_2} values ranging between 165 p.p.m. (by volume) and 425 p.p.m. compared with a pre-industrial value of 280 p.p.m. Another such mechanism involves the separation between regeneration of alkalinity and total carbon that occurs in the oceans because of the fact that organic carbon is regenerated primarily in the upper ocean whereas $CaCO_3$ is dissolved primarily in the deep ocean. The extent of separation depends on the rate of $CaCO_3$ formation at the surface against the rate of upward mixing of deep waters. This mechanism can lead to atmospheric values in excess of 20000 p.p.m., although values greater than 1100 p.p.m. are unlikely because calcareous organisms would have difficulty surviving in the undersaturated surface waters that develop at this point. A three-dimensional model that is being developed to further study these and other problems provides illustrations of them and also suggests the possibility that there is a long-lived form of non-sinking carbon playing a major role in carbon cycling.

1. Introduction

The biologically mediated redistribution of carbon and alkalinity within the oceans plays a major role in controlling atmospheric p_{CO_2} and is the most likely cause of changes in this quantity that occur on timescales of a few hundred to a few thousand years. The observation in trapped air bubbles from ice cores that atmospheric p_{CO_2} increased from approximately 200 p.p.m. (by volume) during the last ice age to its pre-industrial value of 280 p.p.m. on a timescale of order 1000 years (Neftel *et al.* 1982) has led to considerable interest in the relevant processes by showing that the system is far more susceptible to rapid change than had been previously thought. Understanding the redistribution processes is of considerable relevance not only to the interpretation of the ice age termination event and other natural events, but also in predicting the global response to the fossil-fuel carbon dioxide transient.

In this paper we discuss our present understanding of the processes determining the carbon and alkalinity distribution, with a particular focus on those that can cause the largest changes in atmospheric p_{CO_2}; we discuss progress towards the development of predictive models; and we discuss some of the important questions concerning our understanding of the carbon and alkalinity cycles in the oceans that are being raised by this work. A first section provides a background discussion, the section following discusses box models that have been developed to help in identifying those aspects of the carbon cycle to which the atmospheric p_{CO_2} is most sensitive, and a final section preceding the conclusions discusses progress towards the development of three-dimensional models.

2. Background

The partial pressure of p_{CO_2} in the oceans is a function of the total carbon concentration, $[\sum CO_2]$ and total alkalinity, [ALK]. The key to understanding how biological cycling in the oceans controls atmospheric p_{CO_2} is the fact that vertical profiles of $\sum CO_2$ and ALK, such as those in figure 1, show a very evident segregation between surface and deep-water masses. The $\sum CO_2$ and ALK are reduced by organism uptake in the upper euphotic layers of the ocean and are more abundant in the deeper layers where they regenerate. The chemical segregation of these constituents in the oceans leads to an atmospheric p_{CO_2} level that is 3 to 4 times lower than it would be if the oceans were uniformly mixed. In the classical geochemical view, chemical gradients between surface and deep waters are maintained by a balance between the downward flux of sinking biogenic particles and the upward mixing and advection of nutrient, carbon, and alkalinity-enriched deep waters. Sinking particles dissolve or are remineralized in the deep sea, thus closing the cycle. Differences in the regeneration depths of the constituents, which are reflected in the different shapes of the profiles in figure 1, are an important part of the discussion in the next section so are introduced here.

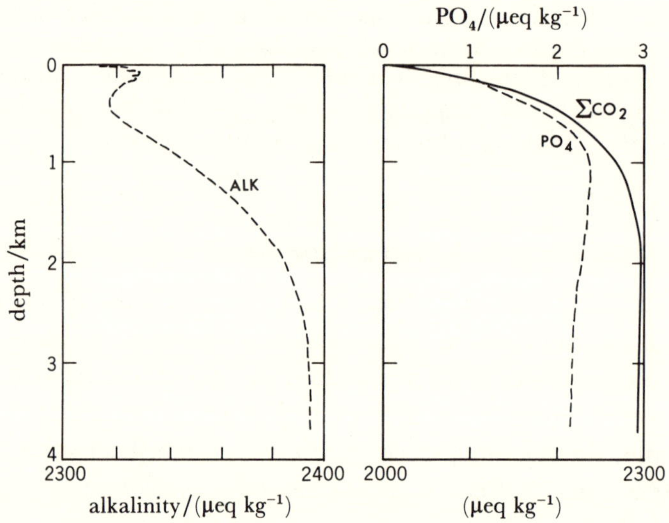

Figure 1. Global average profiles of GEOSECS data. Note the deeper regeneration of the alkalinity relative to the total carbon.

In our modelling approach we treat PO_4 interchangeably with nitrate (NO_3) as the nutrient that is limiting to photosynthesis in most regions of the ocean. This is reflected in the fact that the PO_4 concentration is close to zero at the surface (see figure 1). The remineralization in the water column of phosphorus, which is removed from the surface in organic matter, occurs at shallow depths and combines with the circulation pattern to give a maximum in PO_4 concentration at a depth of *ca.* 1 km.

ALK, on the other hand, varies over a range of less than 5%, and has a broad maximum at about 3 km depth. Its distribution is affected to a small extent by regeneration of organic NO_3, but is dominantly controlled by the cycling of $CaCO_3$ by plankton. Typically, for every mole of PO_4 that sinks out of the surface ocean, there are 15 moles of NO_3, the removal of which increases surface ALK by 15 equivalents, and 32.5 moles of $CaCO_3$ tests, which decrease surface ALK by 65 equivalents. The net decrease in surface ALK is thus 50 equivalents (Takahashi *et al.* 1985). The regeneration of the NO_3 below the surface in the upper part of the water column decreases ALK by 15 equivalents. The dissolution of the $CaCO_3$, which increases ALK by 65 equivalents, occurs at much greater depths in the undersaturated deep ocean by relatively slow processes that are thought to occur primarily after the tests sink to the ocean floor. The effect of biological cycling on ALK is thus to drive its concentration down in the upper ocean both by loss of biogenic material from the surface and regeneration of NO_3 below, and to increase it at depth by dissolution of $CaCO_3$.

The ΣCO_2 has a distribution intermediate to ALK and PO_4 because it is present in both organic matter (130 moles of carbon for each mole of PO_4) and $CaCO_3$ (32.5 moles of carbon for each mole of PO_4).

In our modelling work we use the treatment of Takahashi *et al.* (1980) to find the p_{CO_2} at the surface as a function of ALK and ΣCO_2. Takahashi's treatment includes dissociation of carbonic acid, boric acid and water. In the range of properties found at the surface, Takahashi's equations predict that the removal of organic matter alone decreases the p_{CO_2}, whereas removal of $CaCO_3$ increases the p_{CO_2}. The net effect of removing $CaCO_3$ on the p_{CO_2} of typical low-latitude water is to increase it by of order 45 p.p.m. (Broecker & Peng 1982). In our treatment discussed below we show that the difference in the regeneration functions of organic carbon and $CaCO_3$ with its consequence of putting ALK deeper down in the water column than ΣCO_2, leads to a further increment in surface p_{CO_2} of order 15 p.p.m. A further increase in the separation between these two properties, such as might occur if the deep ocean stagnated, can lead to very large increases in atmospheric p_{CO_2}, as shall be shown below.

3. Box models

Box models are a powerful technique for gaining intuition about how a system functions, and identifying the parameters to which various properties of interest are most sensitive. In doing this it is important that the models developed be simple enough to be readily constrained and interpreted, and that the assumptions made in developing the models and their possible impact be clearly understood. The results from such studies must be viewed with considerable care, as will be shown below by contrasting the results from the two-box model in figure 2 with the models in figures 3 and 5. The model in figure 2 led to an understanding of the carbon cycle in the oceans that suggested only a minimal role for ocean biological cycling and a non-existent role for ocean circulation in leading to changes in ΣCO_2 and ALK through time. These conclusions dominated the thinking of many geochemists until the evidence of rapid changes in atmospheric p_{CO_2} at the end of the last ice age forced a major reconsideration and the development of new models.

In developing the box models, our interest is in gaining an understanding of what controls the atmospheric p_{CO_2} level. The upper ocean equilibrates with the atmosphere on a timescale of approximately one year. We therefore seek in every case to find a solution for the surface

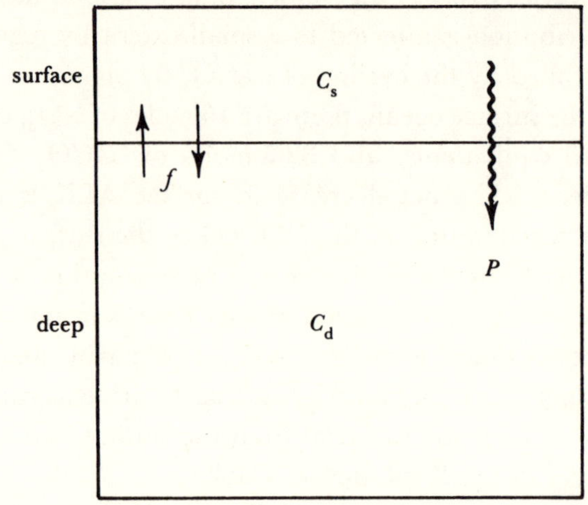

FIGURE 2. A two-box model of the oceans. The f represents physical exchange between the surface and deep box, C_s and C_d represent average surface and deep concentrations, and P represents the particulate rain of phosphate in organic matter. (Taken from Broecker & Peng 1982.)

ocean $\sum CO_2$ and ALK in terms of the minimum number of parameters and concentrations of oceanic constituents. The approach we take is illustrated in detail for the two-box model.

3.1. *Two-box model*

The two-box model approach is based primarily on the observation that PO_4 is very close to zero over most of the surface ocean. Given this, it does not seem unreasonable to assume that one can treat the surface ocean as a single nutrient depleted box. For the sake of simplicity the deep ocean is treated as another single box (see figure 2), although it is possible, as has been done (Oeschger *et al.* 1975) to treat it as a stack of boxes. Assuming steady state, one can put down for the deep box, equations of conservation for $\sum CO_2$ and ALK and solve them for the surface values

$$\sum CO_{2_s} = \sum CO_{2_d} - r_{\Sigma CO_2/PO_4}(P/f), \tag{1}$$

$$ALK_s = ALK_d - r_{ALK/PO_4}(P/f). \tag{2}$$

P is the surface new production of PO_4 and the rs are the Redfield ratios of uptake by organisms for the constituents indicated in the subscript. The values of the Redfield ratios have been discussed in §2. A solution for P can be obtained from the PO_4 balance for the surface box

$$P = f(PO_{4_d} - PO_{4_s}). \tag{3}$$

Combining (1) and (2) with (3) and setting PO_{4_s} to 0 gives

$$\sum CO_{2_s} = \sum CO_{2_d} - r_{\Sigma CO_2/PO_4} PO_{4_d}, \tag{4}$$

$$ALK_s = ALK_d - r_{ALK/PO_4} PO_{4_d}. \tag{5}$$

Equations (4) and (5) show that the surface ocean $\sum CO_2$ and ALK, which determine the atmospheric p_{CO_2}, are themselves determined by the deep $\sum CO_2$, ALK, PO_4, and the Redfield ratio of carbon and alkalinity to phosphate in organisms. Broecker (1982) used this model in his remarkable attack on the problem of the increase of atmospheric pCO_2 that occurred at the end of the ice age. He first collected all the measurements that could provide constraints on the problem, such as the changes in $\delta^{13}C$ that occurred because of loss of carbon in organic form from the oceans, and the temperature and ocean volume changes. The combined effect of all these changes on atmospheric p_{CO_2} is negligible. He then proposed that the oceanic PO_4 level was reduced by more than one third through deposition onto shelf areas, which were flooded by water when the ice melted. One can readily see from (4) and (5) that such a loss would be associated with an increase in surface $\sum CO_2$ and ALK, in effect allowing some of the excess carbon stored in the deep ocean to escape to the surface ocean and atmosphere. The net effect on atmospheric p_{CO_2} is to increase it by the required amount. This mechanism requires several thousand years to come to a completion and would be expected to follow the ice melting, rather than preceding it.

The subsequent discovery that changes in the atmospheric p_{CO_2} occurred not once, but apparently many times during the ice ages and that the changes occurred on timescales of a few hundred years (Siegenthaler & Wenk 1984), eliminated the shelf phosphate storage hypothesis from contention. More recent work has suggested that the atmospheric p_{CO_2} changes actually preceded the ice volume changes, rather than being associated with them, as required by the shelf storage hypothesis (Shackleton & Pisias 1985). All of this suggested the possibility of a faster response, and also one that is more sensitive to climate, than any that can occur in connection with major changes in the oceanic reservoirs of the relevant constituents in equations (4) and (5).

A further difficulty with the two-box model is that it is unable to meet some of the most fundamental constraints, such as predicting the deep-ocean oxygen correctly (Sarmiento 1986). The latter problem arises from the fact that deep waters in the ocean originate in high-latitude regions where the higher oxygen solubilities associated with lower temperatures allow the water to carry a heavier oxygen load into the abyss than average surface-ocean water can; and where the nutrients are not depleted. Nutrients that are utilized by organisms sink to the abyss in reduced form and create an oxygen demand. High-latitude outcrop regions send large amounts of oxygen into the abyss without the oxygen demand that would accompany the nutrient depletion that is typical of lower-latitude regions. These observations led to the development of the high latitude outcrop three-box model illustrated in figure 3 and discussed in the following section.

3.2. *High-latitude outcrop three-box model*

The three-box high-latitude outcrop model shown in figure 3 is the next step up in complexity from the two-box model. This model, or simple variants of it, was proposed simultaneously by three groups working independently to explain the ice-age observations (Knox & McElroy 1984; Sarmiento & Toggweiler 1984; Siegenthaler & Wenk 1984). A simple solution for the high-latitude surface $\sum CO_2$ and ALK can be obtained in a manner analogous to that used in the two-box model:

$$\sum CO_{2_h} = \sum CO_{2_d} - r_{\Sigma CO_2/PO_4}(PO_{4_d} - PO_{4_h}), \tag{6}$$

$$ALK_h = ALK_d - r_{ALK/PO_4}(PO_{4_d} - PO_{4_h}). \tag{7}$$

A full solution of the model requires equations relating the high-latitude $\sum CO_2$ and ALK to the low-latitude surface concentrations and atmosphere, but the most important equations are

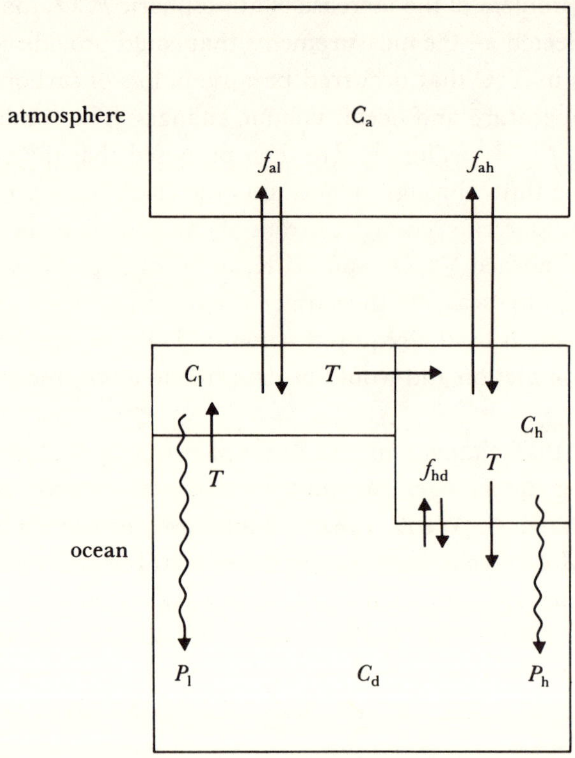

FIGURE 3. A three-box model of the oceans similar to the two-box model in figure 2, only the surface ocean has now been separated into a high-latitude and low-latitude component indicated by the subscripts h and l, respectively. The f_{hd} is exchange between the high-latitude and deep boxes. T represents thermohaline overturning. The Ps are as defined in figure 2. (Taken from Sarmiento & Toggweiler 1984.)

those shown above because the rapid exchange between the deep-ocean and high-latitude box through the f_{hd} exchange term fixes the high-latitude concentrations solidly and allows them to steer the atmospheric and low-latitude concentrations. The full solution, including a description of how radiocarbon and atmospheric p_{CO_2} are used to constrain the parameters, is given in Toggweiler & Sarmiento (1985).

Equations (6) and (7) are virtually identical to (4) and (5) on the right-hand side except for the addition of the PO_{4_h} term. PO_{4_h} can in principle change quite easily. What controls it is a balance between the rate of supply of nutrients to the high-latitude regions by exchange with the deep ocean, and the removal of these nutrients by photosynthetic uptake and sinking. This can be shown by considering the balance equation for the high-latitude box as

$$PO_{4_h} = \frac{PO_{4_d} f_{hd} - P_h}{(f_{hd} + T)}, \tag{8}$$

where we see that an increase in f_{hd} will increase PO_{4_h} and vice versa, and an increase in P_h will decrease PO_{4_h} and vice versa. These changes in PO_{4_h} are linked directly to changes in high latitude $\sum CO_2$ and ALK through equations (6) and (7), and therefore to atmospheric p_{CO_2}.

In simplest terms what occurs is that the high-latitude box is a window to the deep-ocean carbon excess, with PO_{4_h} as the diagnostic of how open the window is. When PO_{4_h} is high, the window is open wide and atmospheric p_{CO_2} can increase, as high as 425 p.p.m. from a pre-industrial value of 280 p.p.m. When PO_{4_h} goes down, the window is closed and the atmospheric p_{CO_2} decreases, as low as 160 p.p.m. Figure 4, taken from Sarmiento & Toggweiler (1984) shows the sensitivity to the forcing parameters in more detail.

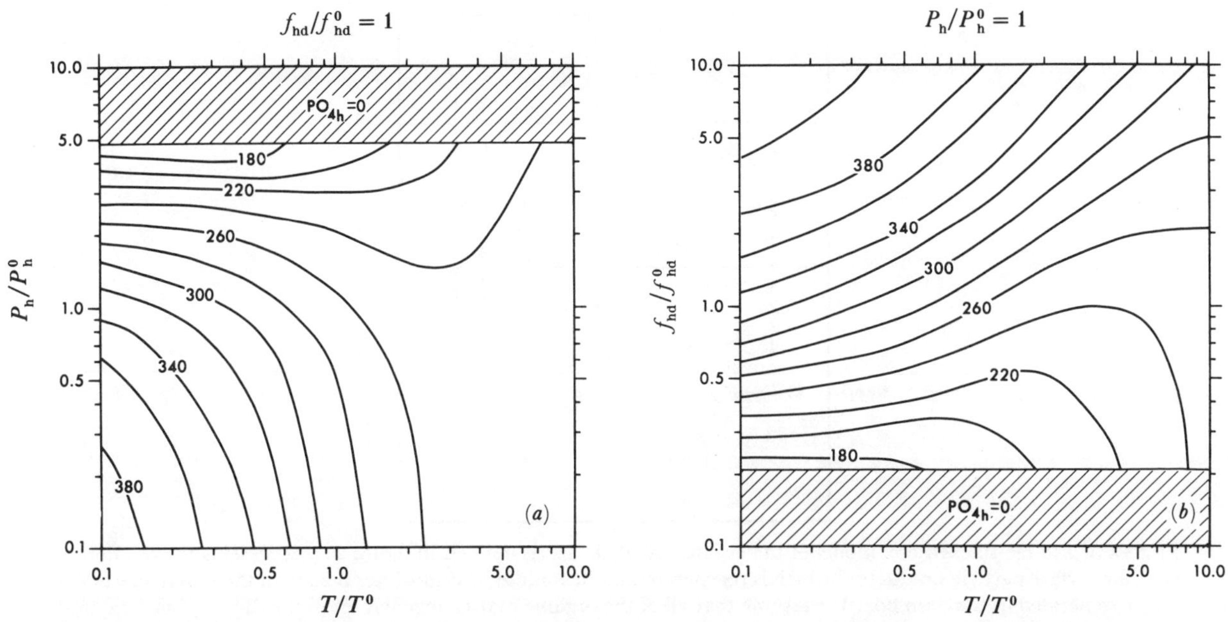

FIGURE 4. Atmospheric p_{CO_2} as predicted by the three-box model for various values of the parameters f_{hd}, P_h, and T. Values $f^0_{hd} = 38.1 \times 10^6\ m^3\ s^{-1}$, $P^0_h = 2.31 \times 10^6\ m^3\ s^{-1}$, and $T^0 = 24.5 \times 10^6\ m^3\ s^{-1}$, are obtained by fitting observations of radiocarbon and atmospheric p_{CO_2}. (Taken from Sarmiento & Toggweiler 1984.)

What this model has shown us is that the high-latitude outcrop regions play a central role in controlling atmospheric p_{CO_2} through the rate of nutrient and carbon supply by deep convection relative to the biological uptake of these constituents and their export to the deep ocean. These are processes that might in principle change quite rapidly. A time-dependent solution of the three-box model shows that after an instantaneous perturbation of one of the controlling parameters it takes of the order of 200 years to reach a new equilibrium (Wenk & Siegenthaler 1985). There is also some sensitivity to the overturning T term, although this sensitivity is reduced as one goes to lower atmospheric p_{CO_2} values. It is possible in principle for T, with its associated low-latitude productivity, to vary over a considerable range with only minimal atmospheric p_{CO_2} changes. There is a way, however, in which the low latitudes may affect the atmospheric p_{CO_2}. To understand how this might occur we must turn to yet a different box model.

3.3. *Three-stacked box model*

One of the first three-dimensional simulations of the carbon cycle that we did, which will be described below, gave surface productivities and atmospheric p_{CO_2} levels that were significantly greater than the present values. From equations (6), (7) and (8) above, we can see that higher productivities should have been associated with lower p_{CO_2}. It quickly became clear that what was occurring in the model was that the high surface productivities, which were caused by the shallow regeneration of PO_4, were giving rise to a large flux of $CaCO_3$ to the deep ocean, which had the net effect of reducing upper-ocean ALK by trapping it in the deep ocean. Dymond & Lyle (1985) have identified the reverse of this mechanism, triggered by lower productivities, as playing a possible role in the lower p_{CO_2} levels of the last ice age. The three-stacked box model in figure 5 was developed to explore this phenomenon further.

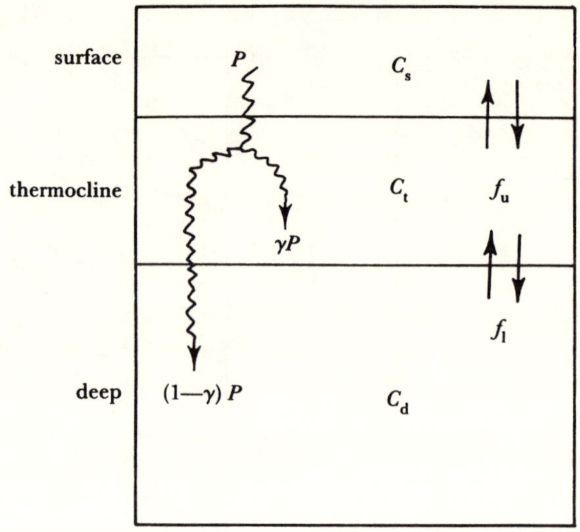

FIGURE 5. A three-stacked box model of the oceans. Symbols are as defined in figure 2; γ represents the fraction of the surface particle production which is regenerated in the middle 'thermocline' box, with the remainder being regenerated in the deep box. We assume that all of the organic matter regenerates in the thermocline box, and that all of the $CaCO_3$ regenerates in the deep box.

Figure 5 shows the three-stacked box model of the ocean. We again take the simplest model possible, this time removing the high-latitude outcrop box (to be returned in the future) and separating the deep ocean into a 'thermocline' box where all of the organic matter produced by organisms is assumed to regenerate, and a deep box where 100% of the $CaCO_3$ is assumed to regenerate. The model is similar to that developed by Dymond & Lyle (1985) in their study, except that instead of f_u they have an overturning term that goes from the surface directly to the deep box and then upwells through the entire water column. A solution for the surface ΣCO_2 and ALK in this model is obtained as above:

$$\Sigma CO_{2_s} = \Sigma CO_{2_d} - r_{\Sigma CO_2/PO_4} PO_{4_d} \left[\frac{1+(f_u/f_l)\,(1-\gamma_{\Sigma CO_2})}{1+(f_u/f_l)\,(1-\gamma_{PO_4})}\right], \tag{9}$$

$$ALK_s = ALK_d - r_{ALK/PO_4} PO_{4_d} \left[\frac{1+(f_u/f_l)\,(1-\gamma_{ALK})}{1+(f_u/f_l)\,(1-\gamma_{PO_4})}\right]. \tag{10}$$

Equations (9) and (10) are again identical to (4) and (5) except for the term in the brackets involving f_u/f_l, the ratio of the upper exchange rate to the lower exchange rate, and the γs. The γs, the values of which can be obtained from the information given in §2, are such that ALK is more sensitive than ΣCO_2 to changes in the exchange rate ratio, f_u/f_l. In simplest terms, these equations change atmospheric p_{CO_2} essentially by sequestering or releasing ALK in the deep ocean. Recall that ALK varies inversely with atmospheric p_{CO_2}, thus an increase in the deep-ocean sequestering efficiency or more vigorous surface mixing will lead to an increase in atmospheric p_{CO_2}, and vice versa.

Figure 6 shows the atmospheric p_{CO_2} predicted by this model for various values of the two exchange-rate terms. The atmospheric p_{CO_2} can be driven up to greater than 20000 p.p.m. before the sequestering of ALK in the deep ocean is great enough to drive the surface ALK to 0! Thus, this mechanism has a far greater range on the high end than does the high-latitude outcrop mechanism, but it has very little flexibility on the lower end (of order 15 p.p.m.) because the present level of deep-ocean sequestering of ALK is quite small. It should be noted that in the realm where ALK gets very low, one might expect other components of seawater than carbonic and boric acids and water to play a role in the carbon system speciation, so that Takahashi's (1980) equations, which we have used to calculate the p_{CO_2}, may not apply. However, it would probably be difficult with this mechanism to drive the atmospheric p_{CO_2} any higher than of order 1100 p.p.m., because at this point the surface ocean becomes undersaturated in $CaCO_3$ and the organisms that use $CaCO_3$ in their tests would have difficulty surviving. It is of interest to consider the possibility that a stagnation of the deep ocean might well be associated with extinction events of calcareous organisms and, conversely, that the development of such organisms would require a relatively rapid exchange of the deep ocean with the upper ocean.

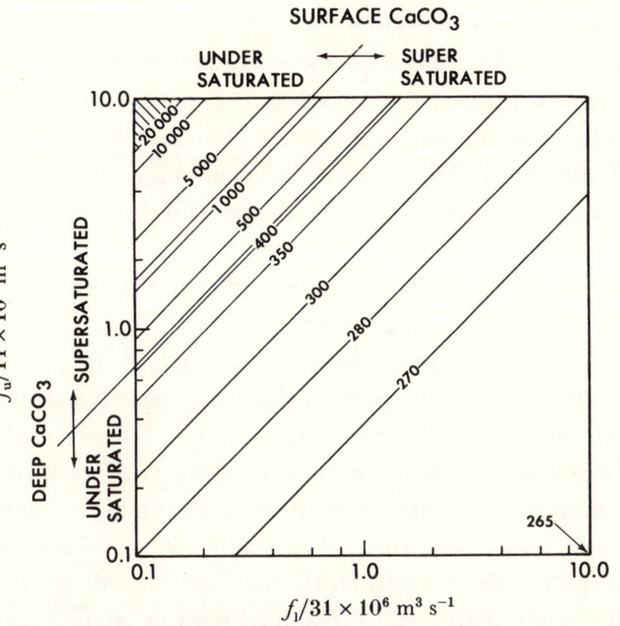

FIGURE 6. Best values for f_u and f_l of 11×10^6 m^3 s^{-1} and 31×10^6 m^3 s^{-1}, respectively, were obtained for the model in figure 5 by forcing the model to fit the pre-industrial value of p_{CO_2} (280 p.p.m.) and the deep-ocean pre-bomb radiocarbon average of -160 p.p.t. These values were then varied over an order of magnitude in each direction, as shown. As the atmospheric p_{CO_2} increases, the deep ocean goes from being undersaturated to being saturated, which would sequester $CaCO_3$ even more efficiently by burying it in the sediments, thus driving the model even more rapidly in the direction of higher p_{CO_2} values. Eventually, however, the upper ocean becomes undersaturated, at which point calcareous organisms would have difficulty surviving.

4. Three-dimensional model

The models discussed in the previous section point towards several ways in which changes in processes internal to the ocean, such as the rate and pattern of ocean circulation and of biological cycling, can cause large changes in atmospheric p_{CO_2}. Such models cannot, however, tell us how those processes would change, nor should the results be entirely trusted unless they can be confirmed with more realistic models or by observations of such changes in the past. In this spirit we are working on the development of integrated models of the ocean carbon system with the major emphasis on understanding how the various processes are related to the physical forcing at the ocean's surface (i.e. the supply of nutrient elements by circulation and mixing, the availability of light, and the dispersion of newly produced material by local turbulence and sinking), as well as chemical and biological regeneration coupled with circulation and mixing in the ocean interior. Such models will eventually be used in conjunction with models of the coupled atmospheric and oceanic circulation to predict the overall response to external perturbations.

At present we have separately developed, or are in the midst of developing, the main components of our series of integrated models. These are: first, the physical-circulation models themselves; second, a one-dimensional model of upper-ocean ecology; third, a model for water-column transformations; and fourth, a sediment-diagenesis model for $CaCO_3$ dissolution to which we will soon add organic carbon oxidation. Our progress in this work is discussed in detail in a technical report (Toggweiler *et al.* 1987).

The production of global integrated models allows us to identify critical unknowns in our knowledge of the ocean's chemical system. When we originally began thinking about modelling chemical fluxes about four years ago, our recipe for making a model was very simple: take a circulation model, add nutrients, convert the surface nutrients into a pool of sinking particles, and remineralize the particles at depth following some scaling law. We found out very quickly, however, that an approach such as this misses important elements of chemical cycling that have much greater consequences for the chemistry of the ocean as well as the atmosphere than we had originally expected.

Our studies have led us to focus on two major aspects of upper-ocean biology that do not fit this simple picture and that may be pressure-sensitive 'points', such as those that have been discussed above. The first, which has already been introduced, is the presence of unutilized nutrients in high-latitude surface water. In these regions the organic carbon production and sinking fluxes are not high enough to deplete nutrients, but the net exchange of O_2 and CO_2 between the atmosphere and deep ocean is greatly enhanced. The work described above has shown us that these regions play a central role in determining atmospheric CO_2 as well as deep-ocean O_2 (Sarmiento & Toggweiler 1984; Toggweiler & Sarmiento 1985).

The second aspect is the production of refractory dissolved organic compounds by marine organisms. We see evidence accumulating that dissolved organic matter (DOM) may be more effective in transporting organic carbon and nutrients out of the upper ocean than particles. We find that the inclusion of DOM in our models can change their behaviour quite dramatically.

In the deep ocean our preliminary studies have led us to focus on the problem of regeneration of $CaCO_3$ and how this affects the distribution of alkalinity. The three-stacked box model discussed above was a direct outcome of some of the results of the work that will be described below.

In what follows we will discuss some results from our first version of a three-dimensional simulation and a two-dimensional version of this. In this model the carbon cycle is parametrized in a very simple fashion, our intention being primarily to begin exploring how the large-scale distributions of $\sum CO_2$, ALK, and PO_4 can be used to constrain our understanding of the present ocean, as well as what the role of DOM and of the rate of deep-water formation are in controlling the distribution of these constituents.

4.1. *Model description*

The ocean circulation is predicted by using the primitive equation approach developed by Bryan (1969) for a sector model basin with idealized geometry that is 60° wide and goes from 70° N to 70° S. It has vertical sides and a flat bottom at 4 km with 25 levels in the vertical and 2° horizontal resolution. The wind forcing is the zonally averaged world ocean mean. Temperature and salinity are predicted by using a newtonian damping boundary condition at the surface in which the temperature and salinity are restored towards the zonally averaged Atlantic Ocean data. A similar version of this model is described by Bryan (1986).

The zonally averaged meridional overturning predicted by the model is shown in figure 7. The interior circulation has a much stronger vertical component to it than is obtained with models that have more realistic topography (Bryan & Lewis 1979) or than is generally expected to obtain, although the total deep-water formation rate of order $24 \times 10^6\ m^3\ s^{-1}$ is comparable to the Atlantic. The predicted upper-ocean transport has realistic values of Ekman pumping, including strong equatorial upwelling, which is balanced by equatorward transport at depth. The choice of using the model in its present form for these first experiments was made because it is simple and quick to run and because the many experiments carried out by Bryan (1986) have given us considerable familiarity with the behaviour of the model. One of the

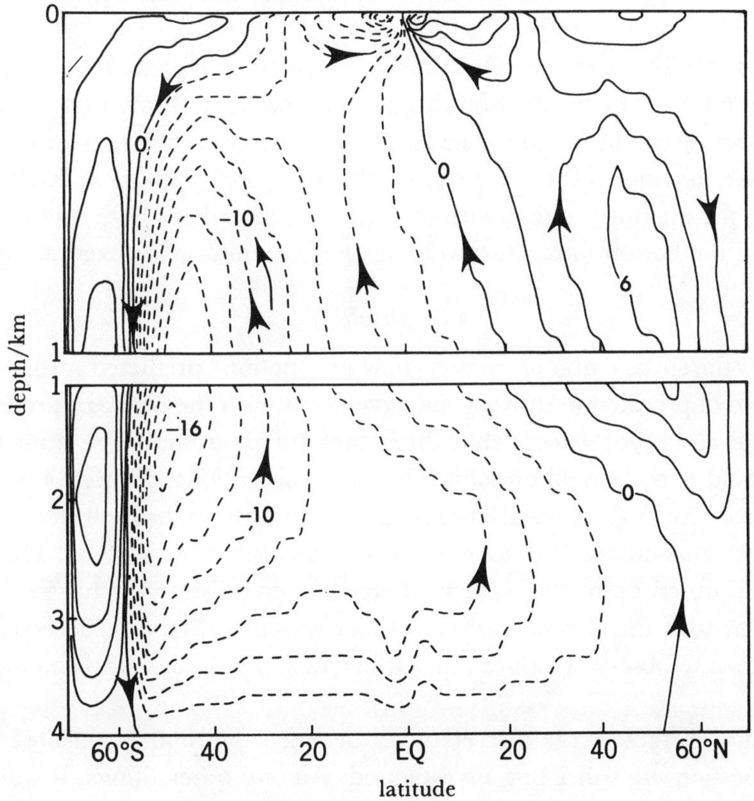

FIGURE 7. The zonally integrated meridional overturning predicted by the three-dimensional sector model described in the text. Units given are $10^6\ m^3\ s^{-1}$.

experiments that we have carried out involves changing the rate of meridional circulation shown in figure 7. The experiments of Bryan (1986) show how this can be done by changing the vertical diffusivity.

The approach we follow in predicting the surface nutrient concentration is essentially the same as for temperature and salinity: the full equation of conservation is solved except for the inclusion of a newtonian damping term of the form $-\gamma([PO_4]-[PO_4^*])$. This term forces the model-predicted PO_4 back towards the observed PO_4^*, which is obtained from GEOSECS observations, thus removing PO_4 from the surface in the process. The model-predicted PO_4 is generally greater than the PO_4^* data because it is continuously supplied from below; γ is the reciprocal of a time constant, which in this case was taken to be 5 days after several experiments.

The PO_4 that is removed from the surface is regenerated in the water column by using a simple scaling law obtained from sediment trap observations (Martin *et al.* 1987), namely:

$$\partial[PO_4]/\partial t = \text{advection} + \text{diffusion} + A\,z^{-1.858}. \tag{11}$$

A is specified for each timestep such that the vertically integrated regeneration for a given geographic location is exactly equal to the nutrient removal at the surface at that location. The assumption implicit in this approach is that the removal of organic matter from the surface occurs primarily by the large particles that are caught in sediment traps. We considered that this was a reasonable initial approach to take, partly because this is the flux term we know most about, and partly because, until we obtained the model results shown below and became aware of some of the most recent work in this area, we considered that this flux term was indeed the major one.

$CaCO_3$, which is produced at the same time as nutrients, is regenerated entirely within the deepest box of the model. The production rate is obtained by multiplying the PO_4 production rate by the appropriate Redfield ratio. The $\Sigma\,CO_2$ and ALK are predicted by removing them from the surface as a function of the PO_4 removal, by using the appropriate Redfield ratios, and then regenerating the organic fraction with the same depth scale as PO_4, and regenerating the $CaCO_3$ fraction in the bottom box. Gas exchange with a uniformly mixed atmospheric box is also included.

4.2. *Results*

Figure 8*a* shows the surface flux of carbon (new production) predicted at the surface by the model. The pattern of production shows good agreement with the primary production map of Koblentz-Mishke *et al.* (1970), except that the values are about an order of magnitude larger than what one would expect based on sediment trap observations. Figure 8*b* of PO_4 at 180 m shows why this is so. The PO_4 values are too high, particularly in the eastern equatorial region where they achieve values almost 3 times greater than any that are observed. These high values are advected and diffused up to the surface where they give rise to the high new production. In later simulations with the same ocean-circulation model, where we succeeded in bringing the PO_4 values down to observed values, the surface new production also came down into the observed range.

What is the cause of these large discrepancies between the model and observations? The answers to this question are still being investigated. Among other things, it will be necessary to carry out simulations with ocean-circulation models that have more realistic meridional overturning. However, our first conclusions are that the increased surface production and

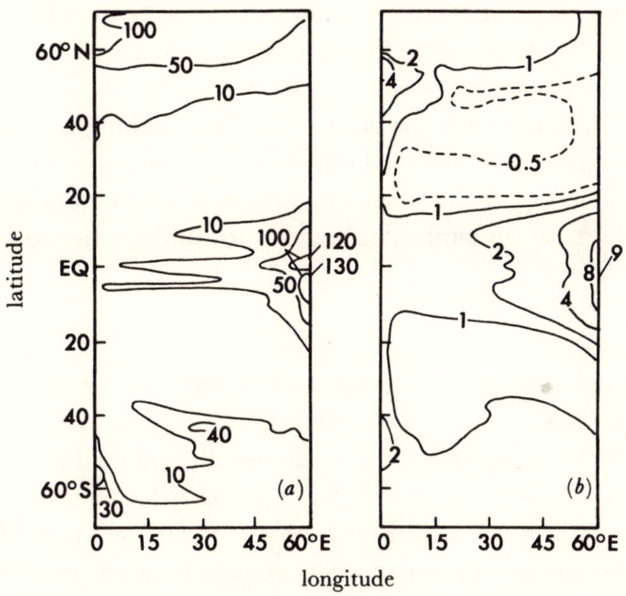

FIGURE 8. (*a*) Flux of organic carbon predicted by the model in figure 7 at the base of the euphotic zone at 100 m. Units are mole per square metre per annum. The average is 13.4 mol m^{-2} a^{-1}. (*b*) Phosphate at 180 m as predicted by the model in units of μmole per kilogram.

PO_4 result from an incorrect characterization of the cycling of organic matter in the oceans. This explanation is explored in what follows.

The major differences between the PO_4 field predicted by the model and the observed PO_4 field are that the model concentrations are too high, and that the zonally averaged nutrient maximum, which is at approximately 400–500 m at the Equator in the model, is too shallow by about 500 m. Our model is very efficiently trapping nutrients in the upper part of the water column, particularly in the tropics. Such trapping is a normal characteristic of regions where the waters transported away from the trapping region have their nutrients removed very efficiently before they can escape the region, and where the waters flowing in to the region trap along the way the nutrients that are removed from the outflowing waters. In a steady state, such as we have assumed, the inflow of nutrients must balance the outflow, but it may take very large concentrations within the trapping region to provide enough nutrients to the outflowing waters in order to balance the inflowing nutrient flux. The equatorial upwelling regions have these characteristics, in that the outflowing water is at the surface, where organisms can very effectively remove nutrients, whereas the inflowing waters are in the thermocline where most of the nutrients are regenerated. Our model is an example of this naturally observed phenomenon that has gone to an extreme.

The regeneration function that was used for organic matter in our first experiment (equation (11)) is obtained from observations of the decrease with increasing depth of material caught in sediment traps (Martin *et al.* 1987). The flux decreases to 1/e at a distance of 220 m below the 100 m base of the euphotic zone. This can be compared to the 1/e scale depth of 340 m obtained by Jenkins (1982) from his oxygen-utilization rate estimates, and 400 m obtained also from oxygen observations by Wyrtki (1962). Not only is the sediment trap scale depth smaller but, as Martin *et al.* point out, their power-law function gives a more rapid drop off than the exponential fits of the other investigators such that the latter have much more regeneration occurring in deeper waters. A likely explanation for this difference, which we explore below, is that organic carbon is being removed from the surface in forms that are regenerated at greater depths than the large particles caught in sediment traps.

Sediment traps typically catch large particles which have sinking rates of order 100 m d^{-1}. Such materials sink too rapidly to be affected significantly by the oceanic advection field. This is implicit in our modelling approach, which requires that all the organic matter (as well as the $CaCO_3$) that is produced at the surface be regenerated in the water column directly below. Note that the rapid sinking rates and short regeneration scale depths of the large particles caught in sediment traps imply an extremely short timescale for regeneration of these large particles.

Other materials that may account for removal of organic matter produced in the surface, such as small particles, or dissolved organic matter (DOM), either sink very slowly or not at all and would therefore be more susceptible to being carried around by the advection field, if their regeneration timescales are longer than those for large particles. The recent observations by Suzuki *et al.* (1985) and Sugimura & Suzuki (1988), that the amount of DOM in the water column appears to be far greater than had been previously been measured, have led us to investigate the possibility that DOM transport and regeneration may account for the difference between the model and observations.

To carry out a larger number of investigations, our three-dimensional model was zonally averaged and reincarnated as a two-dimensional model. This model predicted the PO_4 and surface productivities shown in figure 9*a*. The problem of high PO_4 in the equatorial region is worse in the two-dimensional model than in the three-dimensional model primarily because of the absence of the inflow of low PO_4 waters in the Equatorial Undercurrent. The PO_4 maximum is thus shallower and larger than the zonal average of the three-dimensional model. This model predicts a very high p_{CO_2} of 620 p.p.m. because of the strong driving, by the high surface production predicted by the model, of the alkalinity pump mechanism explored above with the three-stacked box model.

Figure 9*b* shows the result of a two-dimensional simulation in which half the surface production of organic matter was put into DOM, which was then allowed to diffuse and advect. The regeneration time of the DOM was set at 166 years by requiring that the DOM inventory match the observations of Sugimura & Suzuki (1988). The results in figure 9*b* are greatly improved. The nutrient maximum has moved down to a more appropriate depth and has been reduced to a more realistic value. The production of DOM with a long regeneration time allows the nutrients brought into the equatorial region by upwelling and taken up by organisms to escape from the region by surface Ekman transport and participate more fully in the large-scale circulation patterns before being regenerated.

Another way in which the results of the model are improved by including DOM is that the atmospheric p_{CO_2} now falls to a value of 308 p.p.m. from the value of 620 p.p.m. predicted by the non-DOM case. This is because of the cycling of $CaCO_3$. In the model without DOM, the high upper-ocean productivities are equivalent to increasing f_u relative to f_l in the three-stacked box model described above. The high surface productivities shuttle unusually large amounts of $CaCO_3$ to the deep ocean, thus reducing the surface alkalinity and giving rise to the exceptionally high atmospheric p_{CO_2} levels obtained. With normal upper-ocean productivities the ocean is able to bring the deep waters up with sufficient efficiency that the atmospheric p_{CO_2} can be kept within normal levels.

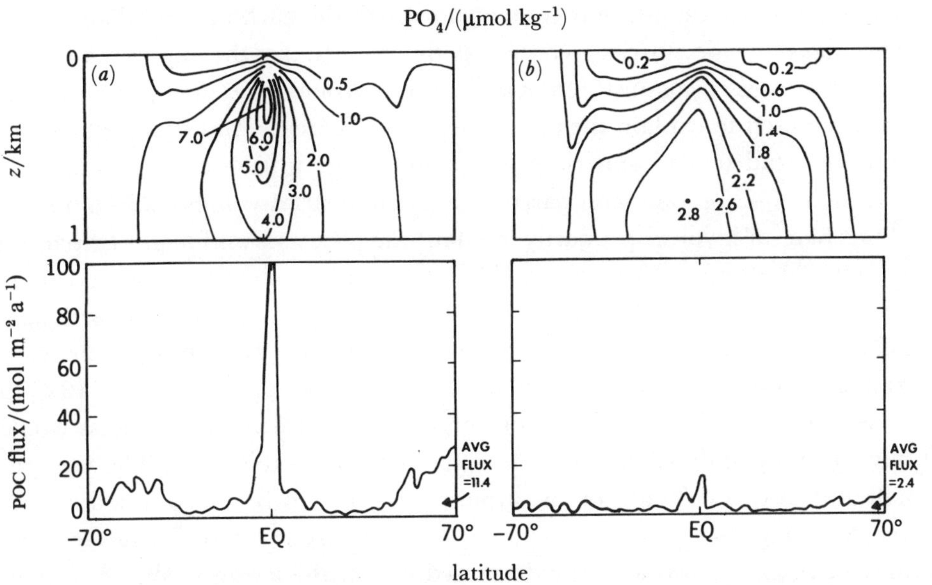

FIGURE 9. (*a*) Phosphate in the upper 1000 m, and particulate organic carbon flux (POC) predicted by a two-dimensional model obtained by zonally averaging the three-dimensional model in figure 7. (*b*) The same figures for a version of the two-dimensional model in which half of the organic carbon production is put into the form of dissolved organic matter with a regeneration timescale of 166 years.

There are other possible explanations for our nutrient trapping problem. The role of the circulation, which has been alluded to above, needs to be investigated. Simply increasing the regeneration scale depth of the sinking material helps as well, but it needs to be kept in mind that the scale depths we use are observationally based. We prefer to take as a working assumption that the scale depth used is appropriate for the large particles and to consider, as we have, what other constituents may be involved in the removal of organic matter from the surface.

Finally, another aspect of the problem that has not been discussed here is the rôle of oxygen. At PO_4 levels much above about 3 μmol kg^{-1} the cumulative oxygen demand is great enough that most waters are at or near anoxia. Under such circumstances the oxidation of organic matter ceases. When we include a prediction of oxygen in our model the entire region under the Equator goes anoxic and the organic matter falls below the anoxic region before it is regenerated, thus in effect lengthening the regeneration scale depth. The PO_4 concentrations predicted in this case are comparable to those observed in the oceans, but the oxygen is not, in that the regions where anoxia is present in the oceans are more limited than those predicted by the model. It is apparent from this that oxygen also provides a strong constraint on the model that needs to be considered.

5. Conclusion

The studies we have carried out show that the oceanic carbon cycle is far more dynamic than had previously been thought, responding, as it does, to the pattern and magnitude of ocean circulation as well as to the biological productivity. We have identified two major mechanisms internal to the oceans that may lead to significant changes in atmospheric p_{CO_2} on timescales as short as of order 200 years. One involves the high-latitude processes that govern the exposure of the atmosphere to the carbon-enriched deep waters. The diagnostic of the extent of contact that we use is the high-latitude nutrient content, to which the carbon and alkalinity contents are linked. The high-latitude nutrient content is determined by the supply of nutrients by physical exchange processes against removal of the nutrients by biological uptake. This mechanism has been used to explain the atmospheric p_{CO_2} increase that occurred at the end of the ice age and may lead to atmospheric p_{CO_2} changes ranging from a low of approximately 165 p.p.m. to a high of 425 p.p.m. compared to a pre-industrial value of 280 p.p.m.

The second mechanism involves primarily low-latitude processes and might be referred to as the alkalinity pump. Higher productivities and/or a stagnation of the deep ocean can give rise to an increasing segregation of alkalinity from total carbon in the oceans, with alkalinity trapped in the deeper waters. In such a case the atmospheric p_{CO_2} can, in principle, go as high as greater than 20000 p.p.m., although an increase beyond approximately 1100 p.p.m. is unlikely because of the fact that the calcareous organisms that drive this process would have difficulty surviving in the undersaturated waters that develop at this point. The segregation at present is small and so it is difficult for the atmospheric p_{CO_2} to be taken much below about 265 p.p.m. by reversing the process. Dymond & Lyle (1985) have suggested that this mechanism may have played a role in the decreased p_{CO_2} of the ice ages. We are interested in exploring whether it may have played a role in the more distant past.

Our first experiments with development of two-dimensional and three-dimensional models of the carbon cycle to further study processes such as the above have led to some very interesting results that are consistent with various observational constraints suggesting that the interior recycling of carbon in the oceans may involve significant contributions from refractory materials capable of being carried great distances from their point of formation. DOM is suggested as the culprit, although the work is far from conclusive.

Phil Tunison and Nancy Duprey helped with preparation of the manuscript. Support for the research described here was provided by the Carbon Dioxide Research Division, U.S. Department of Energy, under contract DE-AC05-84OR21400 with Martin Marietta Systems, Inc. (subcontract 19X-27405C with Princeton University) and by the National Science Foundation under grants no. OCE-8711905 and ATM-8414596.

References

Broecker, W. S. 1982 *Geochim. cosmochim. Acta* **46**, 1689–1705.

Broecker, W. S. & Peng, T.-H. 1982 *Tracers in the sea.* (634 pages.) Palisades, New York: Eldigio Press.

Bryan, F. 1986 *Maintenance and variability of the thermohaline circulation.* Ph.D. thesis, Princeton University, Princeton, New Jersey.

Bryan, K. 1969 *J. Computat. Phys.* **4**, 347–376.

Bryan, K. & Lewis, L. J. 1979 *J. geophys. Res.* **84**, 2503–2518.

Dymond, J. & Lyle, M. 1985 *Limnol. Oceanogr.* **30**, 699–712.

Jenkins, W. J. 1982 *Nature, Lond.* **300**, 246–249.

Knox, F. & McElroy, M. B. 1984 *J. geophys. Res.* **89**, 4629–4637.

Koblentz-Mishke, O. J., Volkovinsky, V. V. & Kabanova, J. G. 1970 In *Scientific exploration of the South Pacific* (ed. W. S. Wooster), pp. 183–193. Washington, D.C.: National Academy of Sciences.

Martin, J. H., Knauer, G. A., Karl, D. M. & Broenkow, W. W. 1987 *Deep Sea Res.* **34**, 267–286.

Neftel, A., Oeschger, H., Schwander, J., Stauffer, B. & Zumbrunn, R. 1982 *Nature, Lond.* **295**, 220–223.

Oeschger, H., Siegenthaler, U., Schlotterer, U. & Gugelman, A. 1975 *Tellus* **2**, 168–192.

Sarmiento, J. L. 1986 In *The role of air–sea exchange in geochemical cycling* (ed. P. Buat-Menard), pp. 65–82. New York: D. Reidel Publishing.

Sarmiento, J. L. & Toggweiler, J. R. 1984 *Nature, Lond.* **308**, 621–624.

Shackleton, N. J. & Pisias, N. G. 1985 In *The carbon cycle and atmospheric CO_2: natural variations archean to present* (ed. E. T. Sundquist & W. S. Broecker), vol. 32 (Geophysical Monograph Series), pp. 303–318. Washington, D.C.: American Geophysical Union.
Siegenthaler, U. & Wenk, T. 1984 *Nature, Lond.* **308**, 624–626.
Sugimura, Y. & Suzuki, Y. 1988 *Mar. Chem.* (In the press.)
Suzuki, Y., Sugimura, Y. & Itoh, I. 1985 *Mar. Chem.* **16**, 83–97.
Takahashi, T., Broecker, W. S., Bainbridge, A. E. & Weiss, R. F. 1980 *Carbonate chemistry of the Atlantic, Pacific, and Indian Oceans: the results of the Geosecs expeditions, 1972–1978,* Tech. Rep. 1, CU-1-80, Lamont-Doherty Geol. Obs., Palisades, New York.
Takahashi, T., Broecker, W. S. & Langer, S. 1985 *J. geophys. Res.* **90**, 6907–6924.
Toggweiler, J. R. & Sarmiento, J. L. 1985 In: *The carbon cycle and atmospheric CO_2: natural variations archean to present* (ed. E. T. Sundquist & W. S. Broecker), vol. 32 (Geophysical Monograph Series), pp. 163–184. Washington, D.C.: American Geophysical Union.
Toggweiler, J. R., Sarmiento, J. L., Najjar, R. & Papademetriou, D. 1987 *Models of chemical cycling in the oceans.* Ocean Tracers Laboratory Technical Report no. 4, Princeton University, Princeton, New Jersey.
Wenk, T. & Siegenthaler, U. 1985 In *The carbon cycle and atmospheric CO_2: natural variations archean to present* (ed. E. T. Sundquist & W. S. Broecker), vol. 32 (Geophysical Monograph Series), pp. 185–194. Washington, D.C.: American Geophysical Union.
Wyrtki, K. 1961 *Deep Sea Res.* **8**, 39–64.

Discussion

D. J. Webb (*Institute of Oceanographic Sciences, Wormley, Godalming, Surrey, U.K.*). The three-box model showed that both the thermohaline circulation and the vertical mixing at high latitudes affected the flux of CO_2 into the deep ocean. Sedimentary evidence indicates that the thermohaline circulation changed significantly during recent ice ages. Can Professor Sarmiento therefore say what levels of CO_2 are predicted by the model as being possible caused by changes in just this one term?

J. L. Sarmiento. The three-box model parameter T, which we have referred to as the thermohaline overturning term, can affect the CO_2 level by greater than 100 p.p.m., although the sensitivity to this term diminishes greatly as one approaches the low CO_2 levels of the last ice ages (see figure 4). The 'vertical mixing' term, f, which we have characterized as being in some ways equivalent to Antarctic Bottom Water Formation, is thus probably the most important one to study. It should be pointed out that f must also involve significant lateral exchange within the ocean basins. Therefore much of the evidence that exists for changes in water-mass distribution is relevant to this term as well as to T.

W. J. Jenkins (*Woods Hole Oceanographic Institution, Woods Hole, U.S.A.*). Could not the unrealistically large 'new production' and subsequent anoxia and anomalous phosphate distributions in Professor Sarmiento's model be ameliorated by a choice of a longer vertical scale length for his remineralization function? In particular, I have in mind the kind of vertical attenuation rates observed by myself in subtropical gyre oxygen-utilization rates. By depositing the remineralized nutrients in deeper strata that will have characteristically longer return timescales, one should deepen the nutrient maxima and slow down new production rates.

J. L. Sarmiento. We have done several experiments where the vertical scale length of regeneration was increased. The results, although an improvement, were not as satisfactory as those obtained with the dissolved organic matter experiments. In our model it appears to be necessary that a mechanism exist that allows the organic matter to be advected and diffused horizontally away from where it is produced. The difference in vertical scale length between Dr Jenkins's oxygen-utilization rate results and the sediment-trap results provides strong support for the idea that there is some organic matter which is not being caught by the sediment traps, but it does not carry direct information about what form that organic matter should take. Our model provides at least some suggestive evidence that it must be something that can be advected and diffused.

C. Wunsch (*Center for Meteorology and Physical Oceanography, Massachusetts Institute of Technology, U.S.A.*). Would Professor Sarmiento comment on the issue of whether the biogeochemical results of a model should be taken seriously when the model is known to have serious physical shortcomings? This question is stimulated by his reference to Toggweiler's results.

J. L. Sarmiento. 'Toggweiler's results', by which I believe Professor Wunsch means the three-dimensional model study, are actually Najjar's results. Yes, I do believe that the biogeochemical results should be taken seriously, otherwise we would not be taking this approach! We are trying to understand the basic mechanisms underlying the carbon cycle so that we can predict the response of the globe to perturbations in this cycle. The particular model we are working with has the advantage of being inexpensive to run, and of being well understood, in that Bryan (1986) has done a detailed study in a model similar to it of what controls the rate of thermohaline overturning. One of our major goals in working with this particular model is to take advantage of Bryan's work to carry out a series of experiments to test out the ideas we have developed concerning the role of high latitude deep water formation in controlling atmospheric CO_2.

Our serendipitous finding about the importance of the role of some refractory form of organic matter in carbon cycling is a good example of the usefulness of the approach we are taking. We do not feel we can conclude based on the model alone that there is a need for a refractory carbon source, but we do feel that the model points towards the possible major impact of having such a source. More realistic ocean-circulation models, which are far more expensive to run, are available and will be used when we have a better understanding of how to simulate the carbon cycle.

H. Elderfield (*Department of Earth Sciences, University of Cambridge, U.K.*). The vertically stacked three-box model that Professor Sarmiento has described has similarities in principle to the 'rain ratio model' where differences in the ratio of organic to inorganic carbon flux between the deep sea and the margins (principally upwelling régimes) allow changes in coastal productivity to affect the CO_2 system between glacial and interglacial times. Has Professor Sarmiento considered lateral exchange in his models, in particular those that consider the role of ocean margins?

J. L. Sarmiento. No, we have not considered lateral exchange, but clearly this is something of great importance that we need to do.

M. Whitfield (*Marine Biological Association, Plymouth, U.K.*). General circulation models of the atmosphere suggest that increases in atmospheric temperature associated with the CO_2 'greenhouse effect' will be accompanied by a decrease in the temperature gradients from

Equator to Pole. One would naïvely assume that this would result in a slowing down of the ocean circulation. Are we in a position to predict the 'sign' of the associated p_{CO_2} changes? Will a positive feedback effect result?

J. L. SARMIENTO. Interestingly, the work of Manabe & Bryan suggests that the rate of deep-water formation does not change with increasing CO_2. This is explained as being caused by the fact that the north–south ocean density does not change significantly with the warming. The density of seawater is not very sensitive to temperature in the high latitudes and salinity changes tend to compensate for whatever density changes do occur. If the deep-water formation did slow down, the three-box model predicts that the equilibrium atmospheric CO_2 would drop, thus giving a negative feedback. However, in a non-steady state this needs to be balanced against the slow down in uptake of the fossil-fuel CO_2 excess.

Reference

Manabe, S. & Bryan, K. 1985 *J. geophys. Res.* **90**, 11689–11708.

J.-F. MINSTER (*CNES/GRGS, Toulouse, France*).

1. In the three-dimensional model, in addition to possible difficulties related to the circulation pattern and the redissolution function, there might be problems because of the forcing parameter of surface phosphate concentration. This parameter is significantly smaller (time delay larger) in tropical oceans compared with higher latitudes. This could lead to the calculation of too-large productivities.

2. In the stacked three-box model, it is not clear whether carbonate redissolution from the sediments, when the diffusion parameters are changed, is well taken care of. It seems to me that this process should prevent the surface layer from taking very low ΣCO_2 and ALK values.

J. L. SARMIENTO.

1. It is tempting but incorrect to try and relate the time constant in our newtonian damping term to the actual timescale of the biological response to a change in nutrient concentration. By analogy with the use of the equivalent boundary condition often used in predicting temperature in primitive equation models, if one is restoring towards the temperature in the atmosphere above the ocean, then one should use a timescale that relates to the physical processes of heat exchange. However, if one is restoring towards the observed ocean temperature, then the appropriate restoring time is infinitely small, or as small as the timestep allows. Our nutrient model is equivalent to the latter case.

2. The effects of carbonate dissolution have not been simulated explicitly in our models, but our calculations of the carbonate saturation levels supports Dr Minster's point. When atmospheric CO_2 levels near 1100 p.p.m., calcium carbonate becomes undersaturated in the surface box. This has two consequences: firstly, the dissolution of shallow calcareous sediments would provide a strong buffer to any further CO_2 changes; and secondly, calcareous organisms would probably not even survive under such conditions so that the alkalinity pump would most likely cease.

The use of tracers in modelling the oceanic uptake of carbon dioxide

By A. J. Crane

Central Electricity Research Laboratories, Central Electricity Generating Board, Kelvin Avenue, Leatherhead, Surrey KT22 7SE, U.K.

Increasing levels of atmospheric carbon dioxide from the burning of fossil fuels and changes in land use pose a threat of significant global climatic change in the 21st century. Owing to uncertainty in the pre-industrial atmospheric CO_2 concentration and in CO_2 releases from land-use change, direct estimates of the airborne fraction of this man-made CO_2 are not well established. Effort has therefore been devoted to estimating the strengths of the oceanic and biospheric sinks as an alternative route to determining the airborne fraction. This paper reviews the development of oceanic CO_2-uptake models.

One-dimensional box models, with first-order exchange kinetics or vertical diffusion to simulate CO_2 penetration into the deep ocean and calibrated against natural ^{14}C distributions, appear inadequate. Their uptake is too small to be consistent with the recent atmospheric record and most estimated histories of CO_2 release from land use change.

Models incorporating representations of specific oceanic mixing processes important in CO_2 uptake achieve somewhat larger CO_2 uptake, especially when calibrated against 'short timescale' tracers, such as radiocarbon and tritium derived from bomb tests. Despite this general conclusion, substantial differences between such models remain. A comparison between my two-dimensional advection–diffusion model and Siegenthaler's outcrop–diffusion model illustrates how the relative importance of air–sea gas exchange rate and rate of ocean mixing in limiting CO_2 uptake depends critically upon modelling assumptions.

The failure of most models calibrated with a single oceanic tracer to reproduce well the distribution of other tracers has encouraged the development of multi-box, geographically realistic, models whose circulation and mixing patterns are determined by simultaneous inverse solution of a set of conservation equations for a range of tracers. This technique, when augmented with additional dynamical constraints, probably offers the most promise for advancing CO_2-uptake modelling while suitable three-dimensional oceanic general circulation models are being developed. The latter, atmospherically driven, models will eventually play a key role in assessing how any future climatic change may feed back on atmospheric CO_2 levels. Feedback could arise either by alteration of the mixing processes responsible for man-made CO_2 uptake, or more fundamentally if changes in the surface-ocean productivity result from changes in circulation-mediated nutrient supply.

1. Introduction

Man's activities over the past century have caused substantial releases of carbon dioxide into the atmosphere. Since 1957, when continuous accurate measurements began, the atmospheric concentration has risen by about 10%, from 315 to around 348 parts per million by volume (p.p.m.v.). Carbon dioxide, water vapour and several other trace gases in the atmosphere are strong absorbers at the wavelengths near the peak of the Earth's infrared emission spectrum. Their natural concentrations maintain the surface temperature at habitable levels. Any increase in their abundance poses a possible threat of global-scale warming and consequent climatic change. Although estimates of the likely climatic change are highly uncertain, it is probable that an increase in these infrared-absorbing trace gases corresponding to a doubling of the pre-industrial level of CO_2 would produce significant effects. Projections of future emissions suggest that such an increase may occur by, or soon after, the middle of the 21st century.

The rate, and probably the nature, of any climatic change will clearly depend on the rate of increase of CO_2 concentration. This in turn will depend upon the rates of future emissions and upon the fraction of these cumulative emissions that will remain airborne. This airborne fraction is itself a function of CO_2 emission growth rate and thus likely to vary with time.

Scientific research is unlikely to yield improved estimates of future emission rates; these will be governed by political and economic considerations. It should, however, provide improved understanding of the way excess CO_2 emitted to the atmosphere is partitioned between the atmospheric, oceanic and biospheric carbon reservoirs, and hence estimates of the airborne fraction likely to prevail for a given emission scenario.

In as much as past values of the airborne fraction, α, are a useful guide to the future, it is instructive to review how accurately these can be determined from available knowledge of emissions and atmospheric CO_2 concentration changes. Over the period 1860–1980, the main uncertainties in the calculation lie in the pre-industrial concentration of CO_2 and the amount of CO_2 that has been released through deforestation and land-use changes, in addition to that from fossil fuels. Assuming fossil-fuel emissions to have been as reported by Rotty (1981), figure 1 shows that the mean α over this period may have lain anywhere in the range 0.32–0.60 for the indicated ranges of uncertainty in the pre-industrial CO_2 level (275 ± 10 p.p.m.v.) and the forest and soil emissions from deforestation and land-use changes (150 ± 50 Gt C). Over the period 1958–1982, for which accurate measurements of the increase in atmospheric concentration are available, Bolin (1986*a*) adopts annual forest–soil emissions of 1.6 ± 0.8 Gt C, giving α in the range 0.34–0.46.

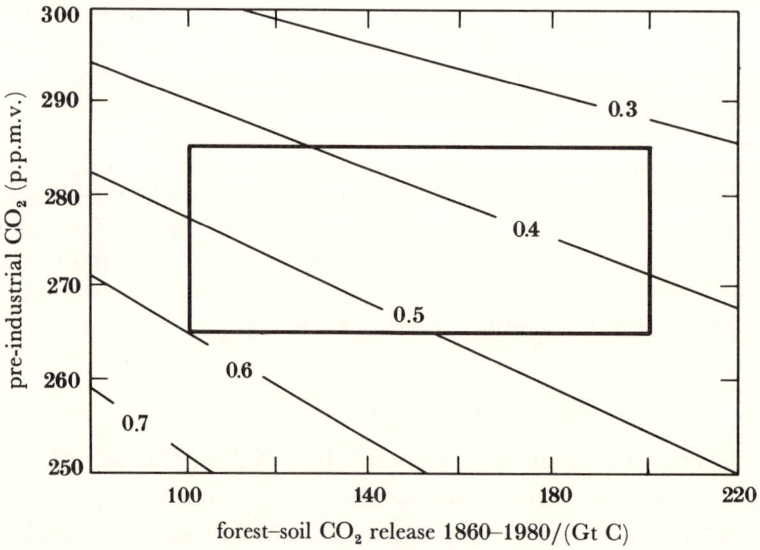

FIGURE 1. The average airborne fraction of CO_2 emissions from 1860–1980 as a function of pre-industrial atmospheric CO_2 concentration and emissions due to deforestation and changing land use, assuming fossil-fuel CO_2 emissions as given by Rotty (1981). The heavy lines indicate the uncertainty estimated by Bolin (1986*a*). (After Bolin 1986*a*.)

Because of isotopic fractionation by plants, the size of the global biosphere at any time is indicated by the $^{13}C:^{12}C$ ratio of atmospheric CO_2. On this basis estimates of past forest–soil emissions have been made by using carbon-cycle models to deconvolve records of $^{13}C:^{12}C$ ratio recorded in tree rings (see, for example, Stuiver *et al.* 1984; Peng 1985). These are subject to the uncertainty associated with climatic and growth effects on the $^{13}C:^{12}C$ ratio and with the accuracy with which carbon-cycle models simulate the carbon system. These studies have tended to calculate similar total non-fossil carbon releases over the industrial period, but with quite different distribution in time, giving much lower releases over the past 30 years than direct estimates. The same technique has been used by Druffel & Benavides (1986) employing measurements of the $^{13}C:^{12}C$ ratios in a banded Jamaican sclerosponge, which is considered a more accurate recorder of prevailing oceanic, and hence atmospheric, $^{13}C:^{12}C$ ratios. They calculate a much smaller overall forest–soil emission, but accord with Stuiver *et al.* (1984) in obtaining a value close to zero for the recent period, indicating α around 0.58.

Deconvolution of the CO_2 record from measurements of air trapped in an Antarctic ice core (Siegenthaler & Oeschger 1987) also indicates a smaller release (0–0.9 Gt C annually for the period 1959–1983) than obtained from direct estimates of deforestation and land-use change, giving α in the range 0.58–0.48. This range results from the use of two alternative carbon-cycle models. The lower value, 0.48, is obtained by using the model whose ocean is able to absorb larger amounts of carbon and that can thus permit a larger non-fossil carbon emission rate (0.9 Gt C a^{-1}) given the same atmospheric CO_2 record. This model is considered to provide an upper limit to the oceanic uptake.

The inconsistency that exists between the carbon-isotope – model-based analyses and the direct estimates of forest–soil emissions, and the consequent wide range of uncertainty in the current airborne fraction, indicate a continuing need to reduce further the uncertainties associated with (i) direct estimates of forest–soil carbon release, (ii) the size of the oceanic sink and (iii) the numerous other minor sinks for man-made carbon that have been proposed in the face of this failure to balance the carbon budget. This paper addresses the second of these requirements, and examines the way in which oceanic tracers have played a key role in the development of oceanic carbon-uptake models.

2. Factors governing the CO_2 uptake by the oceans

The extent to which the oceans take up excess CO_2 is controlled primarily by three factors: the capacity for CO_2 absorption when the ocean is in equilibrium with the atmosphere; the rate of CO_2 exchange between atmosphere and surface ocean; and the rate at which surface water, having absorbed CO_2, is mixed into the deeper layers.

(*a*) CO_2 *capacity*

Dissolved inorganic carbon exists in the ocean in four forms: dissolved carbon dioxide gas, undissociated carbonic acid, carbonate ion and bicarbonate ion. Ignoring the influence of the borate system, their relative concentrations are governed by the reactions

$$CO_2(aq) + H_2O \rightleftharpoons H_2CO_3(aq), \tag{1}$$

$$H_2CO_3(aq) \rightleftharpoons H^+ + HCO_3^-, \tag{2}$$

$$HCO_3^- \rightleftharpoons H^+ + CO_3^{2-}. \tag{3}$$

It is essentially the presence of the CO_3^{2-} ions, which comprise about 10–15% (by mass) of total dissolved inorganic carbon (ΣC), that give seawater its ability to absorb CO_2 in amounts substantially larger than dissolution alone would provide. CO_3^{2-} ions react with H^+ ions, derived from dissociation of H_2CO_3, to form HCO_3^- ions. This effectively removes H^+ ions and the system responds by dissolving more CO_2. With the present atmosphere–ocean carbon system, although dissolved CO_2 constitutes less than 1% of ΣC, reaction with carbonate ensures that at equilibrium ΣC typically increases by 1% for every 10% increase in dissolved CO_2 concentration or atmospheric partial pressure of CO_2, p_{CO_2}. This ratio of the fractional changes

$$\frac{\delta p_{CO_2}/p_{CO_2}}{\delta \Sigma C/\Sigma C}$$

is known as the Revelle factor, R. The value of 10 adopted is an average value of R, which ranges from about 9 in low latitudes to 14 in high latitudes, depending primarily on temperature. As p_{CO_2} rises, R increases, so gradually reducing the capacity of the oceans to take up CO_2.

Many carbon-cycle models do not attempt to handle the different species of dissolved inorganic carbon and therefore rely on the use of the Revelle factor to simulate correctly the response of ΣC to changes in the atmospheric concentration.

(*b*) *Air–sea exchange*

The surface layer of the ocean influenced by wind stress is generally regarded as well mixed in carbon-cycle models. The rate at which its carbon content responds to a change in atmospheric CO_2 concentration is determined by the rate at which CO_2 can cross the air–sea interface. A review of the many theoretical, laboratory and field techniques that have been applied to evaluating air–sea gas-exchange rates is given by Liss & Merlivat (1986). For most carbon-cycle models developed to date, which are highly aggregated geographically, the most appropriate techniques are those that provide a global mean estimate of gas-exchange rates. One such relies on the distribution between atmosphere and surface ocean of natural ^{14}C. Assuming a steady state before man's perturbation of the system, the radioactive decay of ^{14}C within the oceans must have balanced the excess of input over output at the ocean surface. Thus

$$R_{oc}\, N_{oc}\, \lambda = F(R_a - R_s)\, A_s, \tag{4}$$

where F is the exchange flux, A_s the ocean surface area, R_a, R_s and R_{oc} the relative atmospheric, surface ocean and mean ocean ^{14}C concentrations, N_{oc} the total dissolved carbon content of the ocean and λ the radioactive decay constant for ^{14}C. Setting $R_a = 100\%$, $R_s \approx 95\%$ and $R_{oc} \approx 84\%$ as indicated by results of the GEOSECS programme, Siegenthaler (1986) obtains an exchange rate of 18 mol m^{-2} a^{-1} ($\pm 25\%$) corresponding to a transfer velocity of 17.5 cm h^{-1} for a pre-industrial concentration of 280 p.p.m.v. Given that the upper 70 m of the ocean contain about 150 mol m^{-2}, the residence time of carbon atoms in this layer with respect to transfer to the atmosphere is about 8 years. However, re-establishment of equilibrium between atmosphere and ocean following an input to the atmosphere is achieved in a fraction $1/R$ of the residence time, where R is the Revelle factor. Equilibration between atmosphere and mixed layer is thus realized in less than one year (Broecker *et al.* 1971).

The distribution of natural radiocarbon within the ocean indicates that, on average, surface water is replaced by sub-surface ocean water very much less frequently than this, so the surface ocean can be considered close to equilibrium with the atmosphere. The rate of air–sea exchange is thus not expected to constrain the overall uptake of CO_2 by the oceans. This conclusion may not, however, be valid regionally, a point to be considered later in connection with the development of disaggregated models of the ocean.

(c) *Mixing between the surface and deep ocean*

Whereas the surface mixed layer of the ocean contains roughly the same amount of carbon as the atmosphere, the total ocean content is about 60 times this amount. With a Revelle factor of 10, this indicates that if only the surface ocean had been involved in absorbing man-made CO_2, only 9% would have been taken up. Equilibration with the whole ocean, on the other hand, would result in about 85% uptake. This calculation assumes no dissolution of calcium carbonate sediments, which would alter the alkalinity.

It is clear both from the estimates of airborne fraction discussed in the introduction and from the observed distributions of bomb tracers that the volume of ocean that has been affected by man-made CO_2 is considerably greater than the mixed layer but falls far short of the total volume. The circulation and mixing processes that achieve the transfer of CO_2-laden water into the deeper ocean are not well understood. It is these same processes that will be responsible for transferring heat away from the surface ocean during the course of a climatic warming and that will profoundly influence the rate, and perhaps also the geographical distribution, of consequent climatic changes. It will be at least a decade before three-dimensional ocean circulation models with the required resolution, and the data necessary for their validation, are developed to deal with this task adequately. Lacking such tools, a range of less sophisticated atmosphere–ocean models have been developed during the past three decades to address the carbon-uptake problem. These analogue models have adopted simple representations of the physics of mixing and circulation and have relied heavily on observed tracer distributions for their calibration and validation. Subsequent sections of this paper review the important developments that have taken place.

3. One-dimensional box models

The simplest models (see, for example, Craig 1957; Revelle & Suess 1957; Bolin & Eriksson 1959; Keeling 1973) are those in which the atmosphere, surface ocean and deep ocean are treated as three well-mixed reservoirs (figure 2). Transfer of carbon between reservoirs is by first-order kinetics, i.e. fluxes are proportional to the carbon content of the upstream reservoir. With an input to the atmospheric box from man-made emissions, and with allowance made for the Revelle factor in CO_2 exchange at the surface, a set of ordinary differential equations describes how the carbon in each reservoir changes with time.

The constants of proportionality are evaluated from a consideration of the observed steady-state natural ^{14}C balance. For the atmospheric reservoir, for example,

$$\frac{d}{dt}\,{}^{14}N_a = 0 = -k_{as}\,{}^{14}N_a + k_{sa}\,{}^{14}N_s - \lambda\,{}^{14}N_a + {}^{14}\Gamma_a,$$

where ${}^{14}N_a$, ${}^{14}N_s$ are the ^{14}C contents of the atmosphere and surface ocean, k_{as} and k_{sa} are the atmosphere–surface ocean and surface ocean–atmosphere exchange constants, λ is the ^{14}C

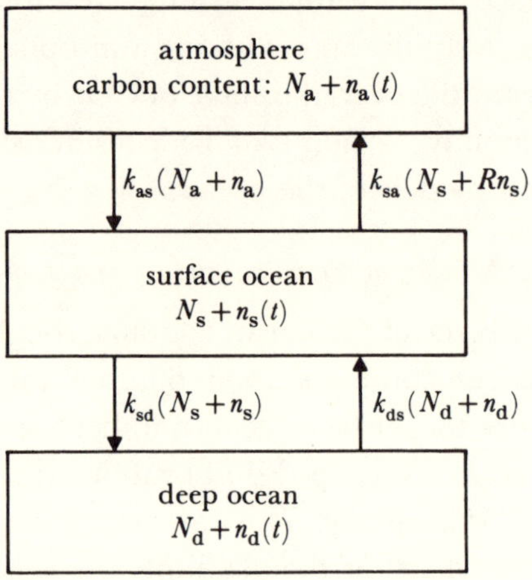

FIGURE 2. Three-box atmosphere–ocean model. N_a, N_s, N_d represent steady-state pre-industrial carbon contents of the atmosphere, surface ocean and deep ocean respectively, and n_a, n_s, n_d man-made departures therefrom. k_{xy} is the first-order exchange coefficient for the carbon flux from box x to box y. R is the Revelle factor.

decay constant and $^{14}\Gamma_a$ the atmospheric ^{14}C source from cosmic rays. Analogous equations for the ocean reservoirs and corresponding equations for the pre-industrial steady-state carbon balance allow all exchange constants to be evaluated.

It is found that if calibrated to match the observed ^{14}C distributions, this model can achieve an airborne fraction of less than 0.60 only if the surface mixed-layer reservoir is more than 500 m deep. For a more realistic depth of 75 m, an airborne fraction of over 0.8 is obtained (Bacastow & Björkström 1981). The assumption that the ocean beneath the surface layer is well mixed prevents the simulation of relatively rapid mixing between the surface ocean and the upper layers of the deep ocean as evidenced by the structure of the ^{14}C distribution.

To simulate this feature, Oeschger *et al.* (1975) developed the box-diffusion model in which the well-mixed deep-ocean reservoir is replaced with a reservoir characterized by vertical diffusion. The diffusion coefficient was chosen to match the mean vertical distribution of ^{14}C. The uptake of CO_2 is considerably enhanced relative to the three-box model but still appears to fall short of that suggested by the atmospheric record. Only a small improvement is obtained by splitting the deep ocean into an 'Atlantic' and 'Pacific' reservoir to reflect the more rapid mixing into the former as revealed in the ^{14}C observations (Crane 1982).

Although early modelling studies had explored the implications of direct CO_2 exchange between the atmosphere and the deep ocean (Craig 1963; Broecker *et al.* 1971), it was the distributions of bomb tracers that graphically illustrated the potential importance for CO_2 uptake of deep convection in high-latitude regions of the Atlantic Ocean (Östlund *et al.* 1974). Subsequent modelling studies have explored various techniques for simulating surface – deep-water exchange (Björkström 1979; Bacastow & Björkström 1981; Killough & Emanuel 1981; Hoffert *et al.* 1981; Crane 1982; Peng *et al.* 1983; Siegenthaler 1983). Generally such modifications to the two-box and box-diffusion models increase their capacity for CO_2 uptake, although the extent depends critically upon the exact structure of the downwelling and return flows adopted (Killough & Emanuel 1981).

The availability of bomb ^{14}C and tritium (^3H) data from the GEOSECS surveys has encouraged recalibration of various versions of the box-diffusion model against these shorter timescale tracers (Broecker *et al.* 1980; Viecelli *et al.* 1981). Whereas the global mean natural ^{14}C profile best matches a vertical diffusivity, K, of 1.25 $cm^2\,s^{-1}$ (Oeschger *et al.* 1975), Broecker *et al.* (1980) calculated $K = 1.7\ cm^2\,s^{-1}$ from ^3H data and $K = 1.9–3.3\ cm^2\,s^{-1}$ from bomb ^{14}C observations. The uncertainty in the latter case arises from difficulties in disentangling bomb-produced and natural ^{14}C in the upper layers of the oceans.

The typical extent of the influence of both the introduction of surface – deep-ocean exchange and recalibration with a bomb tracer is illustrated by Siegenthaler's (1983) outcrop–diffusion model (figure 3), which allows direct exchange between the atmosphere and deep ocean via

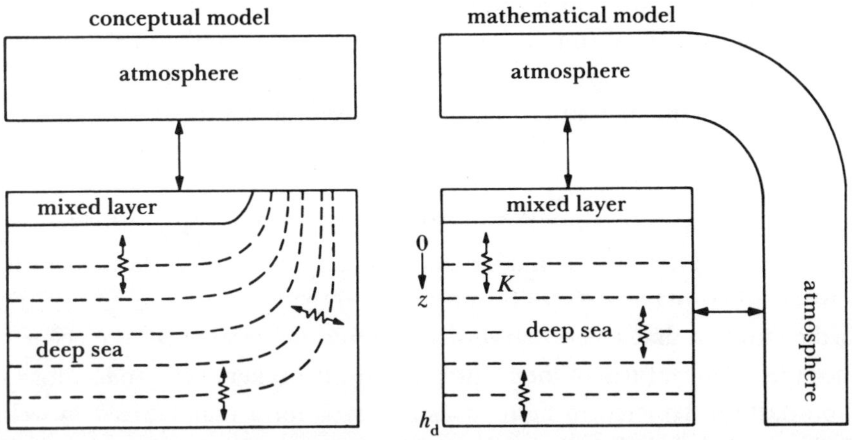

FIGURE 3. The outcrop–diffusion model developed by Siegenthaler (1983). In low latitudes, carbon is exchanged between the atmosphere and a well-mixed surface-ocean layer. At high latitudes, exchange occurs between atmosphere and outcropping isopycnal deep-ocean layers; represented by well-mixed horizontal layers in the mathematical representation. Vertical (cross-isopycnal) eddy diffusion represents the effect of other mixing processes within the ocean. (After Siegenthaler 1983).

outcropping isopycnal surfaces. Calibrated against natural ^{14}C, a change from 0 to 10% outcrop area reduced the airborne fraction, α, from 0.67 to 0.61. When calibrated against bomb ^{14}C, the corresponding figures for α were 0.61 and 0.53 respectively. Not surprisingly, Siegenthaler found that the use of a uniform K value reflecting the overall effects of mixing processes that have operated on the timescale of 10–20 years (bomb ^{14}C) overestimated the rate of mixing into the deep ocean. Generally, low-resolution models calibrated with bomb tracers are unlikely to be appropriate for simulating the long-term response of the oceans to CO_2 input, just as those calibrated against natural ^{14}C appear to underestimate uptake on the short timescale.

To achieve compatibility between short- and long-timescale mixing processes, a much more disaggregated model of the ocean would seem to be required. This in itself, however, poses further potential difficulties with regard to tracer calibration as has been discussed by Peng & Broecker (1985). None of the available tracers provides a perfect analogue for man-made CO_2. Tritium was transferred quickly to the surface ocean in rain and by vapour exchange, so its distribution is strongly influenced by the location of bomb tests. Carbon-14 differs from CO_2 in that it takes approximately 10 times longer for the surface ocean to equilibrate with atmospheric ^{14}C (*ca.* 10 years) than with CO_2 (*ca.* 1 year). Because uptake into the thermocline is thought to occur primarily through mixing along isopycnal surfaces that outcrop at the surface only in winter, very little ^{14}C would enter during the temporary exposure of this water at the surface. However, partial equilibration with atmospheric CO_2 would ensure a CO_2 uptake route not reflected in the bomb ^{14}C distributions. Unlike ^{14}C, ^{85}Kr and chlorofluorocarbons, being non-reactive, equilibrate rapidly between atmosphere and ocean in about one-tenth the time taken for CO_2 equilibration.

The extent to which the differences between the characteristics of CO_2 and these other tracers devalue their individual roles as suitable analogues for calibrating CO_2-uptake models depends in part on the degree to which gas exchange rate is a limiting process. Atlantic circulation models that have recently been developed (Peng & Broecker 1985) (see §4) suggest that mixing rates within the ocean are more likely to limit uptake than air–sea exchange rate. However, these models are not capable of simulating the temporal and spatial variability or the scale of the circulation processes that are likely to control winter-time mixing so resolution of this debate probably awaits the arrival of appropriate ocean general-circulation models.

4. A two-dimensional advection–diffusion model of the ocean

A study of the distribution of oceanic tracers demonstrates that variations in the meridional (depth against latitude) plane are significantly more important than zonal variations. Patterns of tracers further suggest that several large-scale circulation features of the ocean are probably responsible for the main features of their distributions. Broecker *et al.* (1980), for example, used the circulation pattern depicted in figure 4 as the basis for a regionalized upwelling – downwelling-diffusion model. It would seem worthwhile therefore to explore the potential of two-dimensional meridional advection–diffusion models with simple circulation patterns to simulate the uptake of man-made CO_2 emissions. Such a model has been constructed (Crane 1986). The following subsections describe the structure and calibration of the model, and the sensitivity of carbon uptake to key model parameters. Some comparisons are drawn with the model of Siegenthaler (1983).

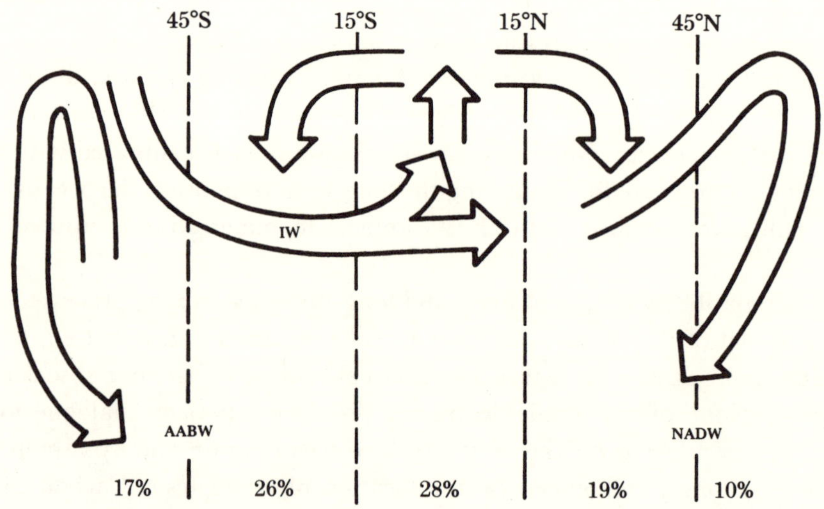

Figure 4. Atlantic circulation system envisaged by Broecker *et al.* (1980) as the basis for a regionalized upwelling–downwelling diffusion model, comprising Antarctic Bottom Water (AABW) and North Atlantic Deep Water (NADW) formation, and northward-penetrating Antarctic Intermediate Water (AIW). Equatorial upwelling feeds water polewards in the upper layers to sink in the subtropical gyres. (After Broecker *et al.* 1980.)

(a) *Model structure*

The two-dimensional advection–diffusion equation

$$\frac{\partial C}{\partial t}+v\frac{\partial C}{\partial y}+w\frac{\partial C}{\partial z}-\frac{\partial}{\partial y}\left[K_y\frac{\partial C}{\partial y}\right]-\frac{\partial}{\partial z}\left[K_z\frac{\partial C}{\partial z}\right]=0$$

is solved by timestepping with an upwind differencing scheme on a finite difference grid. The grid has regular 8° latitude spacing in the horizontal but variable depth spacing ranging from 50 m in the upper ocean (0–200 m), to 200 m at intermediate depths (200–1000 m), to 500 m in the deep ocean (1000–4000 m). The depth is uniform with latitude and the boundaries are vertical. The model is representative of the Atlantic Ocean and the varying width of the ocean is taken into account when water fluxes derived from the specified stream functions are converted into horizontal and vertical velocities. The advective circulation is largely determined by the specification of the stream function for four independent circulation systems: the two sub-tropical gyres, the Antarctic Bottom Water (AABW) system and the Antarctic Intermediate – North Atlantic Deep Water (NADW) system. This idealized circulation pattern was modified during the calibration procedure to permit advective exchange between the rising branches of the AABW and NADW systems in high southern latitudes (figure 5). A simple linear interpolation of stream function to zero at the ocean boundaries and the circulation-system interfaces was adopted. Based on estimates of water mass formation and equatorial upwelling rates (W. S. Broecker & T.-H. Peng, unpublished results) the AABW and NADW circulations were set to $20\times10^6\ m^3\ s^{-1}$ and the subtropical gyres at $5\times10^6\ m^3\ s^{-1}$.

One major difficulty with the model was the choice of diffusivities. Given the uncertainty that exists over the relative role of advective and diffusive mixing processes, token values were adopted initially, and adjusted where necessary during the calibration procedure. For the same reason, no attempt was made to minimize the numerical diffusion generated by the upwind finite differencing by adopting more sophisticated schemes.

(b) *Calibration*

In view of the disaggregated nature of the model of the ocean an attempt was made to calibrate the model against both long- and short-timescale traces. Salinity was chosen as a suitable long-timescale tracer. With timesteps ranging from one thirtieth of a year in the 0–200 m layer to one half of a year in the deep ocean, the model was integrated forward in time until a steady salinity distribution was obtained. Initial conditions comprised observed surface-ocean salinities. Surface salinities were restored to these values after each timestep as a means of representing the salinity sources and sinks at the surface. Salinity was also held fixed in two other regions of the model to represent the highly saline Mediterranean outflow (0–600 m, 28–36° N) and the outflow for the Weddel Sea (200–400 m, 68–76° S). As indicated above, adjustments to the circulation rates centred on the exchange across the rising branches of the AABW and NADW circulations. Figure 6, which compares the modelled and observed salinity distributions, shows that the main features have been reproduced by the model.

The GEOSECS (1972) and TTO (1981) tritium distributions (Östlund *et al.* 1974; Östlund 1983) were used to assess the time-dependent response of the model to short-timescale tracers. The temporal and latitudinal distribution of bomb-tritium input to the surface ocean from 1952 to 1975 was calculated by using the methods outlined by Weiss & Roether (1980) and

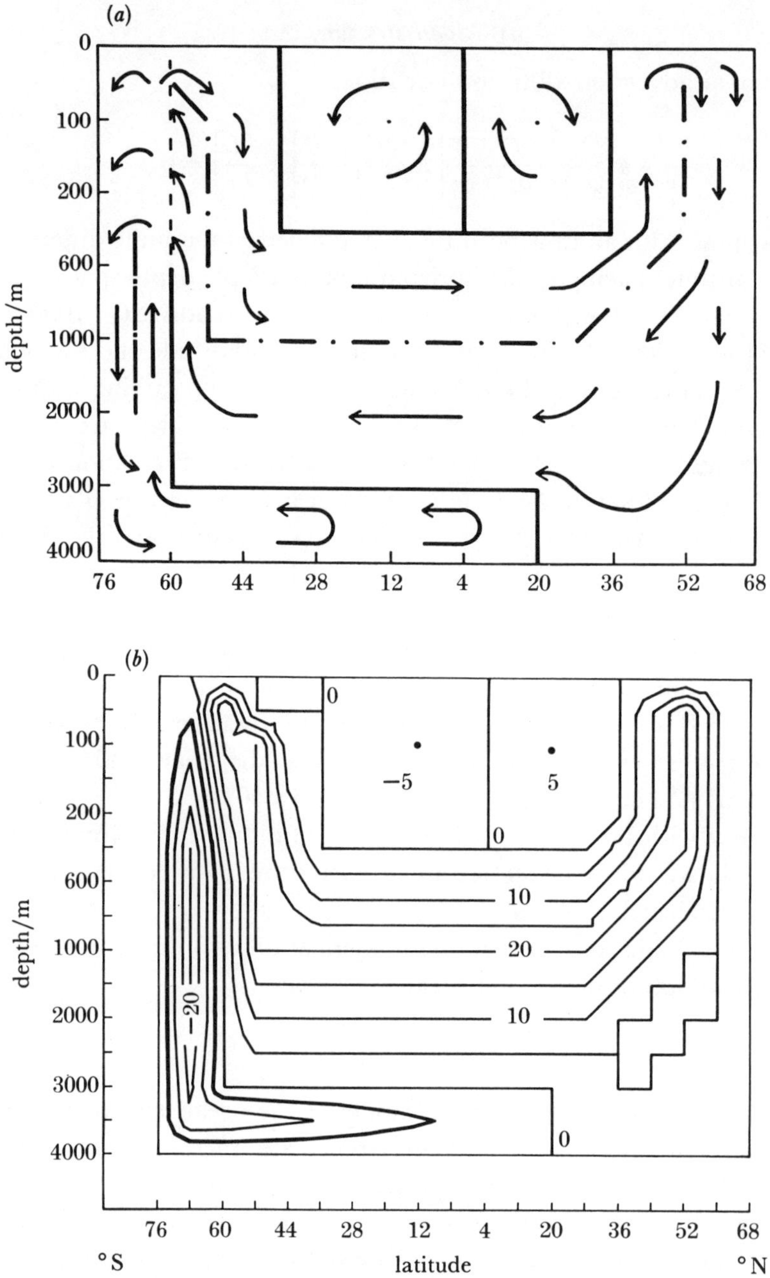

FIGURE 5. (a) Schematic of advective circulation system adopted in the advection–diffusion model of the Atlantic Ocean. (b) Stream function (10^6 m^3 s^{-1}) specified for Atlantic advective circulation depicted in (a). (See text for details.)

used as input to the model into which a tritium-decay term was introduced. In figure 7, a comparison is shown between observations and the modelled tritium distributions in 1972 and 1981, starting from zero concentration in 1952. Although the model contains too many degrees of freedom to ensure that calibration was optimized, the only change in model parameters from those used to simulate salinity that was required to significantly improve the tritium simulations was the enhancement of horizontal diffusivity in northern temperate and polar latitudes at

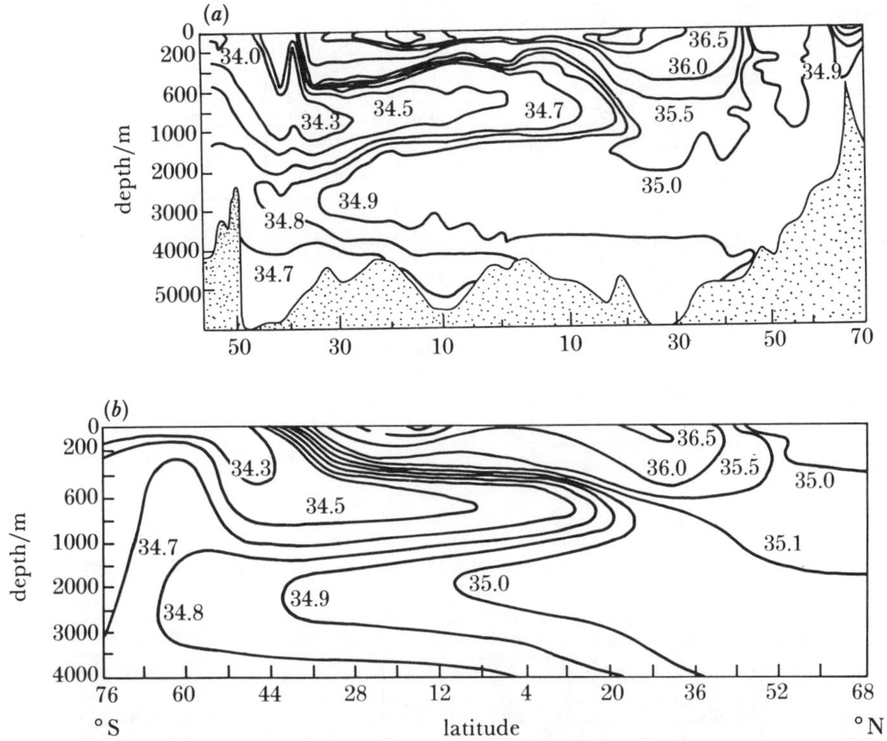

FIGURE 6. Comparison between (*a*) observed (GEOSECS) Atlantic salinity distribution (in units of per mille) and (*b*) modelled salinity in the advective-diffusion model. The latter represents the steady state achieved with a 1200 year integration of the model. Surface salinity is maintained at observed values. Constraints are imposed to represent high-salinity Mediterranean and Weddel Sea outflows. (See text for details.)

intermediate depths. No parameter variation, however, within the bounds dictated by numerical stability considerations was able to simulate the rapid penetration of tritium to the deep ocean in high latitudes.

(*c*) *Response to atmospheric* CO_2 *inputs*

The model was used to simulate the oceanic uptake of man-made atmospheric CO_2 inputs. To represent the world ocean two alternative scalings were adopted. In the first, the surface area of each latitude band was multiplied by 4.5, the ratio of global ocean area to Atlantic area, so the world ocean behaved as a scaled version of the Atlantic. In the second the same factor was applied, but circulation rates in the AABW and NADW systems were then reduced by a factor of 3.3 (the ratio of global to Atlantic Ocean area poleward of 36° latitude) on the basis that these circulations do not exist in the Pacific and Indian Oceans. Of these alternatives the first is likely to overestimate CO_2 uptake on account of too much deep- and bottom-water formation. The second will give rise to a considerable underestimate because it also excludes the Indian and Pacific AIW circulations due to the linked AIW – NADW system in the model.

The airborne fraction, α, over the period 1959–1979 for fossil-fuel CO_2 inputs (1860–1979) given by the model as calibrated was 0.67 from the first method of extrapolation to the world ocean and 0.76 from the second. The sensitivity of airborne fraction to variation in model parameter values was also examined in view of the imprecision of the calibration technique. With circulation rates (i) double and (ii) half the calibration values, α values of 0.60 and 0.73

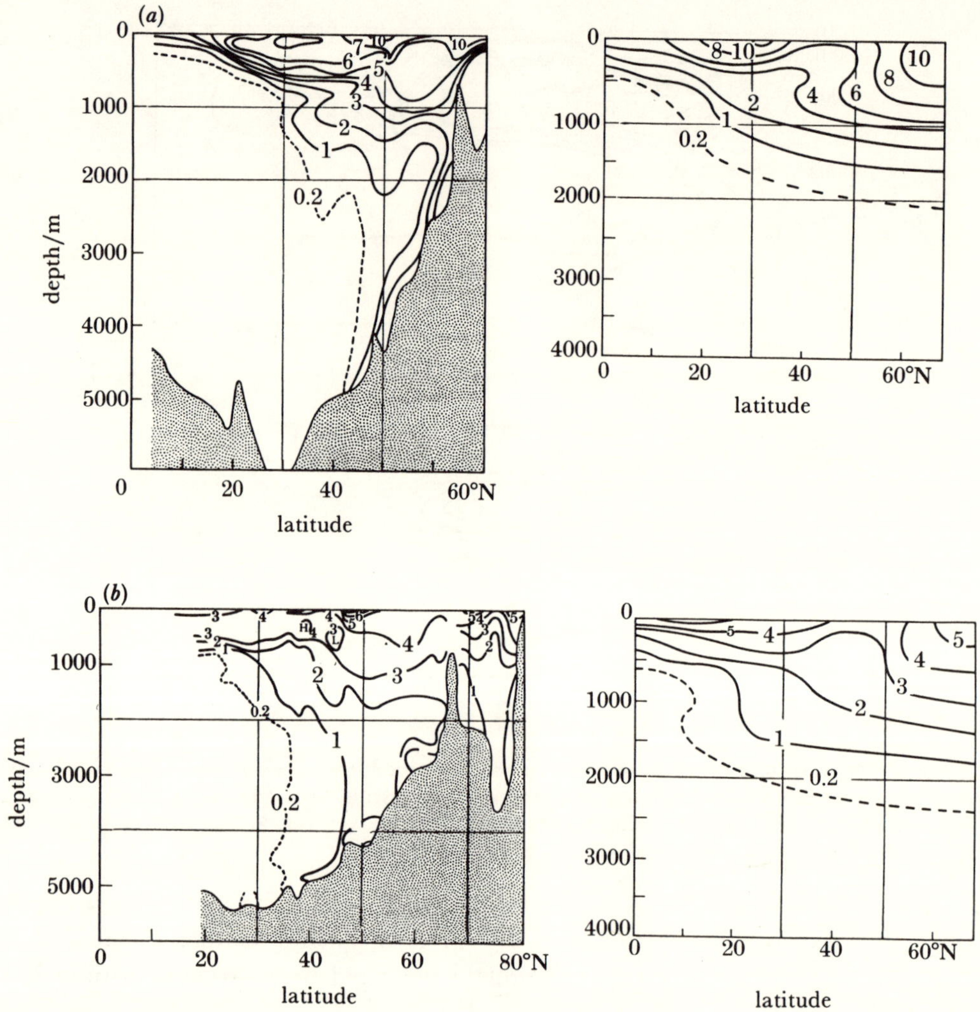

FIGURE 7. (*a*) Comparison between observed (GEOSECS) North Atlantic tritium distribution in 1972 (left) with modelled tritium at 1972 in the advection–diffusion model (tritium units). The latter results from a 21 year integration of the model from the tritium input history 1952–1972 derived by Weiss & Roether (1980). (*b*) As (*a*) but for 1981 (units TU81). Input continues until 1975. (Note the different depth–latitude scaling of the two figures.)

were obtained, by using the first method of extrapolation. Subsequent tritium integrations suggested that the doubled circulation rate continued to give reasonably good agreement with observations, whereas the halved rate led to excessively high surface concentrations.

Airborne fraction was found to be insensitive to realistic variations in the size and latitudinal pattern of air–sea gas exchange rates. An increase at all latitudes from 20 to 30 mol m^{-2} a^{-1} reduced α by only 0.01. It is interesting to relate this feature of the model to the discussion by Siegenthaler (1983) of the importance of air–sea exchange rate in determining the CO_2 uptake of the outcrop–diffusion (O–D) model. The p_{CO_2} of surface water in the outcrop region of the O–D model is far from equilibrium with atmospheric CO_2, with the sea–air flux of CO_2 only 0.1–0.2 times the air–sea flux. This results from the infinitely rapid lateral mixing assumed in the layer of the ocean that outcrops. In the low-latitude region of the O–D model the ratio of the exchange fluxes is about 0.8. Figure 8 shows the latitudinal distribution of the ratio of air–sea to sea–air

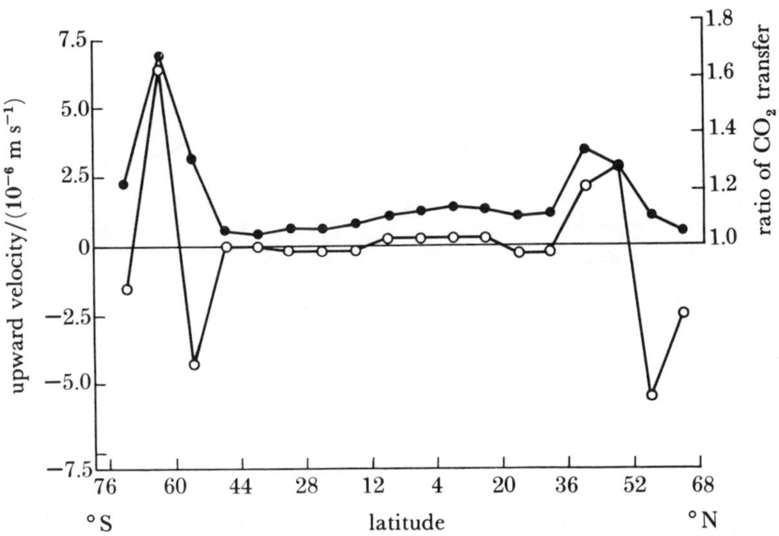

FIGURE 8. Ratio of air–sea to sea–air transfer of carbon (right-hand ordinate, ●) and vertical velocity at 50 m (10^{-6} m s^{-1}) (left-hand ordinate, ○) as a function of latitude in the globally scaled (version 1) Atlantic model. (See text for details.)

transfer obtained in the advection–diffusion model. It is seen that although the net uptake is greater in regions of rapid overturning, the finite rate of mixing into the sub-surface layers restricts the ratio to more modest values. Thus whereas in the O–D model air–sea exchange rate is the limiting factor for high-latitude uptake, it is the rate of circulation that limits uptake in the present model. This would appear to be an important reason why the O–D model generally simulates a rather lower airborne fraction (§3) than the advection–diffusion model.

It is worth reiterating here an important aspect of the CO_2 system which has been noted previously by other authors (Siegenthaler *et al.* 1978; Oeschger & Heimann 1983; Baes & Killough 1985; Siegenthaler 1986; Siegenthaler & Oeschger 1987). The airborne fraction during recent decades not only depends upon the releases of fossil and non-fossil CO_2 releases during this period but also upon the pattern of release in earlier decades. Table 1 illustrates the model response to four CO_2 input histories: fossil-fuel release only and fossil-fuel release combined with the biospheric-release histories calculated by Stuiver *et al.* (1984), Peng *et al.* (1983) (from ^{13}C:^{12}C records in tree rings) and Houghton *et al.* (1983) (from land-use statistics). These three biospheric-release histories are illustrated in figure 9. A large biospheric

TABLE 1. RESPONSE OF ADVECTION–DIFFUSION MODEL TO FOUR CO_2 INPUT FUNCTIONS

(Fossil-fuel emissions (FF) (Clark 1982), and FF plus biospheric release histories estimated by Stuiver *et al.* (1984), Peng *et al.* (1983) and Houghton *et al.* (1983). The Stuiver *et al.* (1984) release was extended from 1970 to 1979 assuming emissions constant at 1970 values. 'Observed airborne fraction' implies (observed CO_2 increase 1959–1979/CO_2 input 1959–1979).)

input function	CO_2 input/(Gt C) 1820–1979	CO_2 input/(Gt C) 1959–1979	airborne fraction 1959–1979 model	airborne fraction 1959–1979 observed	implied 1820 CO_2 level (p.p.m.v.)
FF	160	78	0.67	0.58	292
FF + Stuiver *et al.* (1984)	258	79	0.62	0.58	273
FF + Peng *et al.* (1983)	422	105	0.53	0.43	237
FF + Houghton *et al.* (1983)	357	131	0.61	0.36	254

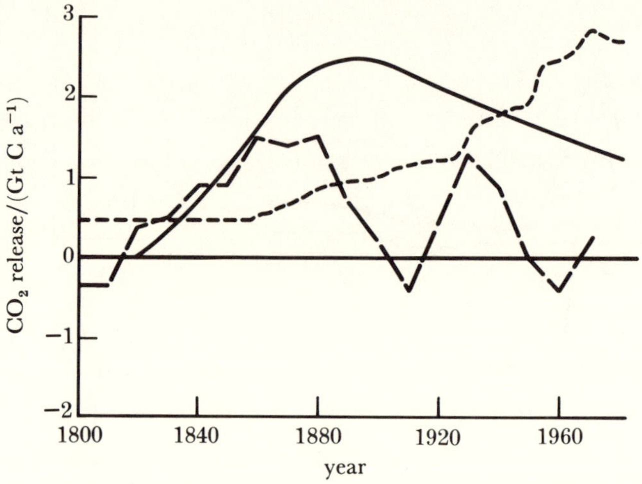

FIGURE 9. Estimates of biospheric CO_2 release to the atmosphere from $^{13}C:^{12}C$ records in tree rings (—, Peng *et al.* 1983; ———, Stuiver *et al.* 1984) and from land-use statistics (-----, Houghton *et al.* 1983).

release during the first half of this century can significantly affect the present CO_2 concentration by providing a decreasing base line while this CO_2 is slowly taken up by the oceans. This is illustrated by the model's response to the release history of Peng *et al.* (1983). Although the total release is the largest of the three, it produces the lowest 1959–1979 model airborne fraction.

It is clear from table 1 that only a biospheric-release history with a moderate pre-1959 release and a near zero post-1959 release, like that of Stuiver *et al.* (1984), allows both a model airborne fraction and a pre-industrial CO_2 concentration roughly consistent with observations. A larger historical release implies too low a pre-industrial CO_2 concentration, whereas significant current biospheric release gives rise to a substantial over estimate of the contemporary airborne fraction.

5. Multi-tracer models and inverse techniques

The advection–diffusion model described in §4 essentially relies upon the use of a single tracer to calibrate an arbitrarily specified circulation pattern. Peng & Broecker (1985) used tritium data to calibrate five alternative 11-reservoir models of Atlantic thermocline mixing. Again, trial-and-error methods were used to determine the flow rates that gave a 'best fit' to the data. Once calibrated, Peng & Broecker (1985) subjected their models to rigorous checks on their ability to reproduce both the temporal and spatial characteristics of the tritium signal, and the time histories, distributions and ratios of other bomb tracers. In this way they were able to judge which of the five alternative mixing patterns gave an overall best fit to all the tracer data employed.

Although multiple tracers were involved in the validation of Peng & Broecker's thermocline models, only one, tritium, was used in calibration. Given the large number of reasonably well-observed oceanic tracers, a more-even allocation between the number of tracers used for calibrating models and the number used for validation would seem to be desirable.

The problem of the simultaneous use of a number of tracers in a three-box ocean model was first studied twenty years ago (Keeling & Bolin 1967, 1968), but was not taken up again until the methodology for multi-reservoir models was developed by Wunsch & Minster (1982) and Bolin *et al.* (1983).

(*a*) *Inverse methodology*

The objective of the inverse method is to find the patterns of advection and turbulent diffusion between model boxes that simultaneously best satisfy the known distributions of a number of tracers. When modelling the carbonate system, the rates of biological production, decomposition and dissolution of carbon, like those of advection and turbulent diffusion, may also be included as unknowns (figure 10). Generally, with M tracers, I ocean boxes and L

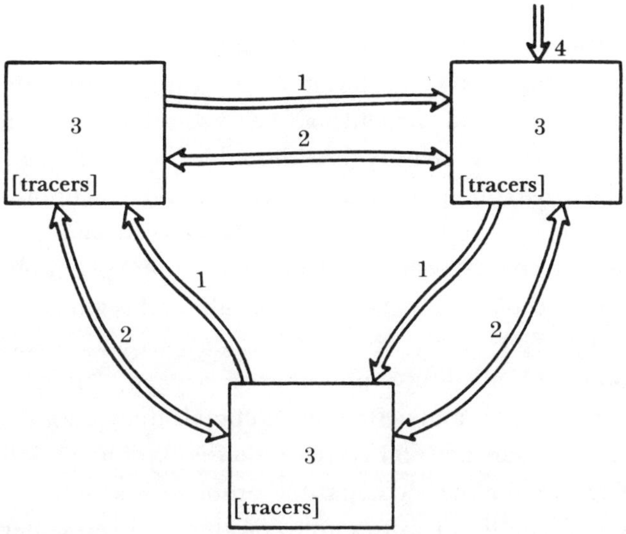

FIGURE 10. The basis of inverse methodology as applied to multi-tracer box models. Arrows represent advective or turbulent diffusive exchanges or exchange at model boundaries. 'Knowns' and 'unknowns' are listed: 1, advective flux across internal boundaries (unknown); 2, turbulent mixing across internal boundaries (unknown); 3, radioactive decay (known) and biochemical processes (unknown) within box; 4, flux across external boundaries (known). (After Moore & Björkström 1986.)

surfaces between boxes, the equation of continuity for each tracer and for water for each box yields a system of $(M+1)\,I$ simultaneous equations in $2L+2I$ unknowns, i.e. velocity and diffusivity at each box interface and source–sink terms for organic carbon and carbonates in each box. In their study, Bolin *et al.* (1983) derive a further $2S$ equations, where S is the number of surface boxes, by requiring that detritus formed in the surface layer decomposes (organic carbon) or dissolves (carbonate) in the layers beneath. In matrix form the problem to be solved is

$$\boldsymbol{Ax} = \boldsymbol{b}$$

where the vector $\boldsymbol{x}$ comprises the $2(L+I)$ unknowns, $\boldsymbol{A}$ is the $r \times s$ matrix of coefficients of the unknowns, where $r = (M+1)\,I+2S$ and $s = 2(L+I)$, and the vector $\boldsymbol{b}$ contains the tracer production and decay terms.

Generally $r \neq s$. If too few tracers are available relative to the number of unknowns $r < s$ and the problem is under-determined or *indeterminate*. If more conservation equations are used than there are unknowns to solve for, $r > s$ and the problem is over-determined or *incompatible*. There is thus no one solution that satisfies all the conservation equations exactly. The methods of solution therefore seek to find a set of values for the unknowns $\boldsymbol{x}$ that either minimize the solution vector $\boldsymbol{x}$ (indeterminate case) or minimize the errors in a least squares sense in the conservation equations $\boldsymbol{Ax} = \boldsymbol{b}$ (incompatible case), subject to any additional constraints that may be imposed, such as the requirement that diffusivities be non-negative. The solution may be influenced by appropriate weightings to stress the role of some tracers relative to others. Water conservation, for example, would be given a high weighting.

Bolin *et al.* (1983) describe how they adapted the solution techniques of Lawson & Hanson (1974) to their particular problem.

(*b*) *The 84-box Atlantic Ocean model*

Bolin *et al.* (1983) tested the inverse methodology on a 12-box world ocean model. When used to simulate the uptake of fossil-fuel CO_2 emissions, the model gave an airborne fraction over the period 1959–1980 of 0.74–0.75. The ^{14}C uptake was likewise smaller then indicated by observation. Although reasonably realistic circulation patterns were derived, they concluded that future studies should explore the benefits of finer resolution by application to a part of the world ocean, that temperature and salinity data should be included in addition to dynamically passive tracers, and that the conservation equations should be supplemented by additional dynamical constraints, such as the requirement of quasi-geostropic flow.

The same group are currently developing an 84-box model of the Atlantic Ocean (Moore & Björkström 1986; Bolin 1986*b*). The ocean is split into 12 regions with up to eight density layers in each region (figure 11). Data from GEOSECS stations within each region define depth profiles for total carbon, ^{14}C, alkalinity, phosphorus, oxygen, salinity and heat. Material is exchanged with the atmosphere and with the Arctic, Indian and Pacific Oceans and the Mediterranean Sea. The system has 696 conservation equations in 540 unknowns so a least squares solution appropriate for the incompatible problem is sought.

A number of realistic circulation features are reproduced: deep-water formation in regions

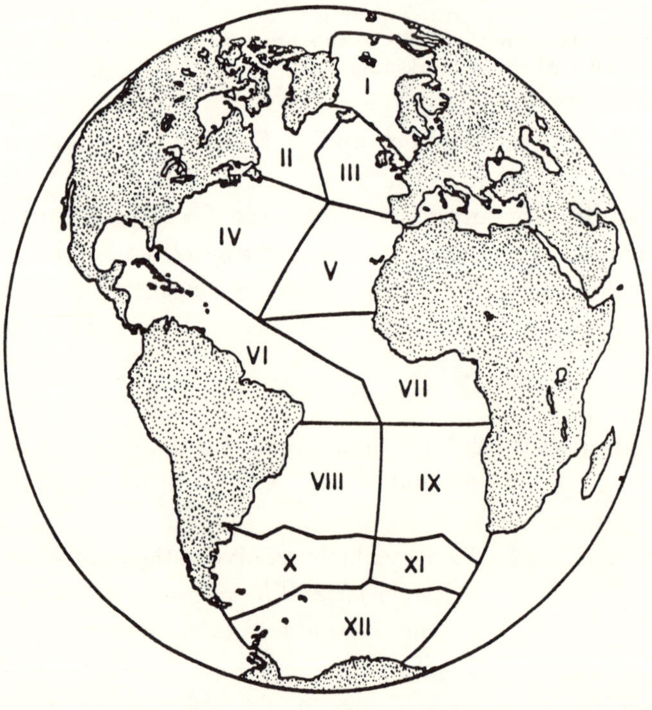

FIGURE 11. The regionalized 84-box model of the Atlantic Ocean of Moore & Björkström (1986). Within each region (depicted by Roman numerals) are up to 8 layers delineated by isopycnal surfaces.

I and III (figure 11), equatorial upwelling, and turbulent exchange between Mediterranean and Atlantic water in regions IV and V. Other features, such as the distribution of biological productivity, are less well reproduced. Preliminary results suggest that the model is more accessible to carbon uptake than the Atlantic component of the 12-box model.

Although the methodology has posed a number of technical problems, there would appear to be potential for further development. Geostrophic-balance equations have been included since the reports published in 1986 were written (A. Björkström, personal communication 1987). Other possible developments include the incorporation of surface wind stress to allow simulation of Ekman pumping. Adoption of even finer resolution may require improved resolution in oceanographic data sets together with a tendency towards under-determined systems of equations, the solution methodologies for which have been less well explored.

6. Climate feedback on the oceans

Within the context of carbon-cycle research, ocean models have so far been primarily devoted to establishing the ability of the oceans to take up fossil-fuel CO_2 and hence determining the airborne fraction of man-made emissions. These models have been used also to simulate future atmospheric CO_2 concentrations from scenarios of future emissions (see, for example, Bacastow & Björkström 1981). As has been shown, the CO_2 uptake by different models varies according to the structure and the rates of mixing processes adopted. The study by Peng & Broecker (1985) nevertheless suggests that, at least for the current timescales of man-induced transients, any models that distribute tritium correctly will take up about the same amount of fossil-fuel CO_2. Moreover, in the context of the overall CO_2 issue, the uncertainty in future concentrations is governed much more strongly by uncertainty in future emissions than uncertainty in airborne fraction. Where, then, does the importance of future ocean-carbon modelling lie?

Recent analyses of ice-core CO_2 data have shown that in the past atmospheric CO_2 concentrations have undergone rapid fluctuations of several tens of parts per million by volume over periods as short as 100 years and have coincided with similar fluctuations in the $^{18}O:^{16}O$ ratio, an indicator of prevailing atmospheric temperature (Oeschger & Stauffer 1986). Any viable explanation would appear to require the redistribution of carbon and nutrients within the ocean, leading to changes in productivity in the surface water. This would alter the dissolved CO_2 concentration in the water and hence the atmospheric CO_2 concentration. Broecker *et al.* (1985) review the various theories that have been proposed on this theme. The fundamental line of reasoning is that the atmospheric CO_2 concentration is strongly affected by the surface-ocean CO_2 partial pressure. This in turn is controlled by photosynthesis, which depletes the carbon content of the upper layers. Photosynthesis is regulated by the supply of nutrients, and this is maintained by the ocean circulation, which recycles nutrients from deep layers where biogenic material decays to surface layers where they are involved in biogenic production.

The possibility of a future CO_2-induced warming operating on the ocean biology in this way has encouraged the development of models incorporating ocean productivity. (Formerly marine biological processes were often excluded on the grounds that they do not represent a significant sink for fossil-fuel CO_2 on account of their not being carbon limited.) Baes & Killough's (1986) two-dimensional model designed to explore the influence of biochemical processes on atmospheric CO_2 levels illustrates the effect of gross changes in the parameters controlling atmospheric CO_2 levels. If in the model biological processes are 'turned off' the atmospheric CO_2 level approximately doubles, whereas a doubling of nutrient concentrations leads to a near halving of atmospheric CO_2 concentration, a result consistent with earlier estimates by Broecker (1982) and Baes (1982). Halved circulation rate reduces atmospheric CO_2 by about 10%, whereas warming raises the CO_2 level by some 4% K^{-1} as expected from the temperature dependence of the distribution of CO_2 between atmosphere and ocean.

In addition to any biologically induced changes in the atmospheric CO_2 level that climate change might cause, direct effects on oceanic CO_2 uptake resulting from circulation changes might also be expected to occur.

It is clear that determining the partitioning of man-made CO_2 emissions between atmosphere and ocean is only the first of a number of increasingly complex and important questions about the atmosphere–ocean system that must be addressed.

Our state of knowledge of ocean circulation, biogeochemical cycling in the oceans, and the likely effects of climatic change on the ocean circulation is at present quite inadequate to address these complex interactions with any predictive capability. Hope of gaining even a qualitative picture of the outcome relies very heavily on continued improvement in modelling capability, which in turn requires the gathering and analysis of the data necessary for their construction and validation. The study of ocean tracers is likely to remain an essential component of this long-term effort.

The work described in §4 and the writing of this paper were carried out at the Central Electricity Research Laboratories and the paper is published with the permission of the Central Electricity Generating Board.

References

Bacastow, R. B. & Björkström, A. 1981 In *Carbon cycle modelling* SCOPE 16 (ed. B. Bolin), pp. 29–80. Chichester: J. Wiley and Sons.

Baes, C. F. 1982 In *Carbon dioxide review 1982* (ed. W. C. Clark), pp. 187–204. Oxford University Press.

Baes, C. F. & Killough, G. G. 1985 A two-dimensional CO_2-ocean model including the biological processes. U.S. Dept. of Energy Report DoE/NBB-0070 TR021.

Baes, C. F. & Killough, G. G. 1986 In *The changing carbon cycle – a global analysis* (ed. J. R. Trabalka & D. E. Reichle), pp. 329–347. New York: Springer-Verlag.

Björkström, A. 1979 In *The global carbon cycle* SCOPE 13 (ed. B. Bolin, E. T. Degens, S. Kempe & P. Ketner), pp. 403–457. Chichester: J. Wiley and Sons.

Bolin, B. 1986*a* In *The greenhouse effect, climatic change and ecosystems* (ed. B. Bolin, B. R. Döös, J. Jäger & R. A. Warrick), pp. 93–155. Chichester: J. Wiley and Sons.

Bolin, B. 1986*b* In *The changing carbon cycle – a global analysis* (ed. J. R. Trabalka & D. E. Reichle), pp. 403–424. New York: Springer-Verlag.

Bolin, B., Björkström, A., Holmen, K. & Moore, B. 1983 *Tellus* **35**B, 206–236.

Bolin, B. & Eriksson, E. 1959 In *Atmosphere and sea in motion* (ed. B. Bolin), pp. 130–142. The Rockerfeller Institute Press.

Broecker, W. S. 1982 *Geochim. cosmochim. Acta* **46**, 1689–1705.

Broecker, W. S., Li, Y.-H. & Peng, T.-H. 1971 In *Impingement of man on the oceans* (ed. W. D. Hood), pp. 287–324. New York: Wiley-Interscience.

Broecker, W. S., Peteet, D. M. & Rind, D. 1985 *Nature, Lond.* **315**, 21–26.

Broecker, W. S., Peng, T.-H. & Engh, R. 1980 *Radiocarbon* **22**, 565–598.

Clark, W. C. (ed.) 1982 *Carbon dioxide review 1982*, pp. 456–460. Oxford University Press.

Craig, H. 1957 *Tellus* **9**, 1–17.

Craig, H. 1963 In *Earth science and meteoritics* (ed. J. Geiss & E. D. Goldberg), pp. 103–144. Amsterdam: North Holland Publishing Company.

Crane, A. J. 1982 *Tellus* **34**, 398–405.

Crane, A. J. 1986 Carbon dioxide uptake by a zonally averaged advection-diffusion model of the ocean. Central Electricity Generating Board Report TPRD/L/2980/R86.

Druffel, E. M. & Benavides, L. M. 1986 *Nature, Lond.* **321**, 58–61.
Hoffert, M. I., Callegari, A. J. & Hsieh, C.-T. 1981 In *Carbon cycle modelling* SCOPE 16 (ed. B. Bolin), pp. 287–306. Chichester: J. Wiley and Sons.
Houghton, R. A., Hobbie, J. E., Mellillo, J. M., Moore, B., Peterson, B. J., Shaver, G. R. & Woodwell, G. M. 1983 *Ecol. Monogr.* **53**, 235–262.
Keeling, C. D. 1973 In *Chemistry of the Lower Atmosphere* (ed. S. I. Rasool), pp. 251–331. New York: Plenum Press.
Keeling, C. D. & Bolin, B. 1967 *Tellus* **19**, 566–581.
Keeling, C. D. & Bolin, B. 1968 *Tellus* **20**, 17–54.
Killough, G. G. & Emanuel, W. E. 1981 *Tellus* **33**, 274–290.
Lawson, C. L. & Hanson, R. J. 1974 *Solving least-squares problems.* New Jersey: Prentice Hall.
Liss, P. S. & Merlivat, L. 1986 In *The role of air–sea exchange in geochemical cycling* (ed. P. Buat-Ménard), NATO ASI Series, vol. 185, pp. 113–127. Dordrecht: D. Reidel Publishing Company.
Moore, B. & Björkström, A. 1986 In *The changing carbon cycle – a global analysis* (ed. J. R. Trabalka & D. E. Reichle), pp. 295–328. New York: Springer-Verlag.
Oeschger, H., Siegenthaler, U., Schotterer, U. & Gugelmann, A. 1975 *Tellus* **27**, 168–192.
Oeschger, H. & Stauffer, B. 1986 In *The changing carbon cycle – a global analysis* (ed. J. R. Trabalka & D. E. Reichle), pp. 89–108. New York: Springer-Verlag.
Östlund, H. G. 1983 Tritium and Radiocarbon: Tritium Laboratory Data Release 83-07, Rosenstiel School of Marine and Atmospheric Sciences, University of Miami.
Östlund, H. G., Dorsey, H. G. & Rooth, C. G. 1974 *Earth planet. Sci. Lett.* **23**, 69–86.
Peng, T.-H. 1985 In *The carbon cycle and atmospheric CO_2: natural variations archean to present* (ed. E. T. Sundquist & W. S. Broecker), Geophys. Monogr. 32, American Geophysical Union, pp. 123–131.
Peng, T.-H. & Broecker, W. S. 1985 *J. geophys. Res.* **90**, 7023–7035.
Peng, T.-H., Broecker, W. S., Freyer, H. D. & Trumbore, S. 1983 *J. geophys. Res.* **88**, 3609–3620.
Revelle, R. & Suess, H. E. 1957 *Tellus* **9**, 18–27.
Rotty, R. M. 1981 In *Carbon cycle modelling* SCOPE 16 (ed. B. Bolin), pp. 121–125. Chichester: J. Wiley and Sons.
Siegenthaler, U. 1983 *J. geophys. Res.* **88**, 3599–3608.
Siegenthaler, U. 1986 In *The role of air–sea exchange in geochemical cycling* (ed. P. Buat-Ménard), NATO ASI Series, vol. 185, pp. 209–247. Dordrecht: D. Reidel Publishing Company.
Siegenthaler, U. & Oeschger, H. 1987 *Tellus* **39**B, 140–154.
Stuiver, M., Burk, R. L. & Quay, P. D. 1984 *J. geophys. Res.* **89**, 11731–11748.
Viecelli, J. A., Ellsaesser, H. W. & Burt, J. E. 1981 *Climatic Change* **3**, 281–301.
Weiss, W. & Roether, W. 1980 *Earth planet. Sci. Lett.* **49**, 435–446.
Wunsch, C. & Minster, J.-F. 1982 *J. geophys. Res.* **87**, 5647–5662.

Discussion

W. Roether (*University of Bremen, F.R.G.*). Dr Crane's two-dimensional circulation model assumed a time-averaged circulation. However, both deep-water formation and surface-water p_{CO_2} are seasonally variable to a large degree. Would Dr Crane expect this fact to affect his rates appreciably?

A. J. Crane. The steady-state cycling of CO_2 between atmosphere and ocean is not treated in the model. The transfers of CO_2 from the atmosphere to the ocean at all latitudes must be interpreted as the departures from steady state arising from increased atmospheric CO_2 levels, namely enhancements in the air–sea fluxes and reductions in the sea–air fluxes when and wherever they occur in the steady state. The influence that seasonality has on the steady-state cycling of CO_2 between atmosphere and ocean is thus not of direct concern. However, as was mentioned in the paper, there is a problem over the extent to which tritium and other tracers are good analogues for CO_2. Because of the differences in surface equilibration rates between CO_2 and other tracers, seasonal variations in p_{CO_2} and rates of surface-water overturning could ensure rates of CO_2 transfer quite different from those estimated on the basis of tracer calibration, depending on the relative rates of uptake and mixing.

P. D. KILLWORTH (*Hooke Institute, University of Oxford, U.K.*). With regard to the part of Dr Crane's talk about box models, even an under-determined set of boxes (merely conserving tracer, say, with no dynamics) may not be able to produce a physically meaningful solution (for example, positive fluxes and diffusions) if the constraints are poor. For example, an interior extremum of tracer cannot be so modelled. Thus we need to choose the boxes with as much physics as possible, and as accurate data as possible.

A. J. CRANE. Dr Killworth is right to emphasize this aspect. The early results from inverse modelling show clearly that physical and dynamical constraints are needed if physically meaningful solutions are to be obtained and if critical oceanic circulation features are to be captured. I think it is generally recognized that progress in this field relies on combining the principles of geophysical fluid dynamics and the properties of tracer distributions, and developing the techniques for solving the increasingly complex models that result.

The use of anthropogenic tritium and helium-3 to study subtropical gyre ventilation and circulation

By W. J. Jenkins

Department of Chemistry, Woods Hole Oceanographic Institution, Woods Hole, Massachusetts 02543, *U.S.A.*

With tritium and helium-3 (^{3}He) data from the Transient Tracers in the Ocean (TTO) expedition and from two other contemporaneous cruises, a synoptic picture of the ventilation and circulation of the subtropical North Atlantic is built. We will see clear evidence of gyre circulation in the tritium–^{3}He age distributions on the shallower isopycnals, permitting estimates of the rates of circulation averaged over timescales from months to decades. The entry points of fluid into the main thermocline and the pathways of exchange with the upper ocean on seasonal to decade timescales appear clearly. It is the time-averaged transport processes on those timescales that are important to the uptake of carbon dioxide by the ocean. The overall relation between tritium and ^{3}He within the subtropics exhibits a systematic hook-like pattern that is consistent with 'strong' ventilation of the gyre thermocline; a fluid parcel entering the gyre thermocline making only about one circuit around the gyre before being ventilated.

Finally, we present a time-series of ^{3}He measurements made over a period of two years near Bermuda. The mixed layer is demonstrably supersaturated in this isotope throughout a large part of the year, requiring a gas-exchange flux of this isotope to the atmosphere. Model results are presented that permit the calculation of the *in situ* solubility isotope ratio anomaly for helium (as affected by bubble injection and gas exchange), and that can be used to estimate the upward flux of this isotope. Because only a small fraction of this flux can be produced in the mixed layer, this helium must be 'mined' from the main thermocline. The computed flux is consistent with the long-term evolution of the inventories of tritium and ^{3}He within the main thermocline. This flux has implications regarding the vertical transport of material within and from the permanent thermocline. A single observation of what may be one of the processes responsible for this upward flux is discussed.

1. On the use of the tritium–^{3}He age

The detonation of nuclear weapons in the atmosphere more than two decades ago produced several hundred kilograms of the radioisotope tritium. By observing the passage of this isotope into and through the ocean, we have an opportunity to characterize the various mechanisms whereby the ocean transports substances. Such information permits us to make inferences about the rates of physical, chemical and biological processes in marine systems. Unfortunately, application of the above strategy is limited by the quantity of data available (both in space and time) that documents the passage of this isotope, and in particular by imperfect knowledge of the precise manner in which tritium has entered the ocean. By simultaneously determining its stable, inert daughter product ^{3}He, however, additional information is gained. This is evident from the distributions of these isotopes in the upper main thermocline of the subtropical North Atlantic, where significant gradients of ^{3}He are observed where the tritium distribution is relatively homogeneous. The timescale of significant radiogenic ^{3}He accumulation is short relative to the recirculation timescale of the upper waters of the gyre, which in turn is short relative to the time since the tritium transient.

A useful concept in discussing the ^{3}He distributions is the tritium–^{3}He age τ, defined as

$$\tau = 17.96 \ln (\xi/\theta), \tag{1}$$

where τ is in years, θ is the tritium concentration, and ξ is the sum of the tritium and the excess ^{3}He concentrations. In a sense, this 'age' may be regarded as the elapsed time since gas equilibration with the atmosphere. It represents the time information implicit in the two tracers while at the same time largely eliminating the dependence on 'initial conditions'. Timescales in the range of 0.1–10 years can be measured.

This simple concept, however, may be complicated by mixing. To quantify the effects of mixing, one can derive an advection–diffusion equation for τ. In the case of two-dimensional isopycnic flow with constant, isotropic turbulent diffusion, one can combine the advection–diffusion equations for tritium and ^{3}He to obtain

$$\tau_t = \kappa \nabla^2 \tau - \boldsymbol{u} \cdot \nabla \tau + 1 - \tilde{v} \cdot \nabla \tau, \tag{2}$$

where κ is the isopycnal diffusivity, $\boldsymbol{u}$ is the isopycnal velocity, and ∇ is the isopycnal gradient operator (Jenkins 1987). The subscript t implies partial differentiation with respect to time; $\tilde{v}$ is a 'mixing pseudo-velocity' arising from mixing of tritium and ^{3}He, defined as

$$\tilde{v} = \kappa \nabla \ln (\xi \theta).$$

One can readily identify the important terms in (2) by comparison with the chronological term (1) associated with the increase of age with time. The magnitude of the mixing terms, on timescales less than a decade, and on surfaces which outcrop to the south of the line of zero windstress curl ($\sigma_\theta \leqslant 27.1$), is of order 10% or less for the eastern subtropical Atlantic (Jenkins 1987). On longer timescales, and where mixing may dominate over advection, the age becomes a much more model-dependent quantity. In general, because of the sense of the tracer gradients, τ underestimates the 'true age'. The sense of this age compression can be seen in figure 1, which is a plot of τ against the advective timescale, $T_a = X/u$ for a variety of one-dimensional advective–diffusive flow simulations. The model was forced by using the theoretical surface tritium time history (from Dreisigacker & Roether 1978), and scaled with the radiotracer Peclet number

$$Pe = u^2/\kappa\lambda, \tag{3}$$

(after Jenkins 1982) where the length scale is given by u/λ.

Typical values of κ and u for the upper main thermocline within the gyre interior are $\kappa = 10^3\ \mathrm{m^2\ s^{-1}}$ and $u = 10^{-2}\ \mathrm{m\ s^{-1}}$, respectively (see, for example, Armi & Stommel 1983; Jenkins 1987), which with a tritium decay probability $\lambda = 1.77 \times 10^{-9}\ \mathrm{s^{-1}}$ lead to $Pe \simeq 50$. Thus one would conclude for areas away from the turbulent boundary regions, that τ is accurate to order 10% for timescales less than a decade. For timescales of a few decades, the bias is of order 20–30%. Moreover, the slope of the relation in figure 1 diverges from the isoline slope in the region of the advancing tritium front (where $\tilde{v}$ is largest). Thus, in frontal regions and where mixing is more important, the interpretation is more complex.

The immediate difficulty in assessing the impact of mixing on τ in such a comparison lies in the choice of the advective age as the correct measure of the timescale. The concept of such a

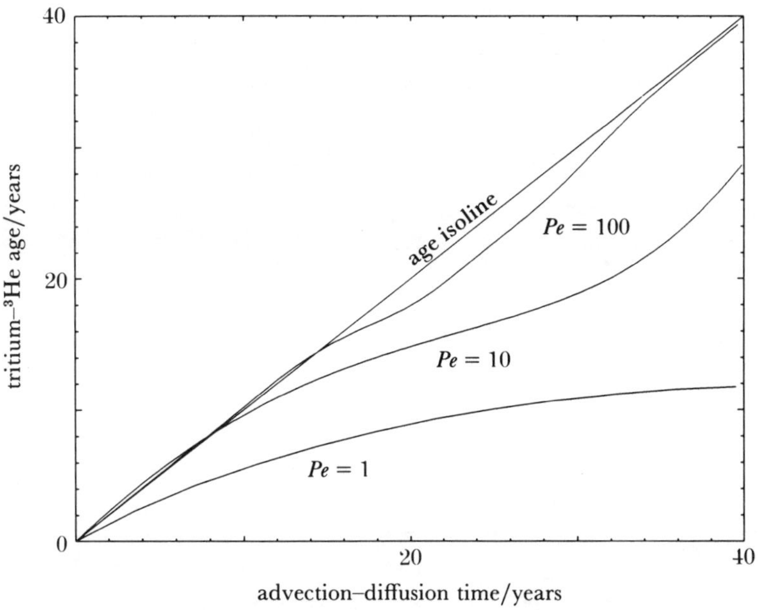

FIGURE 1. Comparison of tritium–^{3}He ages with advection timescale for a one-dimensional model with constant velocity and diffusion (see text). Three values of the Peclet number, $Pe = u^2/\kappa\lambda$, are compared to the ideal (equal age) isoline.

'box-car' time is questionable in the presence of diffusion, and a more appropriate measure may be $T = (1/T_d + 1/T_a)^{-1}$, where $T_d = X^2/\kappa$ is the diffusive timescale. This consideration is necessarily crude, but suggests that part of the apparent disparity may lie more in the concept of the physical timescale rather than the nonlinearities in τ. It is apparent, however, that τ is a model-dependent quantity and hence must be used with caution. It is nevertheless a useful quantity for gaining qualitative insight into gyre ventilation and circulation.

2. Isopycnic maps of tritium and the tritium–^{3}He age

The Transient Tracers in the Ocean (TTO) expedition, in combination with some other cruises occurring within the same time frame (1979–81), has provided the first opportunity to map the distribution of the tritium–^{3}He age (τ) on isopycnic horizons over significant portions of an ocean basin. Included here are four maps (figures 2–5) of tritium and τ on representative density horizons. The tritium and ^{3}He data were linearly interpolated onto the appropriate horizons only when sample points existed within a specified density range (i.e. ± 0.125 for $\sigma_\theta = 26.50$ and 26.75, ± 0.100 for $\sigma_\theta = 27.40$, and ± 0.025 for $\sigma_2 = 37.05$). The tritium–^{3}He age was computed from the interpolated data. The tritium data were normalized to a standard date of 1 January 1981, but the tritium–^{3}He age data were not adjusted, on the assumption that the tritium–^{3}He age is in approximately steady state. This is justified for the shallower surfaces where the advective timescale is much less than the time elapsed since the tritium transient (i.e. where tritium is relatively uniform), but is suspect on deeper horizons.

The $\sigma_\theta = 26.50$ surface

Although the tritium distribution (figure 2*a*) on this surface is relatively homogeneous (it varies by only 30% over the entire gyre), the tritium–helium age shows substantial variation.

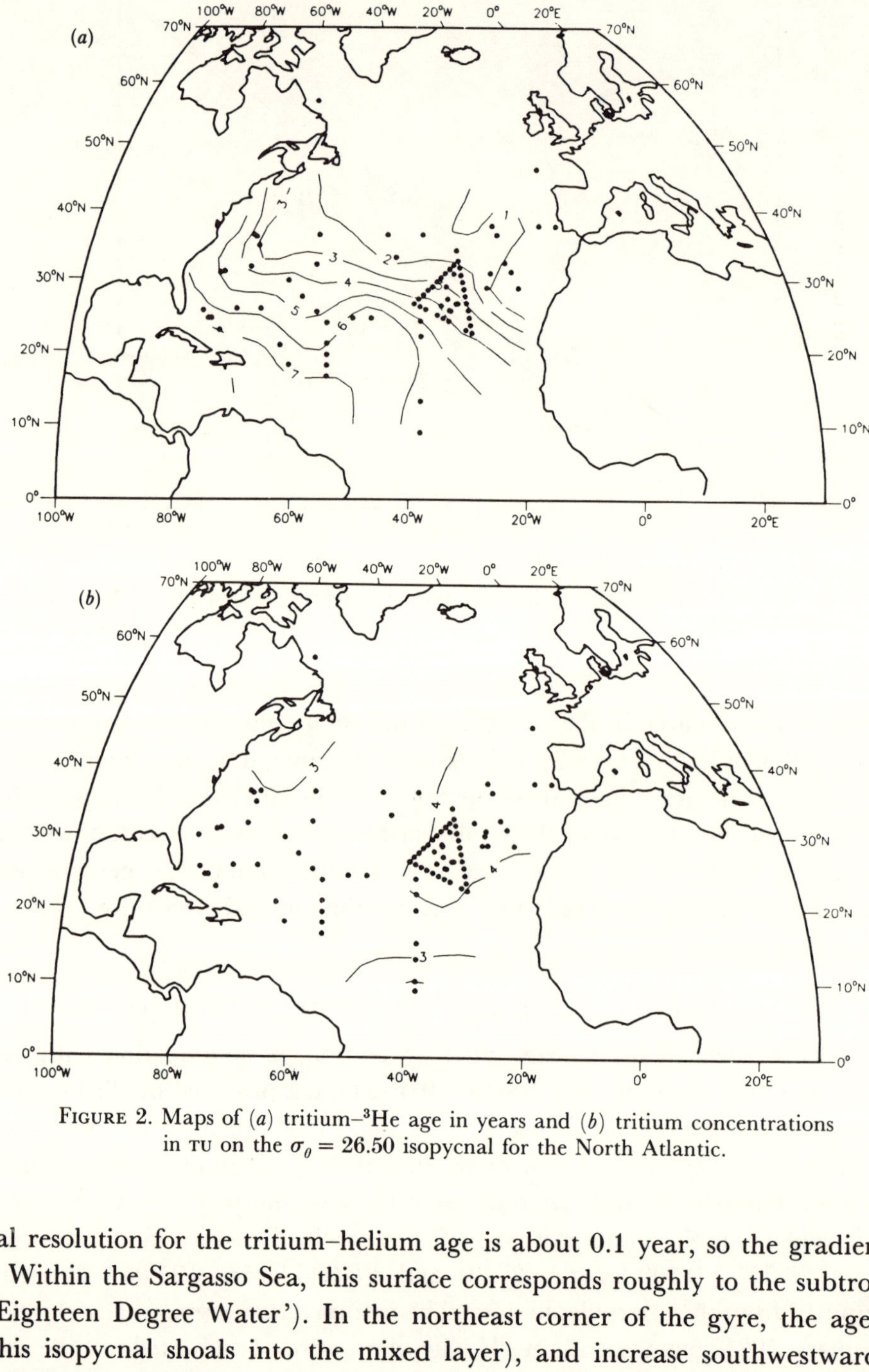

FIGURE 2. Maps of (*a*) tritium–^{3}He age in years and (*b*) tritium concentrations in TU on the $\sigma_\theta = 26.50$ isopycnal for the North Atlantic.

Analytical resolution for the tritium–helium age is about 0.1 year, so the gradients are well resolved. Within the Sargasso Sea, this surface corresponds roughly to the subtropical mode water ('Eighteen Degree Water'). In the northeast corner of the gyre, the ages are small (where this isopycnal shoals into the mixed layer), and increase southwestward at a rate equivalent to a velocity of 1.0 cm s^{-1}. The direction and magnitude of this velocity are close to the β-spiral velocities obtained by Armi & Stommel (1983) for this horizon (cf. Jenkins 1987).

What is more interesting, however, is the larger-scale shape of the isochrons, and what they imply about pathways of ventilation into the interior. The entry of young (ventilated) fluid in the northeast corner of the gyre is a phenomenon readily predicted on the basis of Ekman pumping (see, for example, Leetma & Bunker 1978) and the pattern of wintertime outcropping of the isopycnal (see, for example, Levitus 1982). The overall shape of the isochrons in fact bears some resemblance to the outcrops (cf. Jenkins 1987, figure 1). Specifically, note the general southeastward trend in the central and eastern half of the gyre, and the southwestward bowing in the northwestern region of 'tight recirculation' (the parlance of Worthington 1976; see especially p. 93, figure 41). In the extreme northwest corner of the Sargasso Sea, the 3 and 4 year isochrons turn sharply northward, a sign of the northward flow of older waters in the Gulf Stream. In this region, the isopycnal is at 300–400 m depth. Along the northern rim of the gyre (east of perhaps 75° W), however, the tritium–^{3}He age decreases downstream. This is particularly remarkable considering that the isopycnal is deeper than 500 m in this region. It is possible within the resolution of the station spacing that this decrease is a result of the tight recirculation of waters ventilated further downstream. In any case, considering the large eastward transports (Worthington 1976), one must conclude that ^{3}He is effectively removed, and hence by inference, that all other surface forced properties are reset. This is clear evidence of the strong ventilation proposed by Sarmiento (1983) and by Jenkins (1987).

In the southern part of the gyre, the 6 and 7 year isochrons exhibit a interesting structure. There is evidence of the entry of younger water to the south and east (around 10° N, 40° W), despite the lower tritium concentration. This is the invasion into the gyre of the northward-flowing relatively tritium-impoverished equatorial–southern waters. The confluence of this equatorial water with the broader-scale gyre recirculation component near the Antilles further bends these isochrons southward, causing them to run parallel to the South American coast. It is clear that the simple interpretation of a flow field normal to the isochrons can be misleading in these circumstances, and that mixing must be considered.

The $\sigma_\theta = 26.75$ surface

This isopycnal is close to the one studied by Broecker & Östlund (1979), and lies within the main thermocline at a temperature of approximately 15 °C. During the TTO expedition, it outcropped near 60° N, and deepened southward and westward, reaching a maximum depth of over 700 m in the Sargasso Sea. Although tritium is still more-or-less uniform within the subtropical gyre, the 3 tritium units (TU) isopleth has crept northward by several hundred kilometres on the southern boundary and along the South American coast, and further southward along the North American coast. This is a reflection of the longer ventilation and circulation timescales of this horizon relative to those on the 26.50 isopycnal.

The zero-age region (figure 3) appears in the Irminger Sea beginning a westward march toward the Labrador Sea as one progresses to denser isopycnals. As on the $\sigma_\theta = 26.50$ horizon, the age increases smoothly southwestward in the eastern part of the gyre at a rate and direction in agreement with β-spiral velocities. The tongue-like extension of the 8 year isochron (near 20° N, 40° W) is suggestive of the upper thermocline circulation scheme suggested by Worthington (1976; p. 75, figure 29). Again as on the 26.50 surface, there is evidence of strong 'rim ventilation', seen in the downstream decrease in the tritium–helium age. Along the north-western edge of the Sargasso Sea, the isochrons again show a northerly swing: the slope waters are in fact older than the Sargasso Sea waters on the same horizon. This suggests that on the shallow surfaces, significant ventilation does not occur across the Gulf Stream through the slope waters. This is consistent with float observations and other tracer gradients (Bower *et al.* 1985). However, as on the 26.50 surface, the downstream decrease of the tritium–^{3}He age requires that strong ventilation must occur, perhaps associated with the tight recirculation.

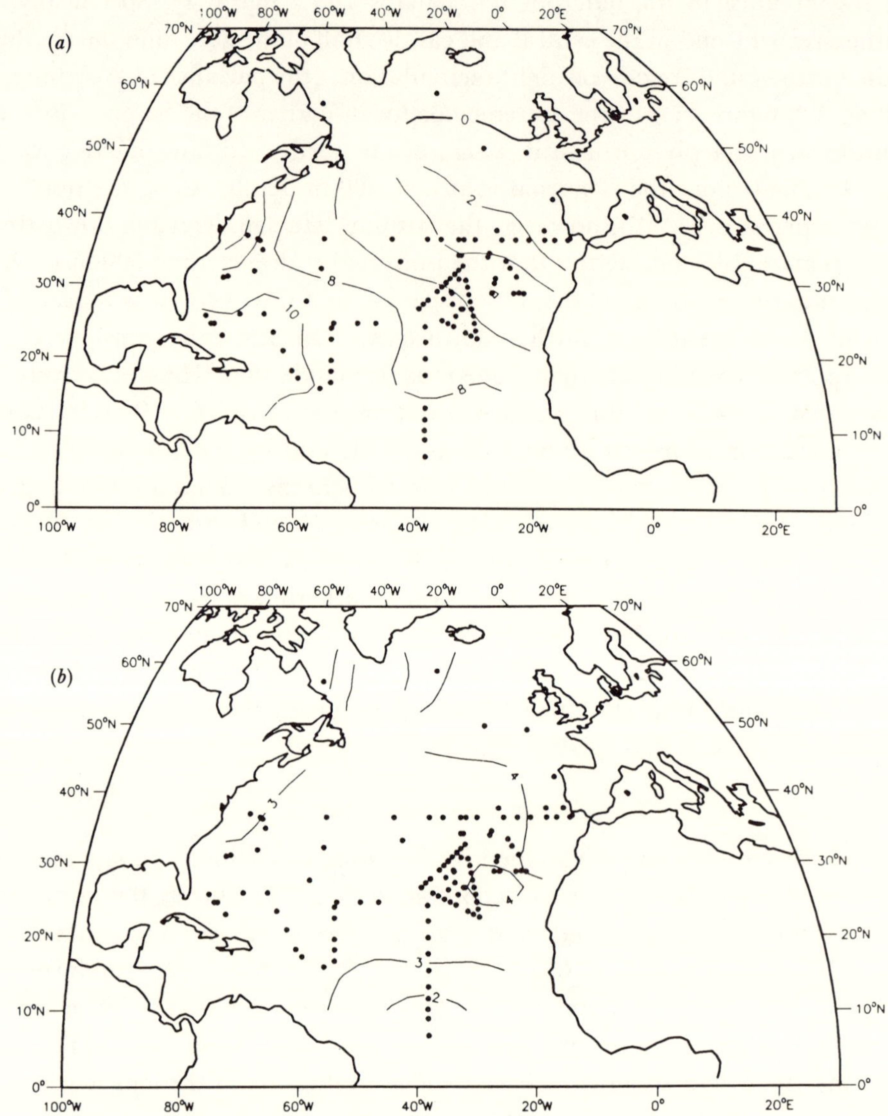

FIGURE 3. Maps of (a) tritium–^{3}He age in years and (b) tritium concentrations in TU on the $\sigma_\theta = 26.75$ isopycnal for the North Atlantic.

The $\sigma_\theta = 27.40$ *surface*

Unlike the previously discussed horizons, this isopycnal (figure 4) outcrops in the wintertime to the north of the line of zero windstress curl (see, for example, Jenkins 1987, figure 1). The zero-age region, and by inference the fluid-entry point, has rotated westward. From a minimum depth and age within the Labrador Sea the layer deepens to 1000 m within the subtropics where it lies generally within the main oxygen minimum. Tritium concentrations are very high near the Labrador and Newfoundland coasts, and decrease away from the Labrador Sea and southward. The 10 year isochron now lies entirely north of the subtropics, whereas the isochrons between 20° and 40° N are zonally aligned. Tritium gradients are similarly aligned within this region, and of lesser concentration than the horizons above. The

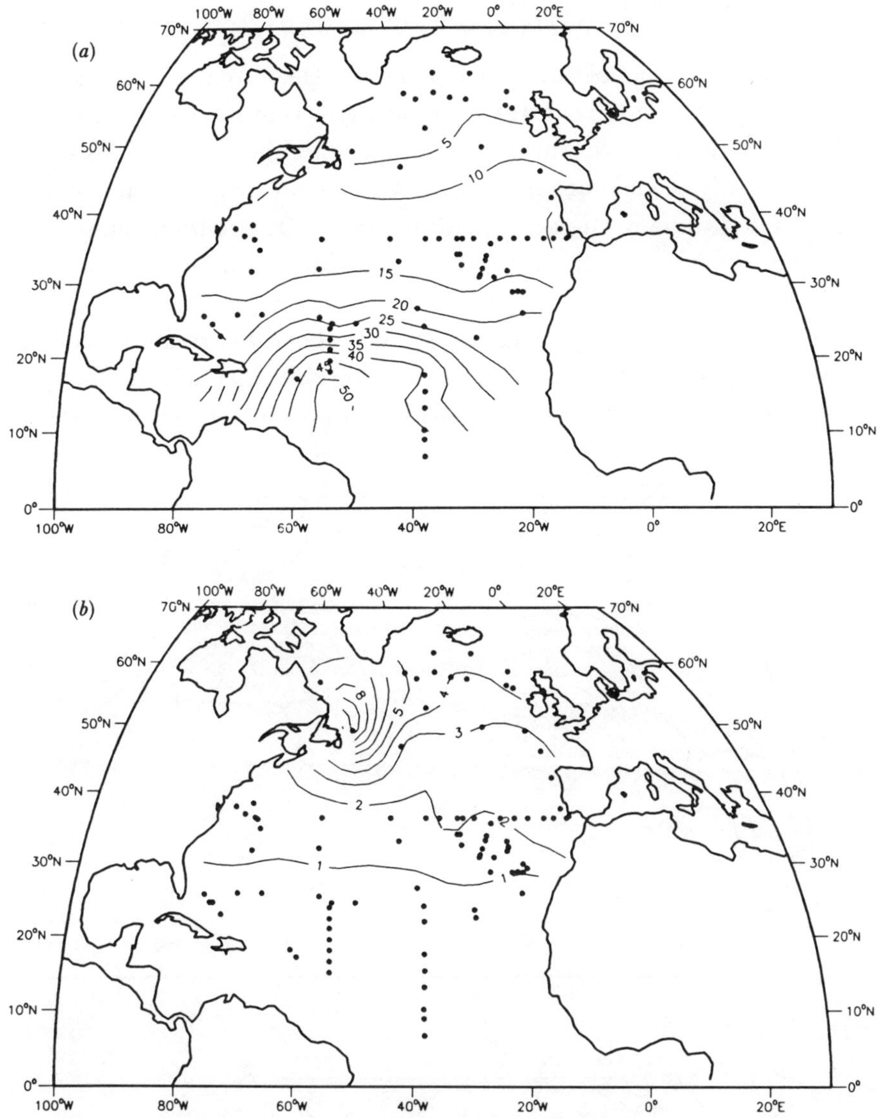

FIGURE 4. Maps of (*a*) tritium–^{3}He age in years and (*b*) tritium concentrations in TU on the $\sigma_\theta = 27.40$ isopycnal for the North Atlantic.

greater age, lower tritium concentrations and lack of significant gyre-like imprint on the isochrons is a reflection of the sluggish ventilation and circulation on this horizon.

To the south, evidence of entry of older water can be seen along the South American coast, both in the isochrons and in the presence of high-silica waters. This is probably Antarctic Intermediate Water. Some of the apparent age may be because of the transport of primordial ^{3}He with this water. From the observed ^{3}He–silicate relation for these waters (Jenkins & Clarke 1976), and the distribution of silicate, it is possible to correct in large part for this influence. This adjustment has been delayed pending compilation of the tritium and ^{3}He data from the TTO–TAS (Tropical Atlantic Study). Clearly, the tritium–^{3}He age in this region should not be interpreted literally, but rather within the framework of model calculations.

The $\sigma_2 = 37.05$ *surface*

This surface lies at the top of the North Atlantic Deep Water complex, at a depth of approximately 3000 m within the subtropics. One sees a rotation from the essentially zonally oriented isochrons of the shallower surfaces to a quasi-meridional slant. This is evident in the 10 year isochron, which extends down from the Greenland Sea and follows the east coast of Greenland (the present data are not adequate to resolve its circumnavigation of Cape Farewell). The 20 year isochron penetrates far down the North American coast, and the 30 year isochron parallels the South American coast, indicating the southward penetration of young water along the western boundary. This surface lies above the boundary current strikingly demarcated by tritium (cf. Jenkins & Rhines 1980; Olson *et al.* 1986), but the tritium isopleths in the figure are strongly suggestive of southward transport along the edge. In the light of the

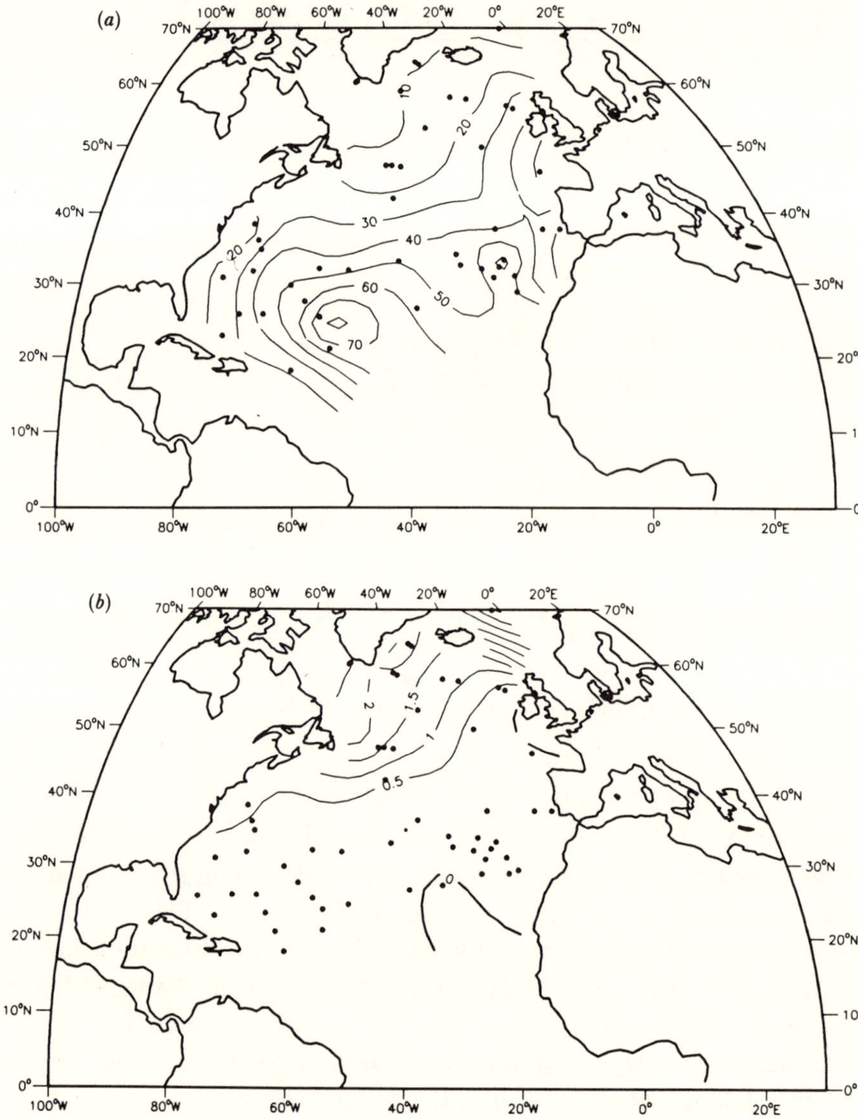

FIGURE 5. Maps of (*a*) tritium–^{3}He age in years and (*b*) tritium concentrations in TU on the $\sigma_2 = 37.05$ isopycnal for the North Atlantic.

large degree of dilution which occurs during the southward journey (Jenkins & Rhines 1980), the ages seen here are not 'box-car' in nature. Current meter measurements suggest smaller transit times. Rather, the age represents some complex mixing age resulting from the ablation and entrainment of interior abyssal waters along the pathway southward. The zonal gradients thus represent the signature of a process of ventilation of the deep interior by shedding of young water from the western boundary. There is a striking similarity in the general shape of the isochrons and the silica isopleths. The large-scale intrusion of apparently older, silica-rich water near 25° N, 50° W needs to be examined more carefully. In particular, one needs to correct for the presence of primordial ^{3}He that has been advected up the South Atlantic from the circumpolar regions.

Summary

The large-scale distributions of tritium and tritium–^{3}He age on a few representative density horizons within the North Atlantic have been presented, and begin to indicate the nature and pathways of ventilation. These distributions are consistent with what we know about gyre circulation, but present us with evidence of very strong ventilation occurring in the north-western segment of the subtropical gyre, as well as the previously supposed entry point of fluid in the northeastern section. We can see the westward march of the outcrop entry point from the Bay of Biscay for the shallowest surface, to the Labrador Sea, and then northward to the Greenland Sea for the deepest horizon. Interior gradients of the tritium–^{3}He age appear to be an important diagnostic feature of ventilation processes in the North Atlantic basin.

3. The tritium–^{3}He scatter plot

Some information can be extracted about the general character of the gyre circulation by examining the overall relation between tritium and ^{3}He for the entire data set. Figure 6*a* is a plot of nearly 2000 data points from the combined TTO–NAS and NATS data sets for the gyre region. The β-triangle data were not included because they were not sufficiently contemporaneous; although the tritium can be normalized for decay, it is not entirely clear how to correct the ^{3}He for the two-year offset between the cruises.

In general, the data fall along a characteristic hook-shaped curve, with the near-surface data lying near zero ^{3}He, and high (3–4 TU) tritium. The thermocline waters rise upward to a maximum of 2–3 TU in ^{3}He, and eventually ramp backward toward zero tritium and ^{3}He in the range 0.5–1.0 TU. It should be noted that the bulk of the data scattering below this curve (in the range 2–3 TU tritium) are in the slope waters. Further, the relation is much tighter on a region-by-region basis.

Figure 6*b* is a plot of results for the one-dimensional model with constant diffusivity (κ) and velocity (u) discussed in §1. The different curves are from varying Peclet number, as defined by (3). This Peclet number must be distinguished from the expected *local* Peclet number of order 100. This can be seen by noting that $L = u/\lambda$, the natural length scale that arises from the non-dimensionalization of the advection–diffusion equations, is of order 5000 km for the upper thermocline. Such a Peclet number thus characterizes the gyre as a whole, and should not be interpreted literally. In a crude sense, it may be regarded as the number of circuits that a parcel of water makes around the gyre before it is ventilated. Comparison with the data indicates that a Peclet number of order one or less provides the best match with the data. This indicates that the gyre is quite strongly ventilated, probably along the northern rim, and that

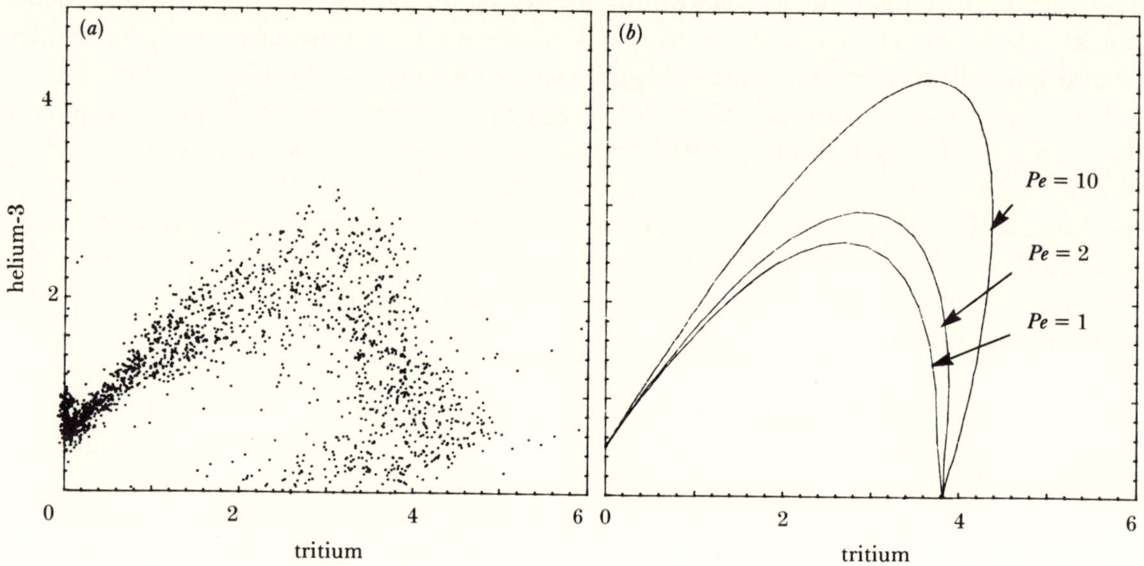

FIGURE 6. The relation between tritium and ^{3}He for (*a*) the North Atlantic TTO and NATS data, and for (*b*) one-dimensional model calculations for various Peclet numbers.

on average, a parcel of water does not make more than one trip around the gyre before it is influenced by ventilation. This is consistent with the high ventilation rates proposed by Sarmiento (1983) and by Jenkins (1987), and with the general character of the τ maps discussed in the previous section.

Another way to regard the scatter plot in figure 6*a* is to recall that the maximum North Atlantic surface tritium in the gyre occurred in the mid-1960s, and was about 18 TU. If a parcel of this water were isolated at that time (i.e. by subduction without excessive subsurface mixing), this water would appear on the plot at 7.8 TU (tritium) and 10.2 TU (^{3}He)! Even allowing for a twofold dilution (to bring the tritium down to the observed range) still gives a ^{3}He value about a factor of two greater than any values observed in the North Atlantic.

4. A TIME-SERIES OF ^{3}He MEASUREMENTS NEAR BERMUDA

The evolution of the transient-tracer distributions in time provides in principle the most useful information about transport processes in the ocean. In addition to quasi-synoptic maps of the tracer distributions obtained at intervals approaching a decade or more, one can obtain relatively frequent sampling at a few sites. Such a time-series is the Hydrostation S near Bermuda, which has been occupied on a nominally bi-weekly basis (with some gaps) for more than 30 years. Over the past decade, sampling has occasionally been done at this station for tritium–^{3}He, and since early 1985, on a regular, more frequent (nearly monthly) basis. The primary goal of this programme is to monitor the longer-term evolution of these two isotopes within the Sargasso Sea and its main thermocline. The more recent effort has also concentrated on the seasonal-timescale processes affecting their distribution. This work is providing important information about the vertical cycling and fluxes of ^{3}He in the upper part of the water column.

Figure 7 is a contour plot of the helium-isotope-ratio anomaly in the upper 1000 m (figure 7*a*) and the upper 250 m (figure 7*b*) for the period January 1985 to January 1987. The isotope-ratio anomaly is defined in the usual fashion as

$$\delta(^3\mathrm{He}) = (R_x/R_a - 1) \times 1000‰, \tag{4}$$

where R_x and R_a are the sample and atmospheric isotopic ratio (^{3}He:^{4}He) respectively. Analytical accuracy is of the order 1.5‰, and sampling depths are indicated in figure 7*b* by small crosses. The lighter isotope is less soluble in water, so that sea water in equilibrium with the atmosphere is characterized by $\delta(^3\mathrm{He}) \approx -16‰$. (This isotope effect has been determined as a function of temperature and salinity to considerable precision by Benson & Krause (1980).)

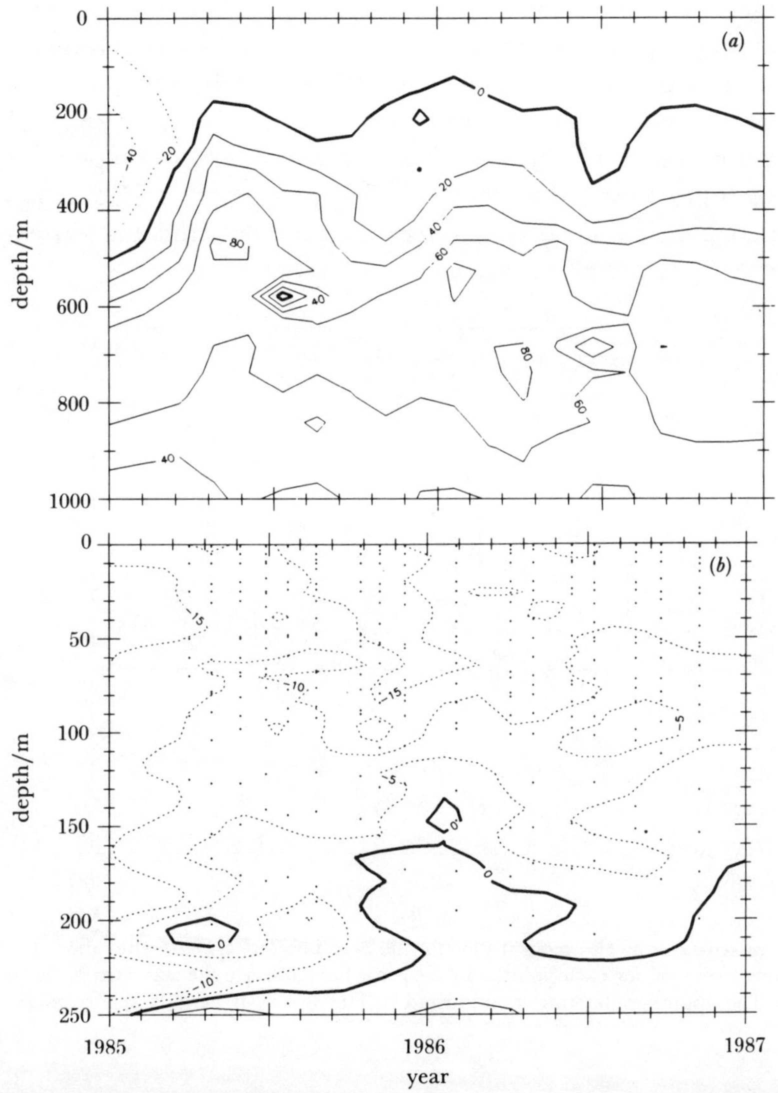

FIGURE 7. A time-series of $\delta(^3\mathrm{He})$ measurements for the years 1985 and 1986 at Station S near Bermuda. The data are contoured in units of per mille for (*a*) the upper 1000 m, and (*b*) the upper 250 m. In (*b*), sample positions are indicated by small + signs. Note the deep maximum at about 700 m depth, and the fact that the static solubility equilibrium value is $\delta(^3\mathrm{He}) \approx -16.5‰$.

The most significant features of the ^{3}He distributions shown in figure 7 are the maximum within the thermocline (common throughout the subtropics) and the vertical excursions of the isopleths on a seasonal basis. The former results from the rollover due to increasing isolation and decreasing tritium with depth. The latter arises in large part from the seasonal changes in the depth of the mixed layer. It is noteworthy that the affected ^{3}He isopleths extend well below the maximum penetration of winter convection for these years, suggesting the significant role of horizontal (or rather isopycnic) processes in the vertical transport of material in the upper ocean.

Examination of figure 7*b* reveals that the $\delta(^3He)$ is supersaturated during a large part of the annual cycle. This is especially evident in figure 8, which is a plot of the mixed layer mean $\delta(^3He)$ as a function of time. The error bars are standard deviation of the mean for each point, and include analytical and systematic error estimates, as well as the random error due to sampling. On the average, about five samples were analysed in the mixed layer for each point. Included for reference is the 'static isotopic equilibrium' $\delta(^3He)$ obtained from Benson & Krause (1980). The supersaturation is greatest in winter, presumably because of the greater flux of this isotope from below resulting from deep convection, and despite expectations of greater gas-exchange rates during this season. In the summer, the $\delta(^3He)$ is lower, because seasonal stratification presents a barrier to vertical transport, and the extremely shallow mixed layer is effectively outgassed despite lower gas-exchange rates.

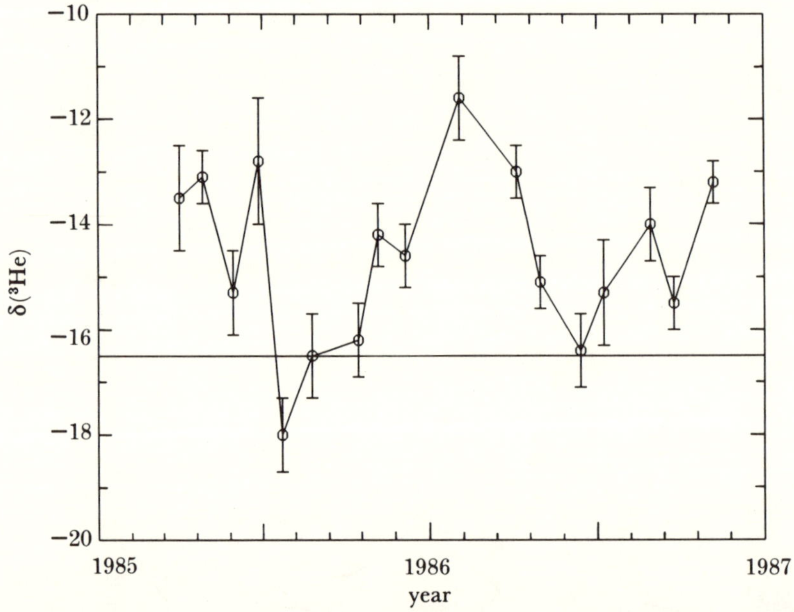

FIGURE 8. The average mixed-layer $\delta(^3He)$ against time for the Bermuda series. Generally, about five samples from the mixed layer were averaged for each point, and the error bars indicate the standard deviation of the mean for each average. The approximate static equilibrium $\delta(^3He)$ is shown here as a horizontal line for visual reference.

The existence of this supersaturation in ^{3}He implies a flux from the sea to the atmosphere, and knowing the gas-exchange rate, it should be possible to estimate the annual average of this flux. From climatological wind data (see, for example, Bunker 1975), and an observed relation between the mean gas-exchange rate and mean wind velocity (Roether 1986), a climatological time history of gas exchange rate can be constructed to an accuracy of order 20%. The more difficult task is to obtain a realistic estimate of the equilibrium isotopic ratio for helium to use as a baseline for estimating the actual ^{3}He supersaturation.

Although seawater equilibrium $\delta(^3\text{He})$ has been determined with extreme accuracy (Benson & Krause 1980), the concentration of helium in the mixed layer is not in equilibrium. In general, air injection maintains North Atlantic surface waters about 5% supersaturated in helium (Jenkins & Clarke 1976). Because, as will be discussed below, the processes responsible are potentially fractionating, the resultant $\delta(^3\text{He})$ may deviate substantially from the equilibrium value determined in the laboratory (see Fuchs *et al.* 1987).

For the purposes of the present calculations one can identify two differing modes of air injection. The first, hereafter referred to as 'trapping', involves the complete dissolution of bubbles forced so far downward by turbulence or wave action that hydrostatic pressure forces collapse and total injection of gases. This process is non-fractionating, but in dynamic balance with gas loss by gas exchange, the resultant supersaturation of two species is given by

$$\Delta_1/\Delta_2 = \alpha_2 D_2^{0.5}/\alpha_1 D_1^{0.5}, \tag{5}$$

where Δ, α, and D are the fractional supersaturation, Ostwald solubility factor and molecular diffusivity respectively (see Fuchs *et al.* 1987). The $\delta(^3\text{He})$ resulting from 5% helium supersaturation at 25 °C would be $-18.8‰$, compared to the 'static equilibrium' value of $-16.3‰$.

The second mode of air injection consists of 'partial injection', which occurs when bubbles undergo only slight dissolution, and escape before significant gas loss (or fractionation of the gas within the bubble) occurs. Such a mode of injection may be important in the swaths of bubbles created by whitecaps. The transport of gas from the bubble to the water is diffusively limited within the boundary layer, introducing an additional fractionation, dependent on the two thirds power of the molecular diffusivity (Levich 1962). Thus for a simple mass balance, one obtains

$$\Delta_1/\Delta_2 = \alpha_2 D_1^{0.167}/\alpha_1 D_2^{0.167}, \tag{6}$$

which differs from the other mode significantly. For the example discussed above, the corresponding $\delta(^3\text{He})$ would be $-14.4‰$. Considering the observations (figure 8), the difference in 'baseline' between the two endmembers would result in factor of three difference in the predicted saturation anomaly in ^{3}He, and hence the flux.

These two modes of air injection are clearly endmembers, and what actually occurs will be a complex spectrum of intermediate degrees of bubble destruction, which will change as a function of wind speed and wave spectra. Within the framework of this discussion, we regard the problem as simply a linear mixture of the two modes, recognizing the limitations of this approach. It remains, however, to discern with adequate accuracy, the relative contributions of these modes. To do this, an additional constraint is needed.

The above equations may be applied to different gases. For neon, the predicted ratios of saturation anomaly (relative to helium) are 1.03 and 0.81 for the trapped and partial components respectively. An additional complication, however, results from radiative heating and mixing induced supersaturations, which are significant for neon. This may be seen (figure 9*a*) in a regression of ΔNe against ΔHe for the GEOSECS Atlantic data (GEOSECS 1987), where the intercept is significantly non-zero ($\Delta\text{Ne}_0 = 1.1 \pm 0.3\%$). To account for these effects, we employ a one-dimensional model similar to the one used successfully to simulate the seasonal

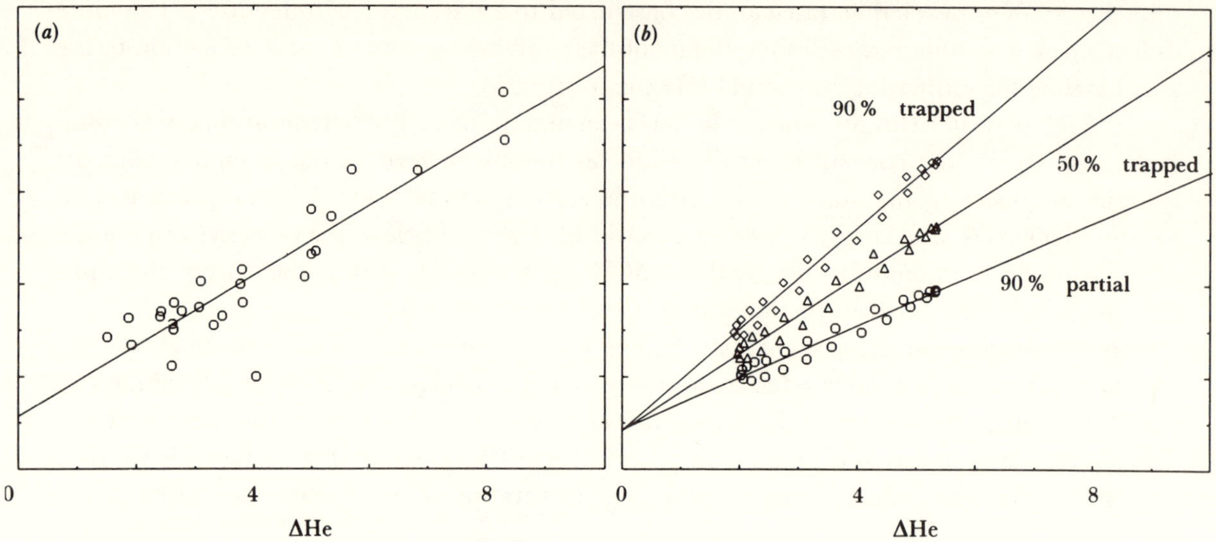

FIGURE 9. The relation between excess neon and excess helium are shown for (*a*) the GEOSECS Atlantic upper ocean data, and (*b*) a numerical seasonal mixed-layer simulation. The regression line in (*a*) corresponds to $\Delta\text{Ne} = (1.1 \pm 0.3) + (0.76 \pm 0.07)\ \Delta\text{He}$, with a linear correlation coefficient of 0.91 for 27 points. The three curves shown in (*b*) are for three representative mixtures of partial and trapped modes of air injection. Note the hysteresis evident in the model data points due to seasonal heating effects on ΔNe.

cycling of oxygen (Musgrave *et al.* 1988) and argon (Chou 1985). The only addition to this model is a wind-dependent air injection component given by

$$W_i = W_0 \cdot u_{10}^{\gamma}(f_t + f_p \cdot \Gamma),$$

where W_i is the injection rate of the gas, W_0 is an adjustable parameter related to the flux of bubbles (and the same for all gases), u_{10} is the wind speed, γ is an adjustable power-law dependence, f_t and f_p are the relative fractions of the trapped and partial components, and is constrained by $f_t + f_p = 1$. Finally, Γ is a normalized diffusivity factor associated with the boundary-layer diffusive limitation for gas transfer from the bubbles, given by

$$\Gamma = D^{0.667}/D_0^{0.667},$$

where D_0 is the molecular diffusivity of helium at 20 °C.

Aside from the relative fractions f_t and f_p, the functional dependence of air injection employed here employs only two unknown parameters. These were adjusted to obtain agreement with observations of the mean helium supersaturation (*ca.* 4.5 %) and its annual range (from *ca.* 2 % in summer to *ca.* 6 % in winter) observed in this time-series. It is encouraging that the optimal case exhibited a cubic dependence on the wind speed, in agreement with acoustic observations of bubble populations (Crawford & Farmer 1988). Within the above constraints, the relation between ΔNe and ΔHe was explored for various mixtures of partial and trapped air injection components. Three regression lines are shown in figure 8*b* for a range of extreme cases. Note that the zero ΔHe intercept for all these is within the errors of that obtained for the GEOSECS observations. From the regression slope and its uncertainty from the GEOSECS data, the optimal mix of partial against trapped components and its uncertainty is determined to be $f_p = 0.63 \pm 0.13$.

This mixture is not inconsistent with the strong attenuation with depth of bubble populations (e-folding depths of order 1 m) observed by Crawford & Farmer (1988), because this would imply a large probability of bubble escape before collapse.

The model can then be used to compute the 'dynamic equilibrium' $\delta(^3He)$ in the mixed layer as a function of time. This, and its uncertainty is shown in figure 10*a*, along with a sinewave fit to the observations. Combined with the climatological gas-exchange rate (see, for example, Musgrave & Jenkins 1987), the annual flux of ^{3}He can be computed (figure 10*b*). The mean gas-exchange flux from the mixed layer is thus computed to be 21 ± 6‰ m d^{-1}. The uncertainty includes contributions due to uncertainty in the dynamic equilibrium $\delta(^3He)$ and a 20% uncertainty in the gas-exchange rate.

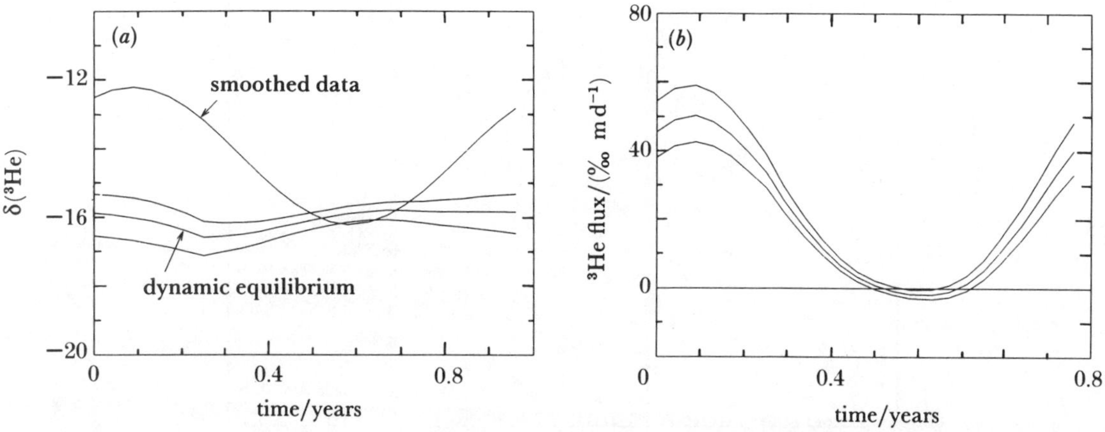

FIGURE 10. (*a*) Comparison of the range of dynamic equilibrium $\delta_0(^3He)$ (lower three curves) predicted by the upper-ocean model, and as restricted by the ΔNe–ΔHe results with a sinusoidal fit to the time-series data. (*b*) Predicted time history of the ^{3}He flux from the ocean to the atmosphere based on the model $\delta_0(^3He)$ and climatological gas-exchange rates obtained from the Roether (1986) windspeed-piston velocity relation and climatological wind data (Bunker 1975).

'Local' production of ^{3}He within the upper ocean can be estimated to be about 10% of the observed flux (in the upper 100 m), so that the upward flux of this isotope from the waters below is 19 ± 6‰ m d^{-1}. This flux represents the 'mining' of tritiugenic ^{3}He from the main thermocline. By comparison of the 1986 profiles with tritium and ^{3}He data obtained from this station in early 1977, it is possible to compare this flux to the long-term changes in the inventory of these isotopes in the main thermocline. The equivalent change rates in the tritium and ^{3}He inventories in the upper 700 m (i.e. above the oxygen minimum) are 12 and 4.5‰ m d^{-1} respectively. The combined flux agrees well with mixed-layer flux estimate.

What does this flux imply about vertical transport? Using the observed vertical gradient of ^{3}He in the upper main thermocline (figure 7*a*) of 0.17‰ m^{-1}, one would require a vertical turbulent diffusivity of approximately 10^{-3} m^2 s^{-1}, a number that is unreasonably large by any standard (see, for example, Jenkins 1980). It is, of course, unreasonable to demand a local balance between the observed mixed layer flux of ^{3}He and the thermocline inventory changes. The tritium–^{3}He age maps discussed earlier, however, point to the Gulf Stream recirculation region as an important window of ventilation for the shallow isopycnals, so that we may indeed be observing part of the process of ventilation.

The data do contain an additional hint as to the nature of one possible mode of return of ^{3}He to the surface layer. Figure 11 is a comparison beween two reoccupations of the Bermuda

Site (stations 593 and 597) in July and August 1986. The earlier station (solid line) shows anomalous structure compared to the later one (broken line), which is more typical of the series. The anomalous data show significant amounts of nitrate in the bottom of the euphotic

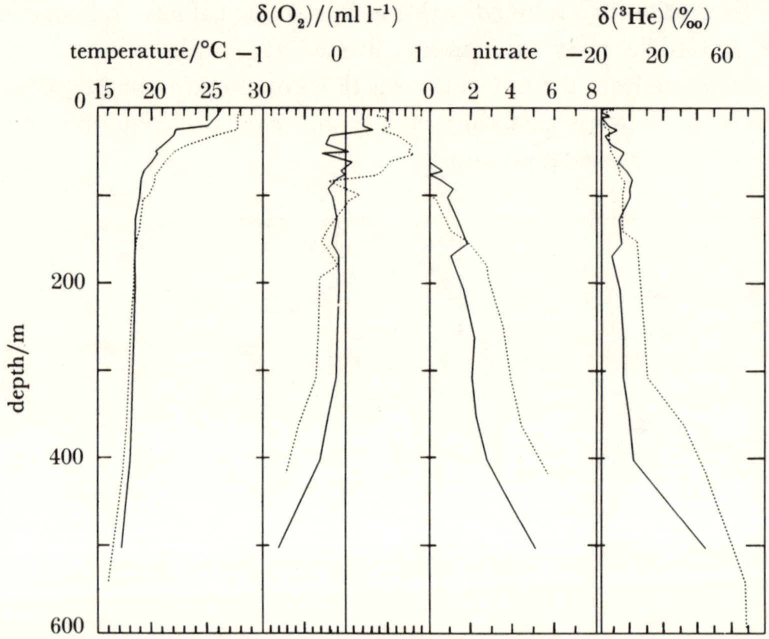

FIGURE 11. Property profiles for two stations from the time-series at Bermuda. Station 593 (7 July 1986, solid line) exhibits anomalous profiles compared with a more typical station (Station 597, 25 August 1986, broken line). Note especially the anomalous spike of nitrate and ^{3}He in the lower part of the euphotic zone (75–100 m), and the comparatively young subtropical mode water below.

zone, accompanied by substantial amounts of excess ^{3}He. The amount of ^{3}He associated with this 'spike' would require in excess of two years of isolation to accrue, indicating that the spike must have come from below. Further, the lifetime of nitrate within the euphotic zone is sufficiently short (less than a day), that the injection must have occurred within the very recent past. Integrating the vertical distribution of the excess ^{3}He in the upper 150 m (the depth accessable by winter convection), one estimates that this spike can account for 20–30% of the annual flux of ^{3}He. Thus one would require only a few of these events to support the observed flux.

It is noteworthy, however, that the peculiar conditions observed at station 593 also are accompanied by a unique structure in the subtropical mode water below. The mode water for this station is characterized by anomalously young tritium–^{3}He age (about half of normal), low nitrate and apparent oxygen utilization (AOU) (again about half of normal), and low potential vorticity. These combine to suggest the presence of a column of freshly ventilated mode water that has been advected around with the gyre-scale circulation. The timing, at least, appears consistent with this. Thus one may be tempted to assert that the introduction of ^{3}He and nutrients into the upper waters may be related to some dynamical feature of this entity. However, in the absence of more complete data (in space and time), this remains rank speculation.

However, by combining large-scale mapping of these isotopes in the thermocline, with the time-series of data that is evolving, we are beginning to observe and quantify the geochemically important processes of ventilation of the ocean.

I am indebted to my highly capable research staff (especially D. Lott, M. Pratt, S. Birdwhistel and M. Davis) for carrying out the demanding tasks associated with the measurements. Samples for TTO were acquired through the hard work of the PACODF group at Scripps Institution of Oceanography, for NATS with the help of C. Wunsch, D. Roemmich and H. Stommel, and for the Bermuda series by the excellent staff of the Bermuda Biological Station (including A. Knap, T. Jickells and R. Sherrif-Dow). Work on the above was funded by NSF grants no. OCE79-21378, OCE81-17998 and OCE85-01171 respectively.

References

Armi, L. & Stommel, H. 1983 Four views of a portion of the North Atlantic subtropical gyre. *J. phys. Oceanogr.* **13**, 828–857.

Benson, B. B. & Krause, D. 1980 Isotopic fractionation of helium during solution: a probe for the liquid state. *J. solution Chem.* **9**, 895–909.

Bower, A. S., Rossby, H. T. & Lillibridge, J. L. 1985 The Gulf Stream – barrier or blender? *J. phys. Oceanogr.* **15**, 24–32.

Broecker, W. S. & Östlund, H. G. 1979 Property distributions along the $\sigma_\theta = 26.8$ isopycnal in the Atlantic Ocean. *J. geophys. Res.* **84**, 1145–1154.

Bunker, A. F. 1975 Energy exchange at the surface of the western North Atlantic Ocean. *W.H.O.I. Tech. Rep. no.* 75–3. (107 pages.)

Chou, J. 1985 Numerical modelling of oxygen cycling in the upper ocean. *W.H.O.I. S.S.F. Report.* (Unpublished.)

Crawford, G. B. & Farmer, D. M. 1988 On the spatial distribution of ocean bubbles. *J. geophys. Res.* (In the press.)

Dreisigacker, E. & Roether, W. 1978 Tritium and ^{90}Sr in North Atlantic surface water. *Earth planet. Sci. Lett.* **38**, 301–312.

Fuchs, G., Roether, W. & Schlosser, P. 1987 Excess ^3He in the ocean surface layer. *J. geophys. Res.* **92**, 6559–6568.

GEOSECS 1987 *GEOSECS Atlas of the Atlantic, Pacific and Indian Ocean Expeditions 7 : shorebased data and graphics.* (200 pages.) Washington: U.S. Government Printing Office.

Jenkins, W. J. & Clarke, W. B. 1976 The distribution of ^3He in the western Atlantic Ocean. *Deep Sea Res.* **23**, 481–494.

Jenkins, W. J. 1980 Tritium and ^3He in the Sargasso Sea. *J. mar. Res.* **38**, 533–569.

Jenkins, W. J. & Rhines, P. B. 1980 Tritium in the deep North Atlantic Ocean. *Nature, Lond.* **286**, 877–880.

Jenkins, W. J. 1982 ^3He results from the Transient Tracers in the Ocean Program: a Peclet number for the North Atlantic subtropical main thermocline. *Eos. Wash.* **63**, 972 (abstract).

Jenkins, W. J. 1987 ^3H and ^3He in the Beta Triangle: observations of gyre ventilation and oxygen utilization rates. *J. phys. Oceanogr.* **17**, 763–783.

Leetma, A. & Bunker, A. F. 1978 Updated charts of the mean annual wind stress, convergences in the Ekman layers and Sverdrup transports in the North Atlantic. *J. mar. Res.* **36**, 311–322.

Levich, V. G. 1962 *Physicochemical hydrodynamics.* Englewood Cliffs, New Jersey: Prentice-Hall.

Levitus, S. 1982 *Climatological atlas of the world ocean.* NOAA prof. paper no. 13. (173 pages.) Washington, D.C.: U.S. Government Printing Office.

Musgrave, D. L., Chou, J. & Jenkins, W. J. 1988 Application of a model of upper ocean physics for studying seasonal cycles of oxygen. *J. geophys. Res.* (Submitted.)

Olson, D. B., Östlund, H. G. & Sarmiento, J. L. 1986 The western boundary undercurrent off the Bahamas. *J. phys. Oceanogr.* **16**, 233–240.

Roether, W. 1986 Field measurements of gas exchange. In *Dynamic processes in the chemistry of the upper ocean* (ed. J. D. Burton), pp. 117–128. New York: Plenum Press.

Sarmiento, J. L. 1983 A tritium box model of the North Atlantic thermocline. *J. phys. Oceanogr.* **13**, 1015–1026.

Worthington, L. V. 1976 *On the North Atlantic circulation.* The Johns Hopkins Oceanographic Studies, vol. 6. (110 pages.) Baltimore: The Johns Hopkins University Press.

Discussion

D. J. Webb (*Institute of Oceanographic Sciences, Godalming, Surrey, U.K.*). The changes produced by the near-surface mixing event discussed are very similar to those produced by events at the edge of a continental shelf. As the site mentioned is near to Bermuda, is it not possible that the mechanism responsible was connected with the island?

W. J. Jenkins. Studies have been made to determine the biological and physical effects of the island, and I do not recall any significant perturbations at this distance, except evidence of dynamic effects in near bottom current meter records. What leads me to suspect that the observed shallow entrainment phenomenon may be an open-ocean process is its apparent association with the deep 'ventilated' structure. Similar structures seem to have been observed in the open ocean (along 55° W) by McCartney *et al.* (1980).

Reference

McCartney, M. S., Worthington, L. V. & Raymer, M. E. 1980 Anomalous water mass distributions at 55 W in the North Atlantic in 1977. *J. mar. Res.* **38**, 147–172.

L. Merlivat (*Université Pierre et Marie Curie, Paris, France*). On the graph of ^{3}He–tritium showing the experimental data of the TTO samples, it appears that for a tritium content equal to zero, the ^{3}He content is not zero. Could Dr Jenkins comment on the origin of this ^{3}He amount, which seems to be associated with zero-tritium waters?

W. J. Jenkins. There are three components of the 'zero-tritium' ^{3}He to which Dr Merlivat refers. A simple extrapolation of the tritium against ^{3}He relation to zero tritium yields an intercept close to 0.5 TU. Part of this (perhaps half or more) may be due to the complete decay of pre-anthropogenic tritium. Work by Roether and co-workers at Heidelberg estimate pre-anthropogenic tritium concentrations in surface waters that are not drastically different from this. The remainder of this ^{3}He, the second component, is likely to be from the invasion of primordial helium via Antarctic intermediate and bottom waters. I interpret the ultimate source of this helium to be at least outside of the North Atlantic because it falls on the mixing curve.

The third component is undoubtedly indigenous injection of primordial ^{3}He from local sources, i.e. the Mid-Atlantic Ridge. We have seen evidence of such injections near the Charlie–Gibbs Fracture Zone (although on an irregular basis), and have observed them directly in the TAG hydrothermal area and the Kane Fracture Zone. Geometric considerations lead me to the conclusion that these sources are not a significant interference with the tritium–^{3}He story in the upper 1500 m in most of the Atlantic, but elsewhere, we must be careful.

W. Roether (*University of Bremen, F.R.G.*). I was very interested to hear that Dr Jenkins has observed mixed-layer excess ^{3}He so consistently. In our own data we seemed to find excesses stochastically in summer observations. Our conclusion was that the finding pointed to transient outcropping of thermocline waters, because the usual time-averaged vertical mixing coefficients were clearly unable to support such excesses. If a large-enough data base could be collected, it might be possible to quantify the outcropping (or whatever it is) effect. Furthermore, I wonder whether Dr Jenkins accounted for the 2‰ depression in the equilibrium $\delta(^3He)$ values that our work suggests?

W. J. Jenkins. My results do not disagree with Professor Roether's observations. In fact they support them. The individual data points in the mixed layer show scatter that is considerably larger than analytical, and that I would attribute (as Professor Roether does) to random injections of tritiugenic ^{3}He from below. The summer data, in fact, are rather close to static equilibrium. The error bars on the mean mixed layer $\delta(^3He)$ include the observed scatter. Where our interpretations diverge, however, is in the modelling of the 'dynamic equilibrium helium isotope effect'. Whereas Professor Roether's work with Fuchs favoured a greater trapped component, and hence projected a 2‰ depression in the $\delta(^3He)$, my work points to a dominance of partial bubble injection, resulting in much less of a depression. However, our differences may well be geographic: Professor Roether's work (as I recall) is from an area where bubble trapping may be dominant. I would not contend that my data and model apply outside of the western subtropics. It is clear, however, that further work would be of considerable value.

Water mass transport and ventilation in the Northeast Atlantic derived from tracer data

By W. Roether† and G. Fuchs

Institut für Umweltphysik, University of Heidelberg, 6900 Heidelberg, F.R.G.

A large ^{3}He and tritium data base collected in the Northeast Atlantic since 1977 allows construction of extensive tracer distributions on isopycnal surfaces in the main thermocline. A distinct southward increase in tritium–^{3}He age is observed that depicts the outcrop boundaries and allows one to quantify the interior transport. This transport is directed towards the south.

1. Introduction

This paper assesses the water mass transport of upper waters in the Northeast Atlantic from the viewpoint of observed ^{3}He and tritium distributions. The underlying method was developed by Jenkins & Clarke (1976) and Jenkins (1980, 1987).

In the North Atlantic, an excess of ^{3}He relative to a solubility equilibrium with atmospheric He is present below the surface layer. This excess results from the radioactive decay of tritium (half life 12.43 years) that has been introduced into the environment by nuclear-weapon testing in the early 1960s. Because of gas exchange with the atmosphere, the oceanic surface layer attains a near equilibrium with the atmosphere (Fuchs *et al.* 1987), whereas in the interior ^{3}He increases while tritium decays. This means that the so-called tritium–^{3}He age, τ, also increases, and according to the decay law is

$$\tau = (T_{\frac{1}{2}}/\ln 2)\ \ln([^3\mathrm{He}^*]/[\mathrm{T}]+1),$$

$T_{\frac{1}{2}}$ being the half life, $[^3\mathrm{He}^*]$ the radiogenic ^{3}He component (excess over the solubility-equilibrium value), and [T] the tritium concentration. One can interpret τ as a formal water age, i.e. the time since the water parcel left the surface layer.

Clearly, the process of ventilation of the oceanic main thermocline, by which is meant its exchange with the surface layer, must be reflected in the distribution of τ. Neglecting mixing processes and assuming that the velocity direction is oriented orthogonally to the isolines of τ and that the distribution of τ is stationary, the velocity $\boldsymbol{v}$ is given by

$$|\boldsymbol{v}| = 1/|\nabla\tau|, \tag{1}$$

where $|\nabla\tau|$ is the magnitude of gradient τ. If mixing is present, the transport direction is no longer necessarily orthogonal to the τ-isolines, and the correct relation becomes (Jenkins 1987)

$$\boldsymbol{v}_{\mathrm{I}}\cdot\nabla_{\mathrm{I}}\tau + w\frac{\partial\tau}{\partial z} = 1 - \frac{\partial\tau}{\partial t} + K_{\mathrm{I}}\nabla_{\mathrm{I}}^2\tau + K_{\mathrm{Z}}\frac{\partial^2\tau}{\partial z^2} + K_{\mathrm{I}}\nabla_{\mathrm{I}}\ln([\mathrm{T_S}][\mathrm{T}])\,\nabla_{\mathrm{I}}\tau + K_{\mathrm{Z}}\frac{\partial}{\partial z}\ln([\mathrm{T_S}][\mathrm{T}])\,\frac{\partial\tau}{\partial t}. \tag{2}$$

† Present address: University of Bremen, FB-1, 2800 Bremen, F.R.G.

In (2) the velocity $\boldsymbol{v}$ is split up into components $\boldsymbol{v}_I$ along the isopycnal surface and w orthogonal to it (i.e. essentially vertical). The two-dimensional gradient operator on the isopycnal surface is ∇_I and z is the orthogonal direction. K_I and K_Z are the isopycnal and diapycnal diffusivities. Here, T is the tritium concentration and $[T_S]$ is 'stable tritium', i.e. the sum of the radiogenic ^3He and tritium concentrations (Jenkins 1987; Fuchs 1987).

In this paper, the tritium–^3He age distribution observed in a certain area of the Northeast Atlantic is described (§2). In §3 a scale analysis is performed to find out which terms in (2) are relevant. The age distribution is then converted into a velocity distribution. Section 4 gives a brief discussion of the results.

2. Tracer data

Samples were collected on ten research cruises between 1977 and 1984. Table 1 summarizes these cruises, giving also the number of samples taken. Station positions are shown in figure 1. Sections along 38° W and the so-called β-triangle (Jenkins *et al.* 1985; Jenkins 1987) have been included. For the Heidelberg data (see table 1), the measuring precision for tritium is between

Table 1. Relevant cruises

cruise	cruise period	number of samples analysed ^3H	T
Knorr-66/2[(a)]	Jun. 1977	60	91
Oceanus-52[(a)]	Oct. 1978	178	282
Meteor-56/6	Apl 1981	395	639
Meteor-64	Apl 1983	84	18
Walther-Herwig-57	Mar. 1983	57	36
Poseidon-104	Nov. 1983	49	—
Valdivia-19	Mar. 1984	173	77
Poseidon-111	Jly 1984	126	29
Meteor-69/2	Aug. 1984	166	87
Meteor-69/6	Nov. 1984	111	37
Atlantic II-107[(b)] / *Oceanus*-78[(b)]	Oct. 1979 / Mar. 1980	467	505

[(a)] Unpublished data from Woods Hole–Heidelberg cooperation (with W. J. Jenkins and W. Weiss).
[(b)] Data from Jenkins *et al.* (1985).
For data from all the other cruises see Fuchs (1987).

0.1 and 0.2 TR and for ^3He between 0.03 and 0.06 TR (1 TR = 1 tritium atom/10^{18} H-atoms or 1 ^3He atom/10^{18} H atoms); hence the precision of τ is usually between a few months and one year (Fuchs *et al.* 1987; Jenkins *et al.* 1985).

Figure 2 shows the age distribution on three isopycnal surfaces ($\sigma_\theta = 27.0$, 27.3, 27.6; the depth range covered by these densities is roughly 500–1000 m over much of the area). The rationale for isopycnal mapping is that, in a first approximation, transport is generally regarded as being adiabatic (Sarmiento *et al.* 1982; Kawase & Sarmiento 1985). To obtain the distributions, individual station profiles were interpolated to the densities in question, the maps being constructed from the interpolated values (Fuchs 1987). In figure 2, the age increases with density as well as southward, the rate of increase becoming larger towards the south. There is also a tendency towards increasing ages towards the African coast. The influence of younger water of Mediterranean origin can be seen on the $\sigma_\theta = 27.6$ surface. Winter outcropping of the

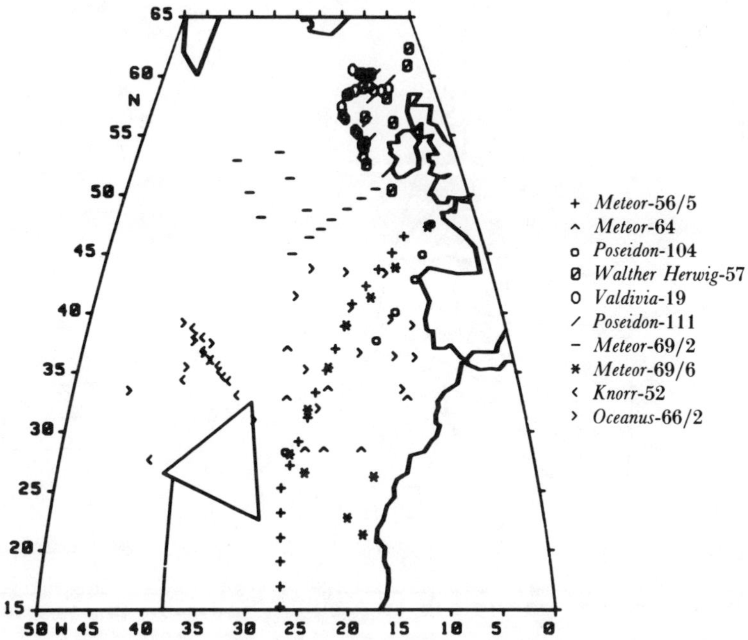

FIGURE 1. Station map including data on the 38° section and β-triangle from Jenkins *et al.* (1985).

surfaces in question occurs far to the north of the area covered in figure 2 so that the age contours are uninfluenced by seasonal effects.

3. Scale analysis and velocity distribution

To derive a velocity distribution from the data given in figure 2, a scale analysis on the basis of (2) was made (Fuchs 1987). The diffusivities were taken to be $K_I = 400\ \text{m}^2\ \text{s}^{-1}$ and $K_Z = 10^{-5}\ \text{m}^2\ \text{s}^{-1}$ (Armi & Stommel 1983; Olbers *et al.* 1985). It was found that north of a certain latitude the diffusion terms contribute no more than about 10%; this is the region where τ gradients are fairly smooth. The diapycnal advection term $w \cdot \partial\tau/\partial z$, with values for w as reported by Olbers *et al.* (1985), was found not to exceed about 5%. Stationarity was checked by comparing profiles at stations close to each other taken a few years apart. The indication was that $\partial\tau/\partial t < 0.15$. It is to be expected that under a stationary velocity field only mixing could make the age distribution vary in time. It is therefore concluded that over much of the area sampled, (2) reduces to

$$\boldsymbol{v}_I \cdot \nabla\tau \approx 1. \tag{3}$$

Assuming the velocity to be orthogonal to the isolines of τ leads back to (1).

Velocities calculated on the basis of (1) are shown in figure 3. Values are given only for positions where the neglected terms contribute no more than 30%. Within this limit, the assumption that the velocity is orthogonal to the age isolines should be valid. It is found in figure 3 that on the $\sigma_\theta = 27.0$ and 27.3 horizons the velocity is directed to the south, with a magnitude around 1 cm s^{-1}. Velocities decrease with increasing density. On the $\sigma_\theta = 27.6$ level a southeastward flow is obtained.

This velocity distribution points to ventilation of the layers in question directly from the winter outcrop regions to the north (Luyten *et al.* 1983; Kawase & Sarmiento 1985; Jenkins

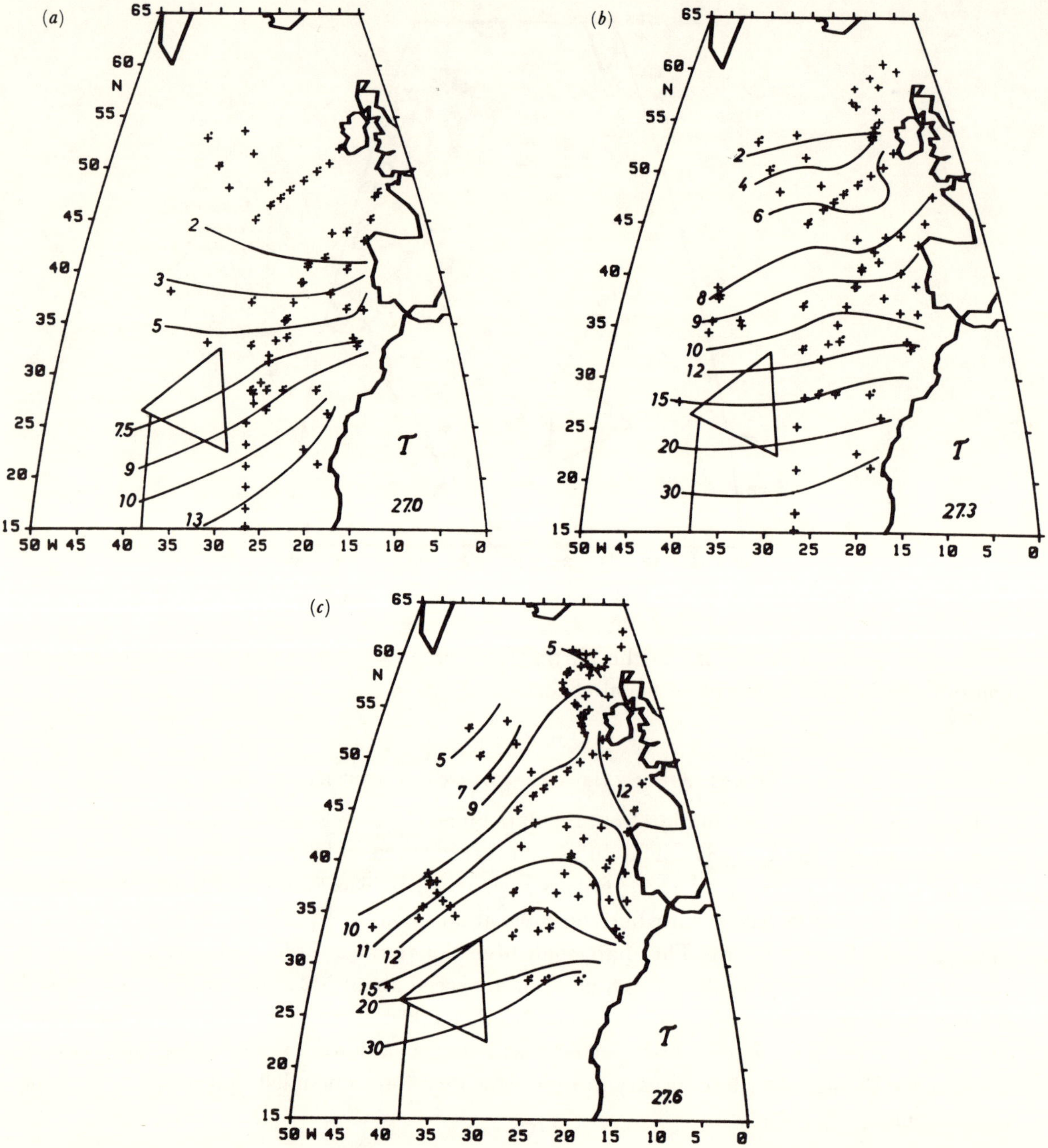

FIGURE 2. Tritium–^{3}He age distribution (units in years) on density surfaces $\sigma_\theta = 27.0$ (*a*), 27.3 (*b*) and 27.6 (*c*).

1987). The total velocities give a southward flow across 45° N of about 5×10^6 m^3 s^{-1} for the layer bounded by $\sigma_\theta = 27.0$ and 27.6 between 10° and 30° W. This is in agreement with an estimate by Krauss (1986). The conclusion that these layers of the Northeast Atlantic are ventilated by advection from the subpolar region is supported by the age distribution in figure 2. An extrapolation of the τ isolines on the isopycnals to zero age leads to boundaries of wintertime ventilation for these isopycnals that are fully consistent with boundaries derived from

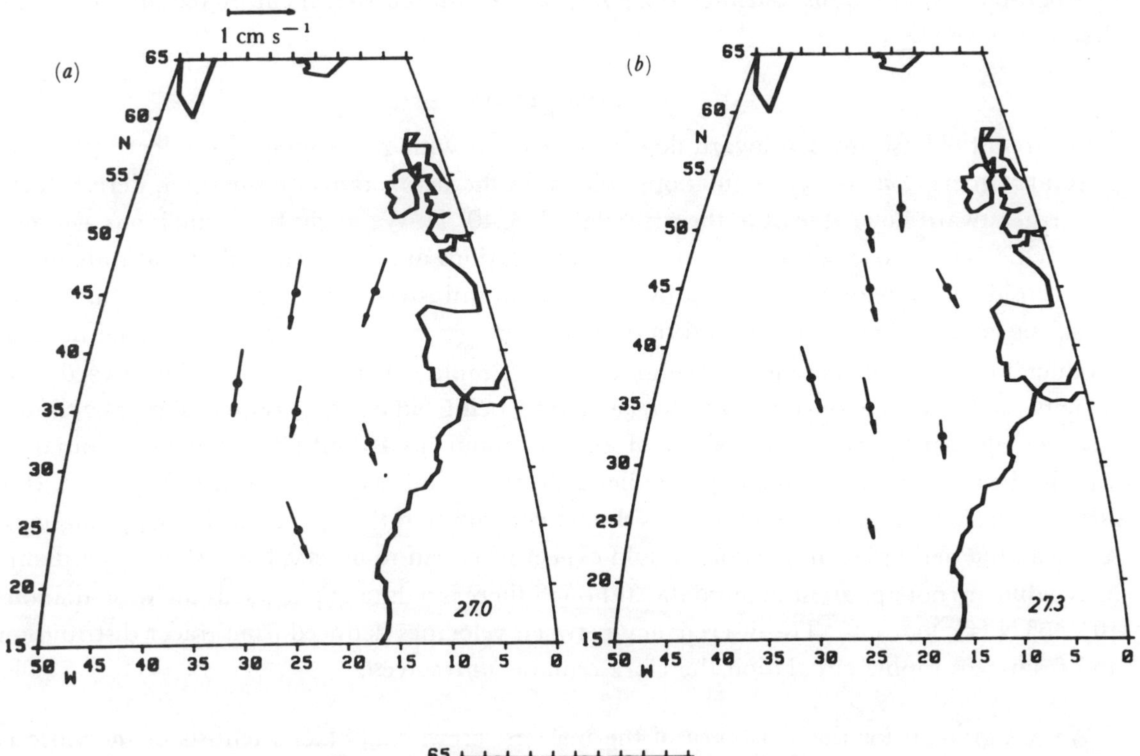

FIGURE 3. Velocity at selected points (centre of the arrows) in the Northeast Atlantic at the density horizons $\sigma_0 = 27.0$ (*a*), 27.3 (*b*) and 27.6 (*c*).

hydrographic observations (Sarmiento *et al.* 1982; see also the assessment by Jenkins (1987) for shallower layers).

4. Discussion

Our result of basically southward flow is at variance with geostrophic (Saunders 1982) and β-spiral (Olbers *et al.* 1985) calculations, which, in the depth range in question, derive north to northeastward flow, at least in the region north of 40° N. We might be wrong if our assumed diffusivities were too small or if the tracer concentrations at some intermediate latitude in the area considered were determined not by advective or diffusive meridional exchange but rather by strong zonal inflow from the west that might then, together with some meridional mixing, produce the observed meridional age gradient. A simple calculation with a northward flow velocity, such as that derived by Olbers *et al.* (1985), being balanced by isopycnal mixing shows, however, that an isopycnal diffusivity of at least about $5 \times 10^4\ m^2\ s^{-1}$ would be required to explain the observed meridional age gradients. Such a value is quite unrealistically large. It is further believed that diapycnal processes should contribute rather less than the isopycnal ones. As for an interfering zonal flow, one would expect it to lead to observable east–west gradients in age that are not apparent in our data, although there is a data gap towards the west at about 40°–45° N (see figure 2). The discrepancy between velocities deduced from tracer distribution and from geostrophic circulation therefore remains unresolved.

We are grateful for the assistance of the masters, crews and chief scientists of the various cruises from which our data originate. We thank especially W. J. Jenkins, P. Schlosser and W. Weiss for allowing us to use their unpublished data. G. Zimmek and G. Bader carried out the helium and tritium measurements at Heidelberg. This study was supported by the Deutsche Forschungsgemeinschaft.

References

Armi, L. & Stommel, H. 1983 Four views of a portion of the North Atlantic subtropical gyre. *J. phys. Oceanogr.* **13**, 828–857.

Fuchs, G. 1987 Ventilation der Warmwassersphäre des Nordostatlantiks abgeleitet aus ^{3}Helium- und Tritium-Verteilungen. Doctoral dissertation, Institut für Umweltphysik, University of Heidelberg.

Fuchs, G., Roether, W. & Schlosser, P. 1987 Excess ^{3}He in the ocean surface layer. *J. geophys. Res.* **92**, 6559–6568.

Jenkins, W. J. 1980 Tritium and ^{3}He in the Sargasso Sea. *J. mar. Res.* **38**, 533–569.

Jenkins, W. J. 1987 ^{3}H and ^{3}He in the beta triangle: observations of gyre ventilation and oxygen utilization rates. *J. phys. Oceanogr.* **17**, 763–783.

Jenkins, W. J. & Clarke, W. B. 1976 The distribution of ^{3}He in the Western Atlantic Ocean. *Deep Sea Res.* **23**, 481–494.

Jenkins, W. J., Lott, D. E., Davis, M. W. & Boudreau, R. D. 1985 W.H.O.I. Helium Isotope Laboratory Data Report no. 2, Tritium and ^{3}He data from the Beta Triangle (AII – 107, Oct. 1979 and OC-78, March 1980) Woods Hole Oceanographic Institution.

Kawase, M. & Sarmiento, J. L. 1985 Nutrients in the Atlantic thermocline. *J. geophys. Res.* **90**, 8961–8979.

Krauss, W. 1986 The North Atlantic Current. *J. geophys. Res.* **91**, 5061–5074.

Luyten, J. R., Pedlosky, J. & Stommel, H. 1983 The ventilated thermocline. *J. phys. Oceanogr.* **13**, 292–309.

Olbers, D. J., Wenzel, M. & Willebrand, J. 1985 The interference of North Atlantic circulation patterns from climatological hydrographic data. *Rev. Geophys.* **23** (4), 313–356.

Sarmiento, J. L., Rooth, C. G. H. & Roether, W. 1982 The North Atlantic tritium distribution in 1972. *J. geophys. Res.* **87**, 8047–8056.

Saunders, P. M. 1982 Circulation in the eastern North Atlantic. *J. mar. Res.* **40** (suppl.), 641–657.

Discussion

W. J. Jenkins (*Woods Hole Oceanographic Institution, Woods Hole, U.S.A.*). I am uncomfortable with the comparison of model or climatologically averaged data that are presented at constant depth with 'isopycnal' velocities. Could part of the discrepancy that Professor Roether reports between the tracer-derived velocities and such flow fields be explained by the northward shoaling of the density surface coupled with vertical shears in the flow field?

Would it not be more appropriate to compare Professor Roether's tracer-derived velocities to geostrophic estimates with the hydrographic data obtained from his cruises?

Further, Professor Roether's comment that one would require substantially larger isopycnic or diapycnic diffusivities (than he has used) to explain the differences is intriguing. Has he done the 'inverse' calculation of estimating how large such diffusivities must be to bring the two flow fields into agreement?

W. Roether. It is true that geostrophic analysis deals with level surfaces in contrast with our isopycnal interpretation and that non-synoptic data may be a problem. Still, these differences in my view do not open a straightforward escape route from our apparent discrepancy, and more work is certainly needed. The isopycnic diffusivity required to override advection is $5 \times 10^4\ m^2\ s^{-1}$. This value has been substantiated also by an inverse-model evaluation of the isopycnal age fields.

World-scale monthly mapping of the CO_2 ocean–atmosphere gas-transfer coefficient

By Fabienne Thomas[1], C. Perigaud[1], L. Merlivat[2] and J.-F. Minster[1]

[1] *UM39 at CNES/GRGS* 18 *avenue Edouard-Belin*, 31055 *Toulouse Cedex, France*

[2] *LODYC, Université Pierre et Marie Curie*, 75230 *Paris Cedex* 05, *France*

Monthly maps of the ocean–atmosphere CO_2 gas-transfer coefficient are derived from meteorological winds for 1982. The seasonal variations are of the order of a factor 2, mostly in the Northern Hemisphere. A strong meridional pattern is observed: the exchange coefficient is larger at high latitudes than in tropical areas. However, the zonal variations within a latitudinal band (generally a westward enhancement) can be as large as the meridional ones. The representativity of the winds is a major concern: interannual variations of the order of 30% are documented by an analysis of four years of the monthly transfer coefficient derived from meteorological winds in the Atlantic ocean. In the future, these maps will be obtainable on a routine basis from satellite data. As an example, a map derived for August 1978 from *Seasat* altimeter winds is compared with the present ones.

0. Introduction

Gas exchange between the ocean and the atmosphere is one of the controlling processes of the CO_2 geochemical cycle. The net flux of gas at this interface can be expressed as

$$F = ks\,\Delta p_{CO_2}, \tag{1}$$

where k is a gas-transfer velocity, s is the solubility of CO_2 in seawater and Δp_{CO_2} is the difference between the CO_2 partial pressure in the surface-ocean water and in the air. The product ks will be called the gas-transfer coefficient. An increasing number of studies provide estimates of Δp_{CO_2} and its variation with season and geography. This requires sampling and analysis of sea water. On the other hand, k and s depend on wind, sea-surface temperature and salinity and can *a priori* be estimated by remote sensing and from climatological data.

The present note analyses the geographical and seasonal variations of k and ks as determined from a meteorological data set for wind speed. Some of the uncertainties in these determinations will be discussed. One important difficulty is the representativity of wind data. This will be illustrated by an analysis of four years of monthly maps of the transfer velocity in the Atlantic ocean, by comparison with the climatological maps of Erickson & Duce (1987) and by comparison with a map deduced for August 1978 from *Seasat* altimeter wind speed measurements.

1. World-scale monthly maps for the gas-transfer velocity

Liss & Merlivat (1986) and Broecker *et al.* (1986) discuss thoroughly the data and models that established the relation between the gas-transfer velocity, wind speed and sea-surface temperature. Here we use the relations proposed by Liss & Merlivat (1986) (see figure 1). The

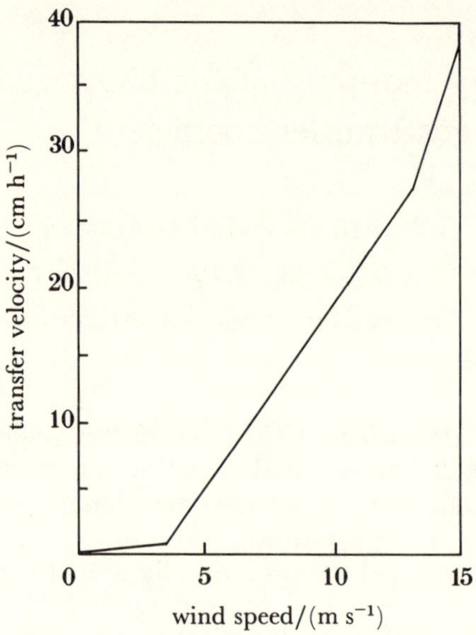

FIGURE 1. Relation between the CO_2 gas-transfer velocity (normalized to a Schmidt number of 600) and the wind speed at 10 m above the sea surface. Adapted from Liss & Merlivat (1986). See text for the corresponding equations.

relation between k and the wind speed is given for three different wind régimes, where the physical processes that control the gas exchange are different: exchange across a smooth surface at low wind speed, across a rough surface at intermediate wind speed, and bubble effects at high wind speed. These relations are

$$\left.\begin{aligned} V \leqslant 3.6\ \mathrm{m\ s^{-1}}, &\quad k = 0.17\ V(Sc(\theta)/Sc(20))^{-\frac{2}{3}}; \\ 3.6\ \mathrm{m\ s^{-1}} < V \leqslant 13\ \mathrm{m\ s^{-1}}, &\quad k = (2.85V - 9.65)\ (Sc(\theta)/Sc(20))^{-\frac{1}{2}}; \\ V > 13\ \mathrm{m\ s^{-1}}, &\quad k = (5.9\ V - 49.3)\ (Sc(\theta)/Sc(20))^{-\frac{1}{2}}; \end{aligned}\right\} \tag{2}$$

where V is the wind speed (in metres per second) at 10 m above the sea surface, k is the transfer velocity (in centimetres per hour), Sc is the Schmidt number ($Sc = v/D$, where v is the kinematic viscosity of seawater and D is the molecular diffusivity of CO_2 in seawater). θ is the sea-surface temperature and 20 is a reference temperature of 20 °C ($Sc(20) = 595$ for CO_2).

In a preliminary study, we derived monthly maps of the gas-transfer velocity (Thomas *et al.* 1986) from the climatological winds of Hellerman & Rosenstein (1983) and from the sea-surface temperatures and salinities of Levitus (1982). To do this we used the relations between s and temperature and salinity of Weiss (1974) and between Sc and temperature of Jahne (1980). By comparison with other estimates, it has become clear that this approach under-estimates k.

Erickson & Duce (1987) recently provided climatological maps of the gas-transfer velocity normalized to a Schmidt number of 600, derived from the *NOAA* $5° \times 5°$ monthly means of measured wind speeds. To overcome difficulties related to the wind-averaging process and to the undersampling of wind measurements in some areas, they introduced an artificial monthly frequency distribution as described in Erickson *et al.* (1986).

To assess the accuracy and representativity of such maps, we derive here monthly values of k and ks from meteorological winds. Such winds are provided as outputs of world-scale meteorological models. The model assimilation is a very efficient cleaning and interpolating process for measured data. In addition, a number of other types of data constrain the models, and thus their various outputs. Finally models generally provide wind fields at intervals of 6 h, so that the estimation of monthly averages should be quite reliable. On the other hand, as winds at the air–sea interface are required, the accuracy depends on the reliability of the meteorological boundary layer parametrization in the models.

The winds that we have used are outputs of the European Meteorological Centre (European Centre for Medium-Range Weather Forecasts; ECMWF). They are 6 h, 1.875° resolution, 10 m height winds. The year 1982 was selected, and k and ks were estimated each 6 h on the grid of the meteorological model and monthly averaged. Figure 2 shows isocontours for k (in units of centimetres per hour) for February, May, August and November. Figure 3 provides numerical values for ks averaged on 7.5° × 7.5° squares for the same months, expressed in units of 10^{-2} mol m^{-2} a^{-1} μatm^{-1}. Figure 3 is mainly provided so that the reader can extract numerical values usable in estimates of CO_2 fluxes from p_{CO_2} measurements. As in Erickson & Duce (1987), our values are not corrected for temperature and salinity; that is, k is normalized to a Schmidt number of 600. Note that the dependence of k on temperature and salinity is small compared to that on wind, so that k could be adjusted by using climatological values of these parameters. As for ks, its dependence on temperature is small (less than 10% for a temperature change of 20 K for winds larger than 3.6 m s^{-1}), so that our values can be used directly for estimating the CO_2-exchange fluxes.

2. Validation

To validate these maps and assess their representativity, we made the four following tests.

The first test is a comparison with the maps of Erickson & Duce (1987). This could only be made qualitatively (i.e. by eye from the isolines). The values agree reasonably well, that is within 30%. This is true even in the tropics where the ECMWF winds are frequently said to be too low (A. Ratier, H. Charnock, personal communications).

The second test is a comparison with gas-transfer velocities derived from wind and sea-surface temperature data provided by a meteorological buoy near 47° N, 17° W (station Romeo off Brittany). Of course, as these data have been assimilated into the model, the comparison for the particular period should be rather good. For this comparison we derived first the monthly mean values of k for the model grid square that contains the buoy, for the years 1981–1984 (heavy line in figure 4). Note that in this data set, the temporal resolution was 12 h. This provides a four-year average of monthly mean values of k. Figure 5 then compares this 'reference' to the values derived from 30 months of three-hourly buoy data during the years 1975–1977. In this figure, the lower diagram shows the monthly mean wind and sea-surface temperature values and their monthly root mean squares (r.m.s.). The upper diagram compares the normalized monthly values for k with the reference curve: the difference is less than 30%, winter 1976 values being consistently lower than those of winter 1977 (this may be related to the 1976 El Niño event and the well-remembered drought in Europe that year). This suggests that, at least at this location, the model winds are reasonably correct.

The heavy broken curve in the upper diagram of figure 5 shows the effect of temperature at this location (that is of the dependence of the Schmidt number on temperature in (2)). The

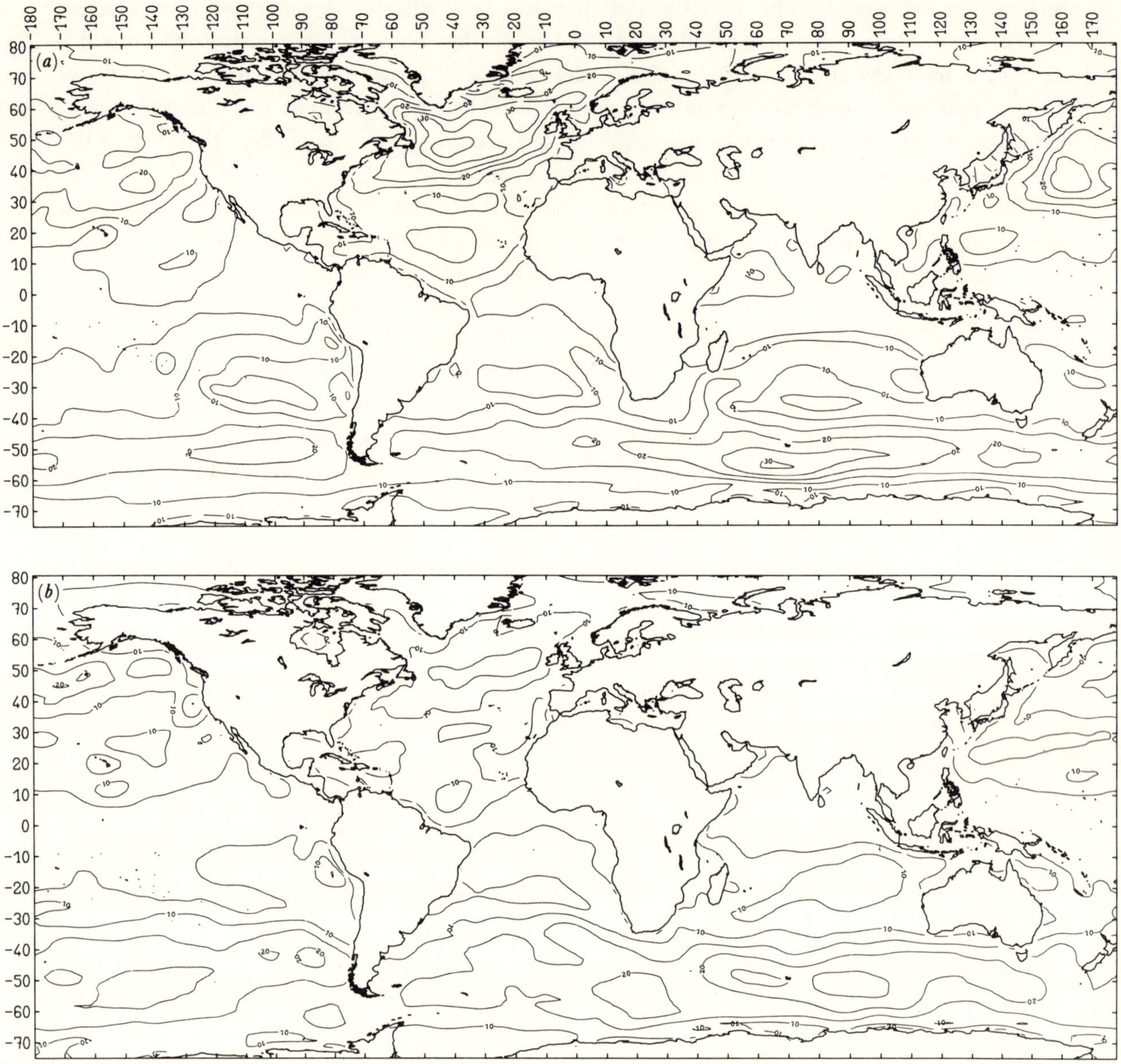

FIGURE 2 (*a*, *b*). For description see opposite.

value of k is decreased by 20% in winter, when the temperature is lower (around 12 °C). Finally the light broken curve illustrates the effects of averaging: wind and temperature are monthly averaged first, before calculating k (contrary to what is done for figures 2–4). Although the wind speed varies by 40% on the average (lower diagram) and its values frequently overlap the 13 m s^{-1}break in the curve of figure 1 (and thus nonlinearity effects are significant), the monthly value of k is generally less than 10% below the heavy broken curve. This would suggest that for estimating monthly mean values for k, monthly averages of wind and temperature are sufficient, provided the sampling gives a good statistical description of the real situation in the area considered.

A third test concerned the effects of sampling frequency on monthly averages of k by using

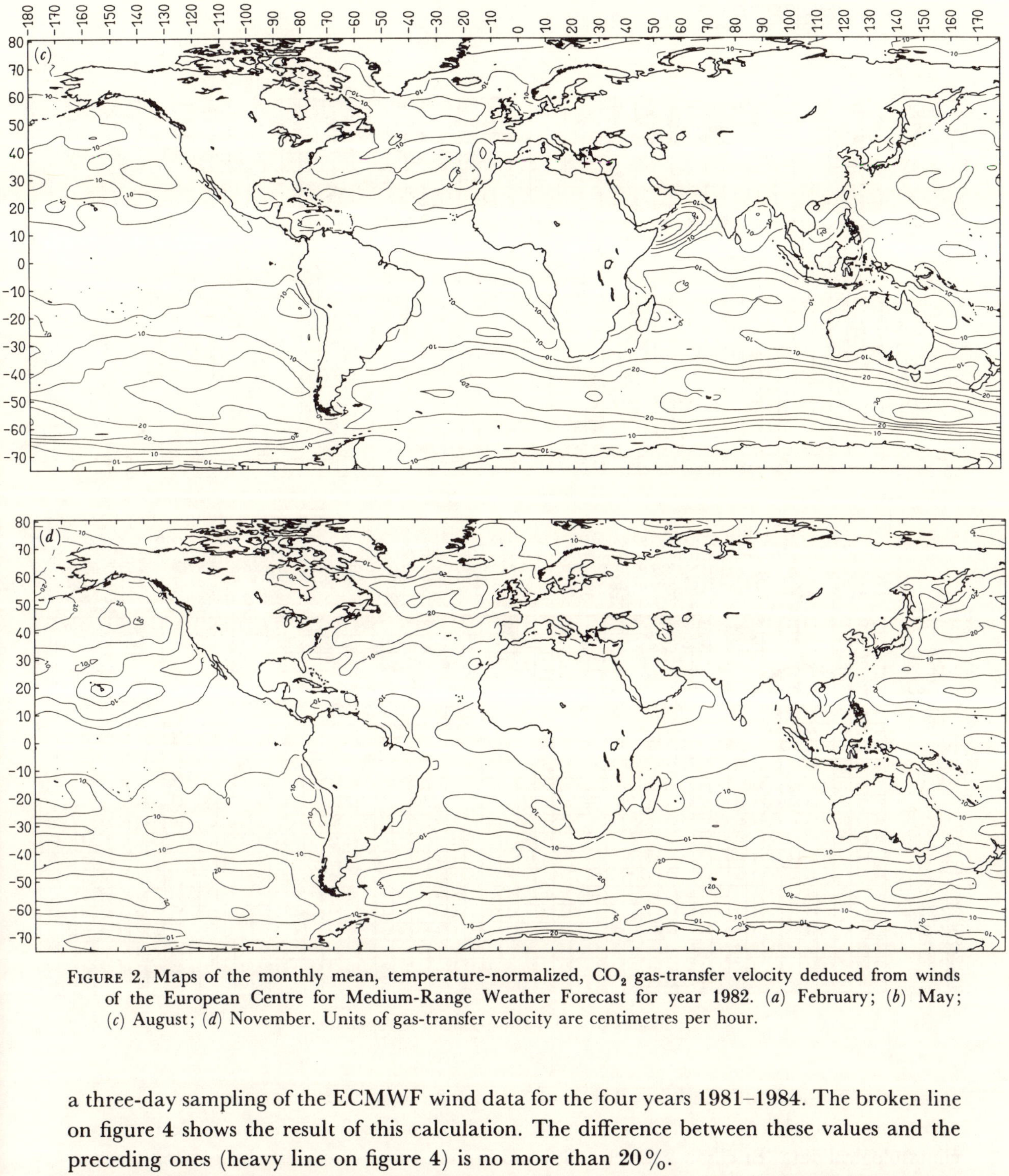

FIGURE 2. Maps of the monthly mean, temperature-normalized, CO_2 gas-transfer velocity deduced from winds of the European Centre for Medium-Range Weather Forecast for year 1982. (*a*) February; (*b*) May; (*c*) August; (*d*) November. Units of gas-transfer velocity are centimetres per hour.

a three-day sampling of the ECMWF wind data for the four years 1981–1984. The broken line on figure 4 shows the result of this calculation. The difference between these values and the preceding ones (heavy line on figure 4) is no more than 20%.

The fourth test evaluates the representativity of the year 1982. This is already indicated in figures 4 and 5, which show interannual variations of up to 30% near 47° N, 17° W. It is further documented by mapping the relative variations of monthly mean values for *k* deduced from the four years of ECMWF winds (years 1981–1984), over the tropical and North Atlantic Ocean (figure 6). These interannual variations can be in excess of 100%, but in well-defined

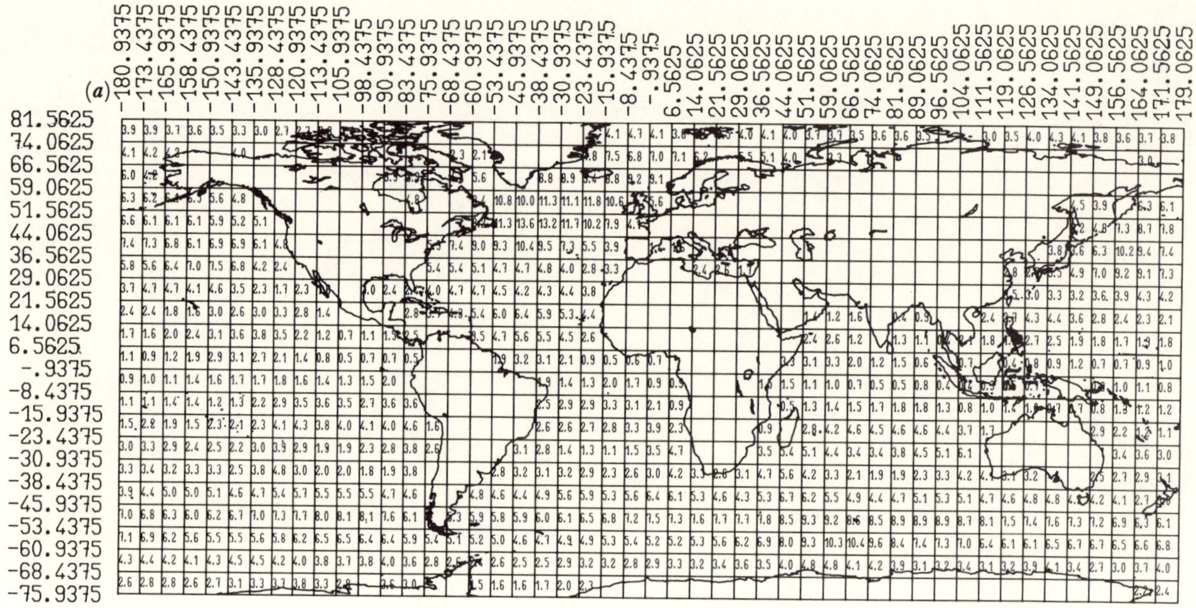

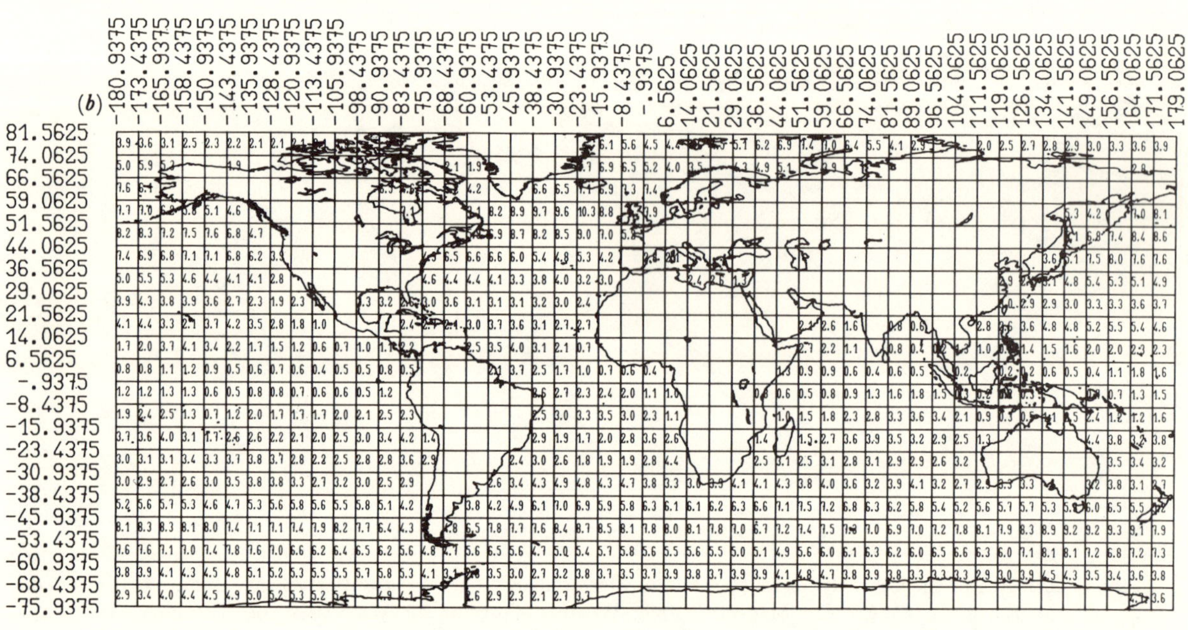

FIGURE 3 (*a*, *b*). For description see opposite.

areas only: in the trade winds at all seasons, and in the westerlies in summer and winter. At other locations, the change is less than 60%. Thus, values from the maps derived from climatological data (Erickson & Duce 1987) and from our maps can differ by a factor 2 for any particular year, because of interannual variability of the wind. However, we suspect that these interannual variations are mainly caused by the varying latitudinal boundary between the trade winds and the westerlies, as shown by the empirical orthogonal function analysis of these ECMWF winds over the North Atlantic Ocean of MacWeigh *et al.* (1988). Most of the large-scale features in figures 2 and 3 should therefore be generally valid.

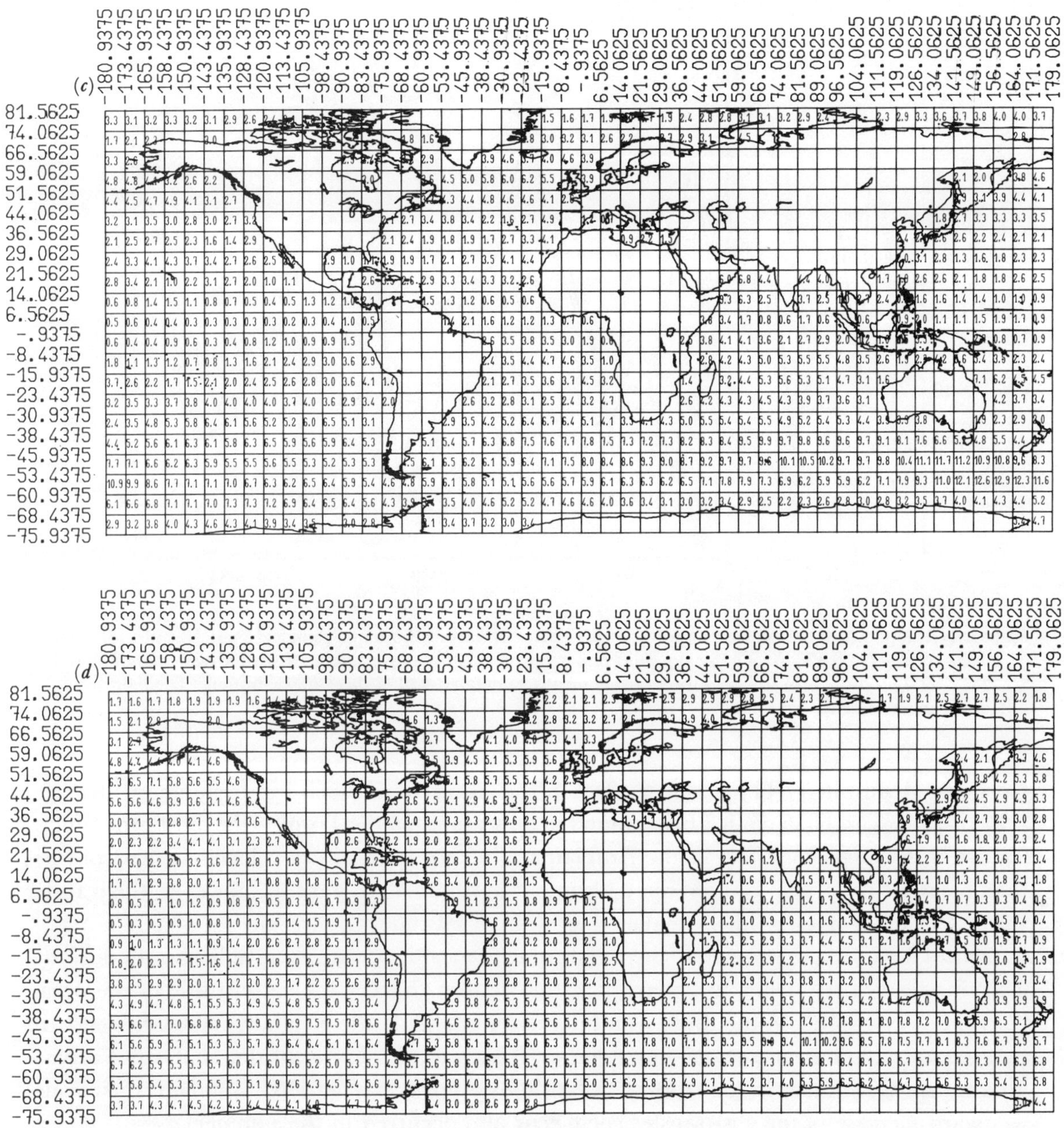

FIGURE 3. Values of the monthly mean CO_2 gas-transfer coefficient (ks) (in units of 10^{-2} mol m^{-2} a^{-1} μatm^{-1}), derived from figure 2 by multiplying k by a constant value for CO_2 solubility at 20 °C (33.37 mol m^{-3} atm^{-1}) and by averaging over 7.5° × 7.5° squares. (*a*) February; (*b*) May; (*c*) August; (*d*) November.

3. Analysis of the results

The k and ks vary from small values ($k < 10$ cm h^{-1} and $ks < 0.05$ mol m^{-2} a^{-1} μatm^{-1}) over most of the ocean, to values in excess of 35 cm h^{-1} and 0.1 mol m^{-2} a^{-1} μatm^{-1} in the Somali area or at high latitudes in winter time. The temperature effect would decrease the actual winter values of k by a factor 2 but the effect on ks would be less than 10%.

The most obvious feature in figures 2 and 3 is the seasonal change (by a factor 2) in the

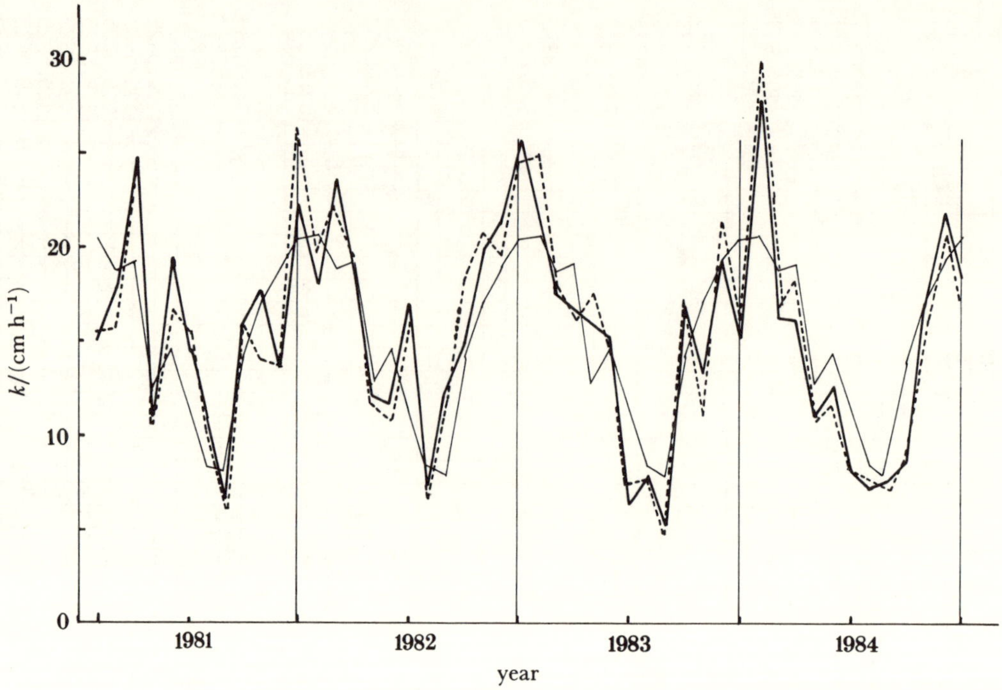

FIGURE 4. Temperature-normalized monthly mean values of k from the meteorological model winds, for 1981 to 1984, near 47° N, 17° W. The heavy solid line is for monthly means calculated with a twelve-hour sampling ($Sc = 600$) and the broken line is for monthly means calculated with a three-day sampling. The light solid line is the average of the four years.

Northern Hemisphere. The seasonal effect in the Antarctic is much less pronounced, transfer velocities in excess of 25 cm h^{-1} being found in all seasons. In the Tropics, the most pronounced feature is the monsoon effect in the Arabian sea, where the transfer velocity varies from less than 5 cm h^{-1} in spring, to more than 30 cm h^{-1} in summer. In general, however, the transfer velocity in the tropics is much smaller than at high latitudes (figure 7).

Figures 2 and 3 also show that there is as much variability with longitude as with latitude. The transfer velocity tends to be larger in the western than in the eastern side of the ocean basins. For example, k in the trade-winds domain in the Atlantic generally increases westward by a factor of 2–5 in all seasons but summer. This was also found in the analysis of ocean–atmosphere CO_2 fluxes in the tropical Atlantic by Andrié *et al.* (1986) (although their zonal westward increase of the gas-transfer coefficient also occurred in the summer).

4. Discussion and conclusion

The world-scale yearly mean is 10 cm h^{-1} or 0.029 mol m^{-2} a^{-1} μatm^{-1}. This is smaller than the global piston velocity deduced from ^{14}C decay model that is of the order of 15–20 cm h^{-1} (see for example, Broecker *et al.* 1986). One could use this to assess the value of our estimates. Uncertainties can be attributed to the wind data set or to the choice of the wind dependence of k (equation (2)).

However, two factors indicate that this is not necessarily the case. First, one should remember

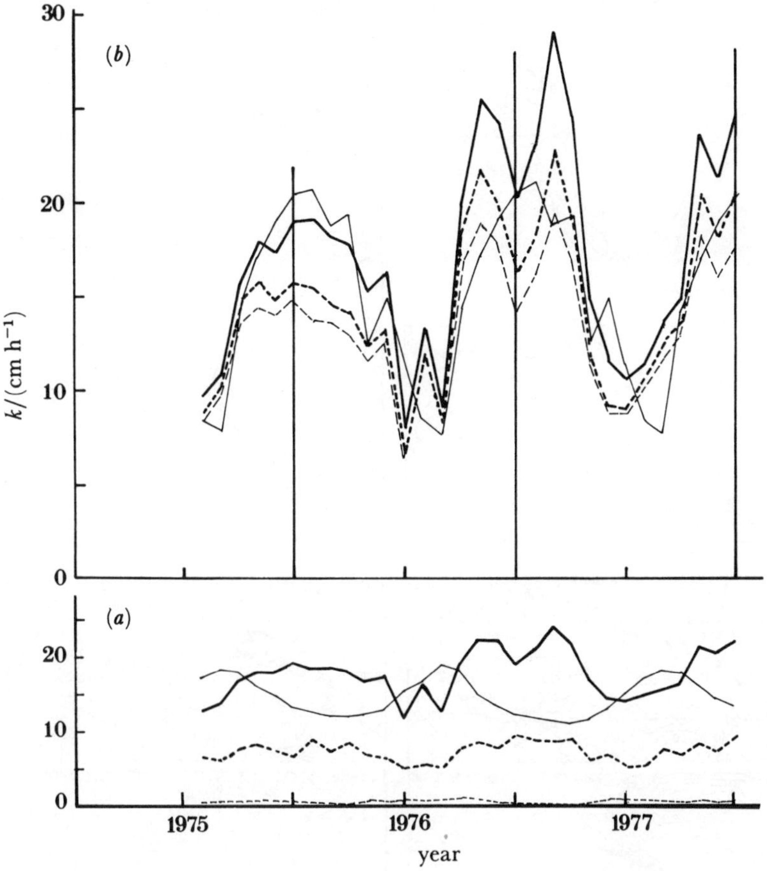

FIGURE 5. (*a*) Monthly mean wind and sea-surface temperature and their monthly r.m.s. values at meteorological buoy Romeo for years 1975–1977 (30 months). Heavy line, wind velocity (knots); light line, temperature (°C); heavy broken line, wind r.m.s.; light broken line, temperature r.m.s. (*b*) Heavy line, temperature normalized, monthly mean CO_2 gas-transfer velocity derived from the buoy data by using equations (2); light line, four-years average from figure 5 (reference); heavy broken line, same as heavy line but corrected for temperature variations; light broken line, monthly transfer velocities deduced from the monthly averaged wind and temperature values by using equations (2).

that the ^{14}C method mostly describes CO_2 invasion at high latitudes and subsequent radioactive decay in the deep ocean. Thus the effective value for k would be much higher than the world average. For example, the yearly mean in the Southern Ocean is of the order of 17 cm h^{-1} and 0.05 mol m^{-2} a^{-1} μatm^{-1}, close to the ^{14}C value.

Second, the observations of strong seasonal and zonal variations of the gas-transfer coefficient suggest that world-scale averages should be made with great care. An example of the effects of averaging in estimating the CO_2 gas flux is given here, by using the data of the North Pacific seasonal programme (Takahashi *et al.* 1986): figure 8 compares the zonal changes of ks from figure 3 with that for Δp_{CO_2} from Takahashi *et al.* In winter, the intensity of degassing due to Δp_{CO_2} strongly decreases from west to east. This is enhanced by 20% by a change of ks. In summer, Δp_{CO_2} is much smaller and of the opposite sign (pumping by the ocean). But the actual exchange is very small, because ks is a factor 2 smaller than in winter.

For estimating the net yearly gas flux to the ocean in this area, one has to average zonally,

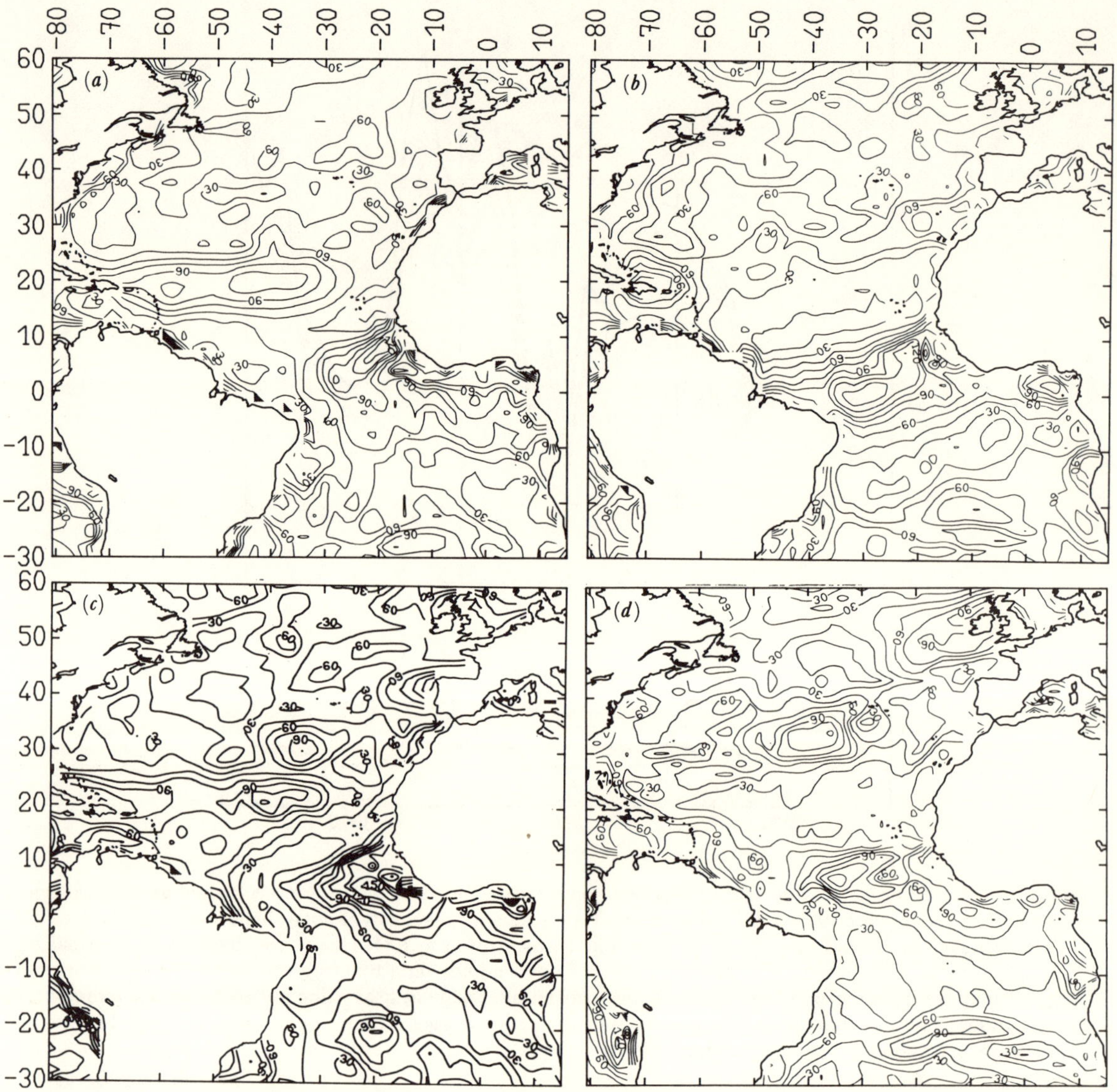

FIGURE 6. Relative variations (percentage, of the temperature-normalized, monthly mean values for the gas-transfer velocity deduced from the ECMWF winds for years 1981–1984. Isococontours each 15%. (*a*) February; (*b*) May; (*c*) August; (*d*) November.

weighting by the surface involved as in Takahashi *et al.* (1986), and seasonally. Denoting the zonal average by an overbar and the seasonal average by an underbar, one gets

$$\underline{\overline{ks}}\;\underline{\overline{\Delta p_{CO_2}}} = 0.81\ \text{mol m}^{-2}\ \text{a}^{-1},$$
$$\overline{\underline{ks}\;\underline{\Delta p_{CO_2}}} = 1.01\ \text{mol m}^{-2}\ \text{a}^{-1},$$
$$\underline{\overline{ks}\;\overline{\Delta p_{CO_2}}} = 3.54\ \text{mol m}^{-2}\ \text{a}^{-1},$$
$$\underline{\overline{ks\;\Delta p_{CO_2}}} = 4.72\ \text{mol m}^{-2}\ \text{a}^{-1}.$$

In the first case, the averages are made on the two terms separately, before making the product. In the last case, the product was made first, then averaged. The result is a factor of 6 larger than in the first case! Of course, the most critical factor is seasonality, as indicated by comparing the second and the third cases. Working with zonal averages of the seasonal values underestimates the flux by 30%.

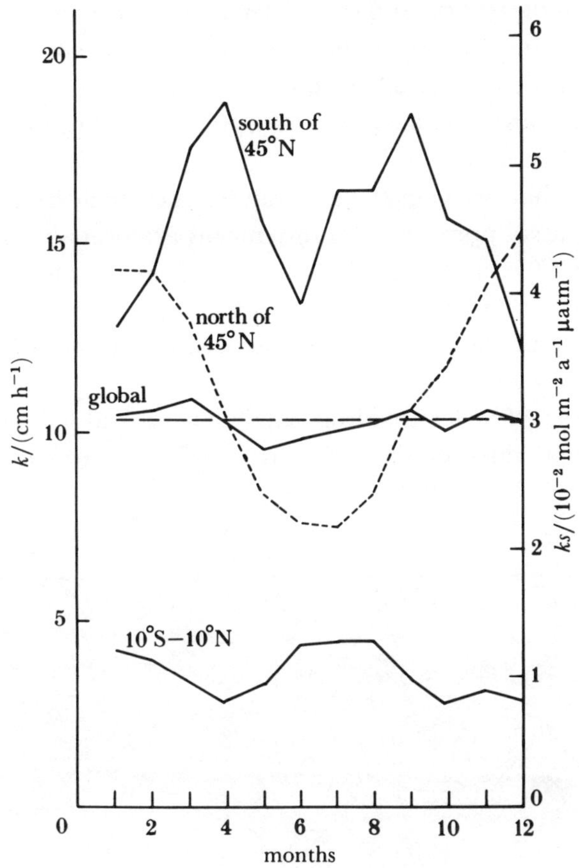

FIGURE 7. Zonal averages of the monthly mean value of the temperature-normalized CO_2 gas-transfer velocity and of the CO_2 gas-transfer coefficient, for high- and low-latitude bands. 'Global' stands for the world-scale averages.

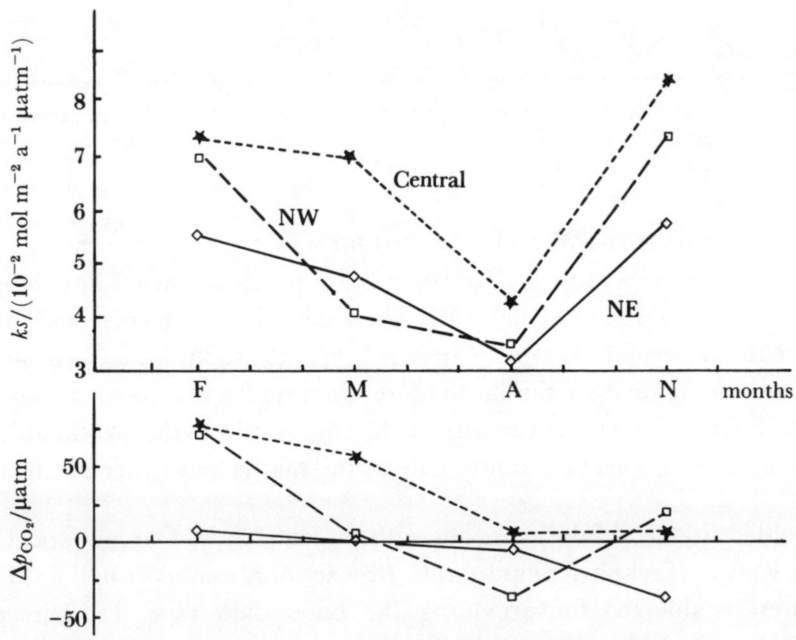

FIGURE 8. Seasonal variations of Δp_{CO_2} (lower diagram) and the gas-transfer coefficient (upper diagram) in the Central, North West (NW) and North East (NE) Pacific; p_{CO_2} data from the North Pacific seasonal program (Takahashi *et al.* 1986). Gas-transfer coefficients are averaged from figure 3.

This very large effect indicates the difficulty in making a correct direct estimate of the net fluxes of CO_2 between various parts of the ocean and the atmosphere. For this, seasonal programmes of Δp_{CO_2} measurements, such as those in the North Pacific (Takahashi *et al.* 1986), in the tropical Atlantic Ocean (Smethie *et al.* 1985; Andrié *et al.* 1986) or in the South Indian Ocean (Goyet 1987) will be necessary.

At present, we believe that deriving the gas-transfer coefficient from meteorological winds with equation (2) is the best approach. Monthly means are probably appropriate in view of the typical timescale for CO_2 equilibration between the ocean and the atmosphere (of the order of one year). From the present analysis, such values are apparently accurate within 30 %, and they are probably more representative than values deduced from wind measurements from ships.

In the future, estimates will be feasible from satellite wind and temperature measurements. As an example, Figure 9 provides a map of the gas-transfer velocity, deduced for August 1978

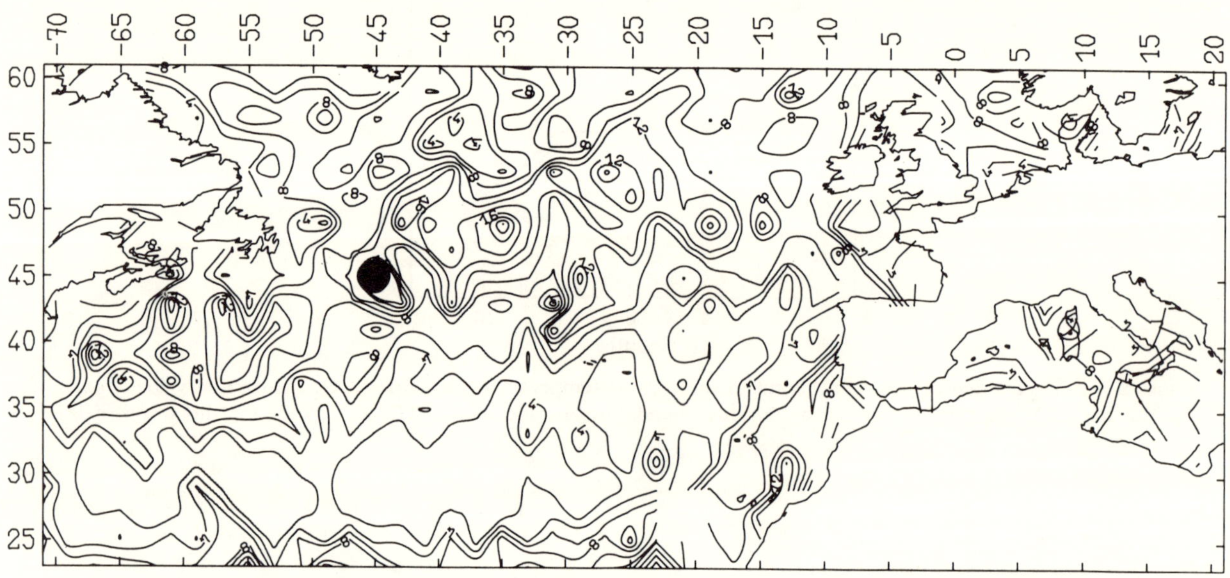

FIGURE 9. Map of the gas-transfer velocity (in units of centimetres per hour, corrected for temperature) derived for August 1987 from the *Seasat* altimeter wind data (see Chelton & Wentz 1986) and the climatological sea-surface temperature of Levitus (1982). Isoline each 2 cm h^{-1}.

from the *Seasat* altimeter wind speeds and corrected for temperature from the Levitus (1982) climatological sea-surface temperatures. These winds are provided each 7 km along the track of the satellite. The wind values, as given in the *Seasat* data files were corrected for their bias at strong wind intensities (see, for example, Chelton & Wentz (1986) for a discussion). Most of the significant features in figure 9 are similar to those of figures 2 and 3 (trade winds, westerlies, intensification near 50° N). Such maps can already be obtained from the Geosat altimeter wind speeds (Cheney *et al.* 1986). Their reliability will be the matter of further studies.

This work benefited from fruitful discussions with C. Jeandel, V. Garçon, and L. Memery, B. Barnier and A. Ratier. Technical help from C. Brossier has been most useful. M. Ollitrault from COB in Brest is thanked for providing the buoy data tape. La Direction de la Méteorologie Nationale provided the ECMWF wind data files.

References

Andrié, C., Oudot, C., Genthon, C. & Merlivat, L. 1986 CO_2 fluxes in the tropical Atlantic during Focal cruises. *J. geophys. Res.* **91**, 11741–11755.

Broecker, W. S., Ledwell, J. R., Takahashi, T., Weiss, R., Merlivat, L., Memery, L., Peng, T. H., Jahne, B. & Munnich, K. O. 1986 Isotopic versus micrometeorologic CO_2 fluxes: a serious conflict. *J. geophys. Res.* **91**, 10517–10527.

Chelton, D. B. & Wentz, F. J. 1986 Further development of an improved altimeter wind algorithm. *J. geophys. Res.* **91**, 14250–14260.

Cheney, R. E., Douglas, B. C., Agreen, R., Miller, L., Milbert, D. & Porter, D. 1986 The Geosat altimeter mission: a milestone in satellite oceanography. *Trans. Am. geophys Un.* **67**, 1354–1355.

Erickson, D. J., Merill, J. T. & Duce, R. A. 1986 Seasonal estimates of global oceanic whitecap coverage. *J. geophys. Res.* **91**, 12975–12977.

Erickson, D. J. & Duce, R. A. 1987 On the global transfer velocity field of gases with a Schmidt number of 600. *Searex Newslett.* **10**, 7–10.

Goyet, C. 1987 Variations saisonnières de pCO_2 dans les eaux de surface du Sud Ouest de l'Ocean Indien. Thèse de Doctorat de l'Université Pierre et Marie Curie, Paris, France.

Hellerman, S. & Rosenstein, M. 1983 Monthly wind stress over the world ocean with error estimates. *J. phys. Oceanogr.* **13**, 1093–1104.

Jahne, B. 1980 Zur parameterisierung des gasaustausches mit hilfe von laborexperimenten. Doctoral Dissertation, University of Heidelberg. (124 pages.)

Levitus, S. 1982 Climatological atlas of the world ocean. *N.O.A.A. Professional Paper* no. 13. Rockville, Maryland. (173 pages.)

Liss, P. S. & Merlivat, L. 1986 Air–sea gas exchange rates: introduction and synthesis. In *The role of air sea gas exchange in geochemical cycling* (ed. P. Buat-Menard). N.A.T.O. A.S.I. series, vol. 185, pp. 113–128.

MacWeigh, J. P., Barnier, B. & Le Provost, C. 1988 Spectral and E.O.F. analysis of four years of E.C.M.W.F. wind stress curl over the North Atlantic ocean. *J. geophys. Res.* (In the press.)

Smethie, W. M., Takahashi, T., Chipman, D. W. & Ledwell, J. R. 1985 Gas exchange and CO_2 fluxes in the tropical Atlantic ocean determined from ^{222}Rn and pCO_2 measurements. *J. geophys. Res.* **90**, 7005–7012.

Takahashi, T., Goddard, J., Sutherland, S., Chipman, D. W. & Breeze, C. C. 1986 Seasonal and geographic variability of carbon dioxide sink/source in the oceanic areas: observations in the North and equatorial Pacific ocean, 1984–1986 and global summary. *Final report to CO_2 research division*, Office of Energy Research U.S.A., Department of Energy. (52 pages.)

Thomas, F., Foubert, J. M., Perigaud, C., Merlivat, L. & Minster, J.-F. 1986 Mapping the CO_2 ocean-atmosphere gas transfer coefficient from climatological wind and sea surface temperature data. *Trans. Am. geophys. Un.* **67**, 1041.

Weiss, R. F. 1974 Carbon dioxide in water and sea water: the solubility of a non ideal gas. *Marine Chem.* **2**, 203–215.

A problem in the application of inverse methods to tracer data

BY J. G. SHEPHERD

Ministry of Agriculture, Fisheries and Food, Directorate of Fisheries Research, Fisheries Laboratory, Lowestoft, Suffolk NR33 0HT, U.K.

Attempts to apply inverse methods to the interpretation of tracer data usually seek some least squares solution of the flux divergence equations

$$AC = S,$$

where A is a matrix of transport coefficients, C a vector of concentrations and S a vector of sources/sinks.

However, what is often really required is a set of values for the elements of A, which will give a satisfactory prediction of the concentrations. This corresponds to finding a least squares solution of

$$C = A^{-1} S.$$

The two problems are not equivalent. The latter corresponds to an extensively re-weighted version of the former, where the weights depend on the solution (the elements of A). The former is linear in the elements of A: the latter is highly nonlinear. In addition, the matrix A is invariably sparse, and required to be so. However, A^{-1} is not, nor is it guaranteed that its inverse will be if its elements are determined freely.

It is not clear whether the standard methods of generalized inverse theory are applicable to the more difficult 'real' problem nor, if they are not, what other methods might be used. It is, however, possible that solutions of the 'real' problem, if they can be found, would be more informative.

1. INTRODUCTION

The application of inverse methods to the interpretation of oceanic geochemical and hydrographic data, with a box model representation of the system, was pioneered by Wunsch & Minster (1982). On the face of it this methodology is ideally suited to the problem. Box models have been used for many years by chemical oceanographers to describe the systems of interest to them (see, for example, Broecker 1979; Keeling & Bolin 1967). The systems of equations involved are not large, and the transport processes are very naturally represented by the set of linear equations that arise.

The success of attempts to use this technique is difficult to judge, because the truth is not usually known. In general, however, those who have tried it seem to have been less than satisfied with their results, either because the solutions contained implausible features, or because the results revealed inconsistencies between the data and the model (Schlitzer 1984; Bolin *et al.* 1987; Mobbs *et al.* 1987). In addition simulation studies (Fiadeiro & Veronis 1984) have revealed that the solutions may be very sensitive to the details of the technique. Although it is true that inverse methodology also permits much information about the indeterminate features of the solution to be obtained (Wunsch 1985), its interpretation is not entirely straightforward. There are several possible reasons why work in this field has not been as productive as had been hoped.

First, it seems likely that we have been attempting to determine too much from tracer data alone. Asking for advection and diffusion to be determined at every interface requires the determination of small-scale features, and may anyway lead to structural indeterminacy. It is after all well known that even in a simple one-dimensional system only the ratio of advective and diffusive transports can be deduced from steady-state tracer data, and not both quantities independently.

Second, obtaining realistic solutions from under-determined systems requires that the auxiliary information used (e.g. choosing the norm to be minimized in constructing a generalized inverse) be well-chosen. This may not be straightforward.

Third, some care is needed to ensure that the mathematical problem is a proper representation of the practical or intellectual problem of interest. It is this aspect that I wish to discuss in this paper.

2. Forward, reverse and inverse problems

The systems of interest are fluids in which the processes of advection and turbulent diffusion interact, and where various tracers are created and destroyed by sources and sinks. The mathematical description is based on the classical differential equation

$$\mathrm{d}C/\mathrm{d}t = \operatorname{div}(K \operatorname{grad} C - uC) - \lambda C + S, \tag{1}$$

where C is the concentration of a tracer, K represents turbulent diffusivity, u the velocity field, λ a decay rate (chemical or radioactive) and S other sources or sinks.

For convenience in fitting to available data this equation is usually applied to a finite representation in the form of a (not very large) number of boxes: this is equivalent to the use of a rather coarse finite-difference or finite-element representation. Concentrating for simplicity only on the steady-state problem, it is easily shown that (1) may be reduced to a set of linear equations

$$\sum_{\mathrm{j}} A_{\mathrm{ij}} C_{\mathrm{j}} = -S_{\mathrm{i}}, \tag{2}$$

where there is one value of C and one of S for each box (i.e. both i and j index boxes). The elements of the (square) matrix A_{ij} are transport (exchange) coefficients, consisting of a diffusive and an advective element. They have the dimensions of volume transport rates; the exact form depends on the discretization process (upstream, weighted or centred diffencing, etc.) adopted. The linear equations (2) may of course be written in matrix form

$$\boldsymbol{A}\boldsymbol{C} = -\boldsymbol{S}. \tag{3}$$

In what follows I shall for simplicity speak as though one were concerned with only one tracer field, and ignore the additional (but very important) constraints imposed by conservation of water. This captures the essence of the problem, because adding additional tracers does not alter the structure of the problem. Conservation of water is in any case probably best ensured by the technique of 'loops' (characteristic circulations) adopted by Bolin *et al.* (1987). It is also simpler (but not essential) to proceed as though the diffusive transports were known or specified.

Mechanistically, the concentration field (approximated by the vector $\boldsymbol{C}$) is produced by the transport processes (represented by the matrix A) working on the field of sources and sinks (the vector $\boldsymbol{S}$). Thus it is natural to regard the solution of (3) for the concentrations, for given A and $\boldsymbol{S}$, to be the forward problem, corresponding to

$$\boldsymbol{C} = -A^{-1}\boldsymbol{S}. \tag{4}$$

Note that (confusingly) the forward problem involves the inverse of A.

In interpreting tracer data, what is usually required is to determine the unknown elements of A, for given values of $\boldsymbol{C}$ and $\boldsymbol{S}$. This is the inverse problem.

The usual procedure for solving this inverse problem has been to seek least squares solutions of (3), i.e. to minimize the residuals of the flux-divergence equations. This is most conveniently done by rewriting the equations in the equivalent form

$$B\boldsymbol{W} = \boldsymbol{S}'. \tag{5}$$

Here B is a sparse matrix whose elements are concentration values, and $\boldsymbol{W}$ is a vector of the unknown transport coefficients. $\boldsymbol{S}'$ is a vector of sources and sinks, modified by decay (and diffusive flux divergence if diffusive exchanges are treated as known). Standard and very efficient methods for the solution of linear equations (e.g. singular-value decomposition) may then be applied, after appropriate attention to row and column scaling (see, for example, Wunsch & Minster 1982; Wunsch 1985). Note that the matrix B is not square, unlike A, and that it may be both underdetermined and inconsistent at the same time (Bolin *et al.* 1987).

This is a simple and efficient way to attack this inverse problem, but it is not the only way, and it may not be the best way. Firstly, the concentration and source/sink data are treated as exact: concentration values in particular are simply fed in as matrix elements, and analysis of the influence and propagation of errors in them is not straightforward. Second, minimization of residuals of the flux-divergence equations may not be the most appropriate goal. For many practical applications, one wishes to determine transport coefficients (i.e. the elements of the vector $\boldsymbol{W}$ or equivalently of the matrix A), to use them in a predictive mode. The goal is not just the interpretation of the data, but the construction of a model (essentially the box structure and a set of transport coefficients) that will have predictive utility. This is the situation when one is aiming to contribute to evaluating the problems of anthropogenic CO_2 releases, or radioactive waste disposal, for example (Bolin *et al.* 1987; Mobbs *et al.* 1987).

In these cases, a more appropriate goal would be to select transport coefficients that would yield solutions of the forward problem that fitted the data as closely as possible, i.e. to minimize the residuals of the forward equations (4), rather than the reverse (flux-divergence) equations (2), (3) or (5).

3. Nonlinearity

These statements of the problem are not equivalent. The forward version corresponds to an extensively reweighted version of the reverse version. In addition it is nonlinear in the unknowns (the non-zero elements of A). Finally, it is not obvious how it may be solved efficiently, especially if one recognizes that the elements of $\boldsymbol{S}$ cannot generally be treated as exact.

This problem was recognized by Wunsch & Minster (1982), who discuss it in §4.2 of their paper, and propose the use of the total inversion method of Tarantola & Valette (1982) (see also Mercier (1986)). This expands the size and complexity of the problem considerably, and may not be the ideal approach. Other methods are conceivable (Bolin *et al.* 1987).

The extent of the problem may be appreciated if one considers an obvious (but very unpromising) method; to solve directly for the individual elements of the inverse matrix A^{-1}. This preserves the linearity of the problem, but increases the number of unknowns to n^2 from (typically) $3n$ (for a three-dimensional configuration of n boxes). In addition, A is required to be sparse (only specified interconnections should exist), but there is no guarantee whatever that the inverse of A^{-1} (once it were determined) would be sparse. Ideally one would like to hold the number of unknowns at the number of non-zero transport coefficients, while guaranteeing the appropriate structure (sparseness and asymmetry) of A.

4. Analogies with linear regression

The problem is quite analogous to that of linear regression where, because the fitted line does not coincide with the data, there is an incidental variable (the fitted value) for each datum. It is possible, as in total inversion, to treat all these incidental variables as unknowns to be determined, subject to the constraints that all must lie on the fitted line. This is unnecessary in the regression problem, because its structure is such that one can determine the usual regression parameters independently, and simply compute the incidental variables afterwards. Alternative approaches to the problem may thus generate both large complex and time-consuming algorithms, or small, elegant and efficient ones.

The relation between inverse problems and regression was noted by Wunsch (1978), but does not seem to have received much attention since then. I suggest that it may be worthwhile to pursue the analogy, because there is a further parallel with regression, in that predictive regression is a linear problem, whereas functional regression (where one allows for errors in all variables) is nonlinear. Indeed, multiple functional regression may be reduced to an eigenvalue problem (Sprent 1969, ch. 6), which of course corresponds to the solution of a polynomial equation, but for which well-established algorithms, with alternative methods, are available. The full box-model inversion problem (allowing for errors in both C and S) may indeed be written as a multiple functional regression problem, and it would be of great interest to pursue this approach to its solution.

A further idea from statistics that may be relevant is that of the loss function. Physical scientists have a tendency to seek the 'true' solution to a problem (corresponding in many cases to a functional regression, for example). However, for practical purposes it may be desirable to use another loss function (i.e. the objective function to be minimized) such as mean square prediction error. This leads (in general) to different solutions, which are often 'shrunk' towards some mean value (see, for example, Copas 1983). In the box-model inversion problem, this might mean seeking models that minimize the prediction error for concentrations, rather than the 'best' physical transport coefficients.

It is therefore still possible that the disappointing results obtained from box-model inversions to date are in part caused by insufficiently developed methodology. As Wunsch has pointed out, the same problems arise in many fields and some methods have been re-invented several times. Work to date has been strongly influenced by geophysical techniques, but there are also results from the statistical theory of regression that may help us to find apt and efficient methods of solution for this most interesting class of problem.

I thank C. Wunsch for his criticism of my abstract, and for drawing my attention both to §4.2 of his 1982 paper, which I had forgotten, and to the work of Mercier.

References

Bolin, B., Björkström, A., Holmen, K. & Moore, B. 1987 On inverse methods for combining chemical and physical oceanographic data: a steady-state analysis of the Atlantic Ocean. In Holmen, K. 1987 Ph.D. thesis, University of Stockholm, Department of Meteorology, ISBN 91 7146 534 0.

Broecker, W. S. 1979 A revised estimate for the radiocarbon age of North Atlantic deep water. *J. geophys. Res.* **84** (C6), 3218–3226.

Copas, J. B. 1983 Regression, prediction and shrinkage (with discussion) *Jl R. statist. Soc.* **B45** (3), 311–354.

Fiadeiro, M. E. & Veronis, G. 1984 Obtaining velocities from tracer distributions. *J. phys. Oceanogr.* **14**, 1734–1746.

Keeling, C. D. & Bolin, B. 1967 The simultaneous use of chemical tracers in ocean studies: Part I: General theory of reservoir models. *Tellus* **19**, 566–581.

Mercier, H. 1986 Determining the general circulation of the ocean: a nonlinear inverse problem. *J. geophys Res.* **91** (C4), 5103–5109.

Mobbs, S. F., Hill, M. D., Gurbutt, P. A., Shepherd, J. G. & Killworth, P. A. 1987 Development of a compartment model of the ocean. In *Oceanic processes in marine pollution*, ser. 2 (ed. T. P. O'Connor, W. V. Burt & I. W. Duedall), pp. 81–97. Florida: Krieger Publishing Co.

Schlitzer, R. 1984 Bestimmung der Tiefwasserzirkulation des Ostatlantiks mittels einer Boxmodell: Auswertung von ^{14}C und anderen Tracerdaten. Thesis, Karl Ruprechts Universitat, Heidelberg. (95 pages.)

Sprent, P. 1969 *Models in regression and related topics*. London: Methuen. (173 pages.)

Tarantola, A. & Valette, B. 1982 Generalised non-linear inverse problems solved using the least squares criterion. *Rev. Geophys. Space Phys.* **20** (2), 219–232.

Wunsch, C. 1978 The general circulation of the N. Atlantic west of 50° W determined from inverse methods. *Rev. Geophys. Space Phys.* **16**, 583–620.

Wunsch, C. 1985 Can a tracer field be inverted for velocity? *J. phys. Oceanogr.* **15**, 1521–1531.

Wunsch, C. & Minster, J.-F. 1982 Methods for box models and ocean circulation tracers: mathematical programming and non-linear inverse theory. *J. geophys. Res.* **87** (C8), 5647–5662.

Discussion

W. J. Jenkins (*Woods Hole Oceanographic Institution, Massachusetts, U.S.A.*). It is my belief that the utility and potential of inverse calculations performed to date is not limited as much by technical issues associated with the computational 'machinery', as it is by the paucity of data and the appropriateness of the model geometry. As to the former, it is clear that there are simply not enough data to confidently answer the questions we are beginning to ask with such calculations. Regarding the latter, I feel that the geometry of the models used in such calculations, and the formulation of the constraint equations themselves have not been as well matched to the questions as they could be. In fairness to the modellers, the data limitations often force painful compromises, but I also think that we as a tracer–modelling community need to gain more experience in such areas. It is by such concerns that we may improve our sampling strategies in the future.

J. Shepherd. There is no doubt that the limitations of the data are an important factor. Anyone trying to interpret tracer distributions must be tempted to ask for full three-dimensional fields of everything on a one-degree spacing (synoptically), which is obviously not feasible and probably not necessary either. Thus the problem is to make the best possible use of data which are (or may become) available. I would argue that one needs to exercise more care over methodology when working with imperfect data, not less. An important aspect of this is allowing for the known imperfections of the data (for example, the existence of errors), which is why fitting techniques (including inverse methods) are such an appropriate tool. But better data will always help.

C. Wunsch (*Massachusetts Institute of Technology*, *Massachusetts*, *U.S.A.*). There are many things to be said (too extensive for this space) about this paper, but perhaps a brief record is possible. It is difficult to understand Dr Shepherd's pessimism. The number of people who have seriously tried to apply inverse methods to tracer data can be counted on the fingers of two hands, if not one. I note that ocean modellers have spent 35 years struggling with forward problems, and all the issues of stability, accuracy, boundary conditions, affordability, etc. Inverse modelling will almost surely require a similar effort, as no one has ever claimed that it was any simpler than for forward problems. Life is hard.

It is worth noting that there are many inverse methods and the most appropriate one to use depends upon the question being asked, the context, convenience, etc. In particular, there is no reason at all why the parameters found through an inverse calculation cannot be required to reproduce the data in the corresponding forward computation. But like any powerful tool, inverse methods can be dangerous to both the user and onlooker. One does not condemn automobiles as useless because they are capable of damage both to driver and bystander.

The intrinsic nonlinearity of most oceanographic inverse problems has been apparent from the beginning. But I would argue that the concentration on linearized versions has been the only reasonable approach in the learning process. The insights gained from linear problems are surely a necessary first step toward ultimate solution of the nonlinear ones. It was known in the nineteenth century that the Navier–Stokes equations are also intrinsically nonlinear. Would one argue that recognition of this nonlinearity should have prevented generations of fluid dynamicists from attempting to first solve linear problems?

J. Shepherd. It is true that the total effort on applying inverse methods has been quite modest. It is still disappointing that a technique that seems so apt turns out to be a little trickier than we thought, but I agree that it is much too soon to become discouraged.

I also agree that the inverse methodology is very powerful but deceptively difficult to drive. Its main virtue is, I think, that, properly applied, it forces one to face up to indeterminacies, the features of the solution that are not determined by the data. Extracting and understanding this information is not always easy, but invariably instructive. The danger with more subjective methods is that one can continue to delude oneself about what is determined and what is not.

The nonlinearity of the problem may have been apparent to Professor Wunsch and his colleagues from the start, but it has not received so much attention as the linear problem and I suspect that it has not been as widely appreciated as it deserves. It is quite possible that the linear methods are giving us most of the answer, and that the main problems lie elsewhere (for example, with the data, or structural indeterminacies, or problems of spatial scale, or box structure). Even so, I would expect everyone to agree that sooner or later we have to allow for errors in the data in a systematic way, and that means that we should not forget about the nonlinear problem even if it is expedient to work for the present with linear systems. Furthermore, if we can find efficient ways to solve the nonlinear problem, it may be desirable to use them, to avoid possible side effects or interactions with other problems.

J.-F. Minster (*CNES/GRGS*, *Toulouse*, *France*). The total inversion technique was used in the paper by Wunsch & Minster (1982) because it was recognized that the coefficients of the matrix were affected by errors and because the 'unknown' parameters cannot, in fact, take any value. The method is certainly useful. It suffers from two difficulties: it is computationally heavy (although this is becoming a less severe problem as the machine capacities are improving) and it is nonlinear. As a consequence, there are possibly several minima to the minimization function, and the solution tends to stay close to the *a priori* estimates. These *a priori* values and their errors have thus to be solved very carefully.

J. SHEPHERD. It does seem to me that the total inversion method is rather a sledgehammer approach to the problem, and I would personally be happier if we could find a scalpel or even a knife that would do the job. In addition, I find those methods that only compute departures from an earlier solution unappealing. Strong dependence on a prior solution implies serious (and maybe structural) indeterminacy in the problem, which should be investigated and pinned down somehow. I am glad to hear that work with total inversion methods is proceeding but I do not think I would encourage the young and innocent to get involved with it.

Tracers of air–sea gas exchange

By P. S. Liss

School of Environmental Sciences, University of East Anglia, Norwich NR4 7TJ, U.K.

The flux of gas across the air–sea interface is determined by the product of the interfacial concentration difference driving the exchange and a rate constant, often termed the transfer velocity. The concentration-difference term is generally obtained by direct measurement, whereas more indirect approaches are required to estimate the transfer velocity and its variation as a function of controlling parameters such as wind and sea state.

Radioactive tracers have proved particularly useful in the estimation of air–sea transfer velocities and, recently, stable purposeful tracers have also started to be used. In this paper the use of the following tracers to determine transfer velocities at the sea surface is discussed: natural and bomb-produced ^{14}C, dissolved oxygen, ^{222}Rn and sulphur hexafluoride. Other topics covered include the relation between transfer velocity and wind speed as deduced from tracer and wind-tunnel studies, and the discrepancy between transfer velocities determined by using tracers and from eddy correlation measurements in the atmosphere.

1. Basic principles of air–sea gas exchange

The net flux (F) of a gas across the sea surface is given by the product of the interfacial concentration (strictly fugacity) difference (ΔC) driving the flux and a factor that quantifies the kinetics of the exchange, often called a transfer velocity (K):

$$F = K.\Delta C. \tag{1}$$

The term ΔC is generally obtained by direct measurement of near-surface air and water gas concentrations and is not discussed further here. There are many papers on the subject, a convenient up-to-date selection of which are the chapters by Andreae (on sulphur compounds), Atlas & Giam (on high molecular mass synthetic organic compounds), Conrad & Seiler (on CO and H_2), Fitzgerald (on mercury), Liss (on low molecular mass halocarbon gases) and Siegenthaler (on CO_2), all in Buat-Ménard (1986).

The transfer velocity in (1) includes contributions from both the air and seawater near to the interface. However, it is possible to split the overall transfer velocity (K_w, where the subscript indicates that ΔC is expressed on a water-phase basis) into its air and water-phase components:

$$\frac{1}{K_w} = \frac{1}{\alpha k_w} + \frac{1}{Hk_a}, \tag{2}$$

where k_w and k_a are the transfer velocities in the water and air, respectively; H is the Henry Law constant of the exchanging gas (dimensionless ratio of air:water concentrations at equilibrium); and α is a factor that accounts for any enhancement of transfer on the water side of the interface from chemical reaction between the exchanging gas and the water. For unreactive gases $\alpha = 1.0$, whereas for those like SO_2 that have rapid aqueous-phase chemistry,

α can have a value of several thousand (Liss 1971). The form of (2) is similar to that for addition of electrical conductances in series and by using the same analogy the equation can be written in terms of resistances to interfacial gas transfer:

$$R_w = r_w + r_a, \tag{3}$$

where $$R_w = (K_w)^{-1},\ r_w = (\alpha k_w)^{-1},\ r_a = (Hk_a)^{-1}.$$

By substituting into (2) and (3) reasonable values for the various quantities it can be shown that for most gases either r_w or r_a is dominant. Thus, for gases that partition strongly into the water (low H) and/or react rapidly in the aqueous phase, $r_a \gg r_w$; gases in this class include H_2O, HCl, SO_2 and HNO_3. On the other hand, for gases that have high values of H and are inactive or only slowly reactive in water, $r_w \gg r_a$. This type includes many of the oceanographically important gases including O_2, N_2, CO_2, the Group 0 elements, SF_6 and the Freons. Thus, for the purpose of this paper it is permissible to confine attention to r_w and the factors controlling it, i.e. α and k_w. A further simplification is also possible because for all of these gases, with the possible exception of CO_2, their slow or non-existent reactivity in seawater means that α is effectively unity. Even for CO_2 the extent of exchange enhancement by chemical reaction is small, with α estimated to be about 1.02–1.03 under typical oceanic conditions (i.e. only a 2–3% increase in k_w compared with an inactive gas). Thus, in the present context, for all practical purposes, it is the value of k_w alone that is important. The discussion of air–sea transfer velocities given here has been kept to a minimum. The interested reader will find a more detailed treatment in, for example, Liss (1983).

2. The use of tracers in the measurement of k_w at the sea surface

A variety of gases have been used as tracers in the estimation of the liquid-phase transfer velocity, k_w, at the air–sea interface. In this section the various methods and the results of their application are reviewed.

(a) *Natural and bomb-produced* ^{14}C

Natural $^{14}CO_2$ in the atmosphere is produced by the bombardment of nitrogen molecules by cosmic rays. Assuming this production is time invariant and that steady state exists in the atmosphere–ocean system, the rate at which $^{14}CO_2$ enters the oceans across the sea surface must balance the rate at which the ^{14}C activity decays in the oceans. As Broecker & Peng (1974) have shown, by substituting measured values for the ^{14}C activity in the atmospheric and oceanic reservoirs into the steady-state equation a global estimate of k_w may be obtained.

Another way radiocarbon enters the atmosphere–ocean system is from the detonation of nuclear bombs. The temporal and spatial variations in bomb ^{14}C in the atmosphere since nuclear-weapons testing started in 1954 are well documented. By integrating the excess ^{14}C (i.e. from bomb detonation) in oceanic depth profiles, the rate of uptake of this radiocarbon and hence its air–sea transfer velocity may be calculated (Broecker & Peng 1974).

Values of k_w, whether obtained from natural or bomb ^{14}C, average about 20 cm h^{-1}, with a probable error of less than 25%. It should be stressed that this is very much a large-scale or global estimate and says nothing about the variability of k_w in space and time, or with possible controlling parameters such as wind speed or wave field. Although lacking in sensitivity, the use of ^{14}C in this context sets important constraints on other methods that may produce values of k_w with greater resolution. For example, a technique that consistently predicts k_w as very different from 20 cm h^{-1} must be suspect (see § 2*e*); unless our understanding of the partitioning of $^{14}CO_2$ in the atmosphere–ocean system is grossly in error, which seems unlikely.

(*b*) *Dissolved-oxygen-balance method*

In this approach a time series of dissolved-oxygen measurements is made at some location. The total change in oxygen concentration between observation times ($\delta O_2{}^t$) is then attributed to a combination of biological production–consumption due to photosynthesis–respiration and decomposition ($\delta O_2{}^b$), and exchange across the sea surface ($\delta O_2{}^e$). Then,

$$\delta O_2{}^t = \delta O_2{}^e + \delta O_2{}^b. \tag{4}$$

The method was first described by Redfield (1984) and he used the so called 'Redfield ratios' to obtain the term $\delta O_2{}^b$, as have all subsequent workers. The Redfield ratio links the change in oxygen to the concomitant change in dissolved phosphate (or other nutrient). Thus, by making measurement of changes in phosphate at the same time as those for dissolved oxygen, it is possible to correct the measured changes in oxygen concentration for biological addition or subtraction in the water column. The residual signal is then attributable to the exchange flux across the sea surface, and knowing the concentration gradient, the transfer velocity for oxygen can be calculated.

Although simple in concept, there are several important assumptions involved in this approach; for example, that the single value of the Redfield ratio ($1PO_4{}^{3-} = 138\ O_2$) applies in every area, and that horizontal and vertical advection of water into or out of the study area brings about no change in oxygen concentration. The method and its associated assumptions are discussed in greater detail in Liss (1983), where the main results of its application in a number of areas are summarized. Values of k_{O_2} measured with this technique in summer range between 5 and 15 cm h^{-1}, with winter values from 40 to 50 cm h^{-1}. The increased values in winter may be because of generally higher wind speeds in that season (see § 3). There would not seem to be any significant discrepancy between the results from this approach and the globally averaged values given by the ^{14}C methods. Some results obtained by Tsunogai & Tanaka (1980) from applying the oxygen-balance method in Funka Bay, Japan, are shown in figure 1. A clear increase in k_{O_2} from summer to winter is apparent, and there is evidence that the transfer velocity increases by more than the single power of the measured wind speed.

(*c*) *Radon-deficiency method*

This technique was first described by Broecker (1965) and relies on measuring the radioactive disequilibrium between the nuclides ^{222}Rn and ^{226}Ra in near-surface seawater. In the deep oceans the two isotopes are found to be generally in radioactive equilibrium. However, near the sea surface the gas ^{222}Rn is lost to the atmosphere because its activity in air is much less than that in seawater. By assuming steady state, for any measured vertical profile the rate of loss of ^{222}Rn to the atmosphere must equal the depth integrated deficiency in ^{222}Rn activity. The following equation shows this equality.

$$k_{Rn}(A_I - A_A) = \lambda \int_0^\infty (A_E - A_z)\, dz, \tag{5}$$

where k_{Rn} is the transfer velocity obtained from the method, A_I is the measured activity of

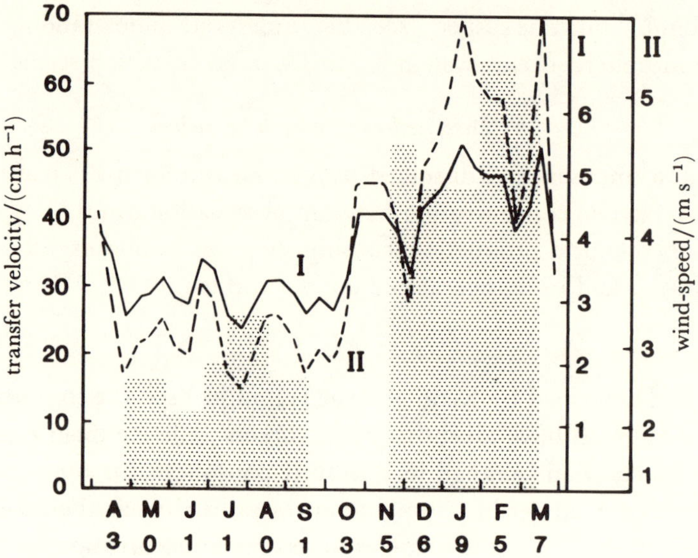

FIGURE 1. Transfer velocity for O_2 (shaded bars) and wind speed (line I) in Funka Bay, Japan, for the period 1979–1980. Curve II is proportional to the square of the wind speed averaged for each period of 10 days. Numbers below the letter for each month indicate the number of days that month when the wind velocity exceeded 10 m s^{-1}. In this study, the meteorological observations were made about 30 km from the water sampling point. (After Tsunogai & Tanaka 1980.)

^{222}Rn in the surface-most seawater, A_A is the activity to be expected in this water if at equilibrium with the amount in air (generally taken as zero in view of the small atmospheric activity and the low solubility of the gas in water), A_E is the activity of ^{222}Rn expected in the water column if equilibrium is achieved with dissolved ^{226}Ra, A_z is the measured activity of ^{222}Rn at depth z, and λ is the decay constant of ^{222}Rn (0.18 d^{-1}). A vertical profile of ^{222}Rn in the surface oceans is shown in figure 2. An important assumption of the method is that the ^{222}Rn profile is not affected by horizontal or vertical water movements.

The method is relatively easy to apply and it has been used quite extensively. Probably the biggest data set is that for the Atlantic and Pacific Oceans obtained as part of the GEOSECS cruises (Peng *et al.* 1979). These results are shown in figure 3: two important conclusions can be drawn. Firstly, the mean value of k_{Rn} for all the data is about 12 cm h^{-1}, which corresponds to an equivalent value for k_{CO_2} between 14 and 16 cm h^{-1}, the range arising from uncertainty over how the conversion from k_{Rn} to k_{CO_2} should be made (for further details see Liss & Merlivat 1986). A more recent set of results obtained by Smethie *et al.* (1985) in the tropical Atlantic gives a mean k_{Rn} of 15.0 cm h^{-1}, corresponding to a k_{CO_2} of approximately 18 cm h^{-1}. Thus, it is reasonable to conclude that there is quite close agreement between transfer-velocity values from the radon-deficiency method and those from radiocarbon.

The second conclusion to be made from the results in figure 3 is a negative one, i.e. that if there is a relation between transfer velocity and wind speed, as might be expected, it is not apparent in this substantial data set. The more recent results from the tropical Atlantic (Smethie *et al.* 1985) do show a better relation but even here the correlation coefficient is only about 0.5 and the data set is much smaller (less than 30 data points compared with about 100 for the GEOSECS results). The reasons for the apparent inability of the method to demonstrate

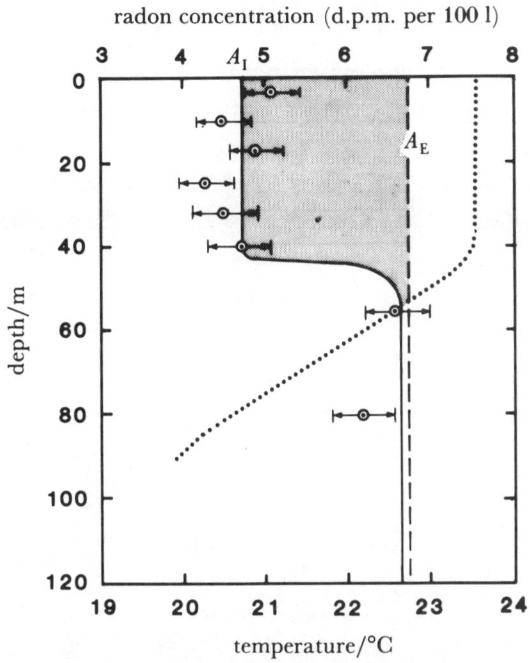

FIGURE 2. Activity of ^{222}Rn as a function of depth at GEOSECS station 57 in the Atlantic (24° S, 35° W). The broken line is the ^{222}Rn activity to be expected if the isotope is in equilibrium with its parent ^{226}Ra. The shaded area represents the amount of ^{222}Rn lost to the air. The dotted line is the temperature profile. (Radon concentration in disintegrations per minute per 100 l.) (After Broecker 1974.)

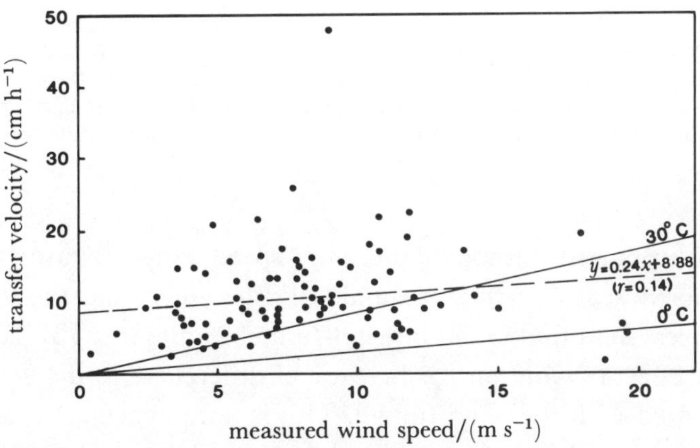

FIGURE 3. Transfer velocity determined on GEOSECS Atlantic and Pacific cruises by Radon Deficiency Method plotted against wind speed measured 0–24 h before sample collection (Peng *et al.* 1979). Wind speeds are at a nominal height of 10 m reduced to 10 cm height by assuming a logarithmic wind profile. Full lines are predictions based on the theoretical approaches of Deacon (1977) and Hasse (1971); broken line is the least squares fit through the data points. (After Hasse & Liss 1980.)

a clear wind-speed dependence are discussed in Liss (1983) and in Roether & Kromer (1984). Basically the vertical radon profile has a response time of a few days to any perturbation, which makes it problematical what value of the wind speed should be used in trying to correlate it to the derived transfer velocity. Further, during the response period the assumption of zero verical water movement may be invalid because of changes in mixed-layer depth and dynamics.

(d) Sulphur hexafluoride

At the time of writing, no gas-exchange experiments with a purposefully added tracer, such as SF_6, have been attempted in the oceans. This situation should change because several ocean releases of SF_6 and similar gaseous tracers are planned over the next few years.

However, sulphur hexafluoride has been used to obtain transfer velocities for a freshwater lake. Wanninkhof *et al.* (1985) added SF_6 to Rockland Lake in New York State and measured its rate of loss to the atmosphere, from which observations transfer velocities were calculated. The results are shown in figure 4 plotted as a function of wind speed. A clear relation between

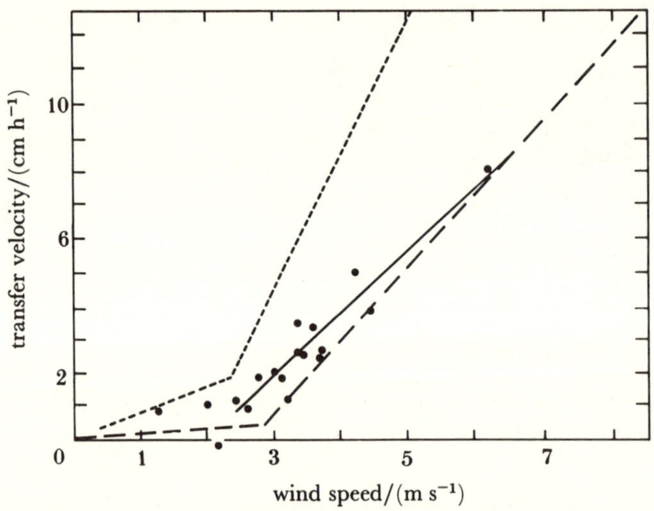

FIGURE 4. Transfer velocity for SF_6 determined on Rockland Lake (New York State, U.S.A.) as a function of wind speed measured 1 m above the lake. The solid line is the least squares fit to the data above a wind speed of 2.4 m s^{-1}. The short-dashed curve is from the wind tunnel study of Broecker *et al.* (1978). The long-dashed curve is an estimate of the relation between the transfer velocity and wind speed if the wind was steady. (After Wanninkhof *et al.* 1985.)

k_{SF_6} and wind speed is apparent, although the wind-speed range is rather limited with no measurements much above 6 m s^{-1} wind. It should also be noted that the slope of the graph is about a factor of two less than that found in wind-tunnel studies (see §3). It seems likely that the apparently smaller effect of wind on k is because of differences in the wind-induced wave field for the lake compared with the wind tunnel. This is not particularly surprising but does indicate that caution is required in extrapolating laboratory results to field situations where, among other differences, much larger wind fetches are the norm.

(e) Micrometeorological techniques

The methods so far discussed for determining k_w all rely on measurements on the seawater side of the air–sea interface. In contrast, micrometeorological techniques require measurements in the atmosphere. They have more usually been used for the determination of 'physical' fluxes of momentum, heat and water vapour, where the controlling resistance is generally on the atmospheric side of the interface. This highlights a basic problem in using micrometeorological methods for measuring k_w. For a gas for which $r_w \gg r_a$ (i.e. k_w is the effective transfer velocity) the fraction of the total concentration gradient in the air phase will be very small. This will inevitably lead to difficulties in measuring the gradient, and high instrumental signal:noise ratios will be important constraints. As an example of the problem, Deacon (1977) has calculated that for air–sea transfer of CO_2 the vertical gradient of the gas in the air between 1 and 10 m height would typically be 0.05 p.p.m. (by volume). This is less than 0.02% of the atmospheric concentration and hence difficult to measure accurately. For this reason the straightforward gradient-measurement approach does not appear to have been used for measuring k_w in the marine environment.

A micrometeorological technique that has been used in this context is eddy correlation. Here gas fluxes, and hence transfer velocities, are obtained from the time-average product of measured gas concentrations in air and the vertical component of the wind. The technique and some of the problems associated with it in estimating gas fluxes are discussed in Liss (1983).

The method has been used to estimate air–sea CO_2 fluxes by two groups (Jones & Smith 1977; Smith & Jones 1985; Wesely *et al.* 1982). The fluxes obtained, and the corresponding values of k_{CO_2}, are up to 100 times greater than those derived from the methods described earlier. This discrepancy, between results from eddy correlation and those from isotopic tracers, is critically discussed by Broecker *et al.* (1986), with replies to the criticisms from Smith & Jones (1986) and Wesely (1986). There are clearly problems with the eddy-correlation method as applied to gas exchange that have yet to be resolved. Fundamentally, as already mentioned, the problem is the poor instrumental signal:noise ratio; for CO_2 exchange, r_a is typically only about 0.6% of the total resistance (R_w).

The eddy-correlation method does appear to give reasonable results for ozone uptake by seawater (data of Lenschow *et al.* 1982, discussed in Liss & Merlivat 1986). The reason why it works for O_3, for which r_w is still the dominant resistance, is probably because the relatively rapid kinetics of the reaction of O_3 with seawater components (α of order 10^1) means that in this case r_a is about 4% of R_w. However, this is not a method for estimating k_w because α_{O_3} is not sufficiently well known.

3. Relation between air–sea transfer velocity and wind speed

In view of the difficulties, discussed in the previous section, of measuring k_w at sea, it would clearly be very useful if it could be related to some easily measured meteorological or oceanographic parameter. Several times already the presence (or absence) of a relation between k_w and wind speed has been mentioned.

Certainly, experiments conducted in laboratory wind tunnels all show clear increase(s) in k_w with wind speed (or, more appropriately, friction velocity). These results are described in Liss (1983), and Liss & Merlivat (1986) give an idealized plot of the wind-tunnel data, which is reproduced here as figure 5.

Liss & Merlivat (1986) identify three régimes each having different k_w against wind-speed relations. In the 'smooth-surface' régime (up to about 5 m s^{-1} wind) the water is largely smooth and undisturbed by capillary waves. At wind speeds between 5 and 10–12 m s^{-1} the surface becomes covered with capillary and larger waves ('rough-surface' régime) and k_w increases with wind speed at a faster rate than for the smooth surface. At higher wind speeds still, bubbles become common because of wave breaking ('breaking-wave (bubble)' régime) and the slope of the k_w against wind-speed relation becomes even greater. As indicated in figure 5, the effect of bubbles depends on the solubility of the exchanging gas in water. This is

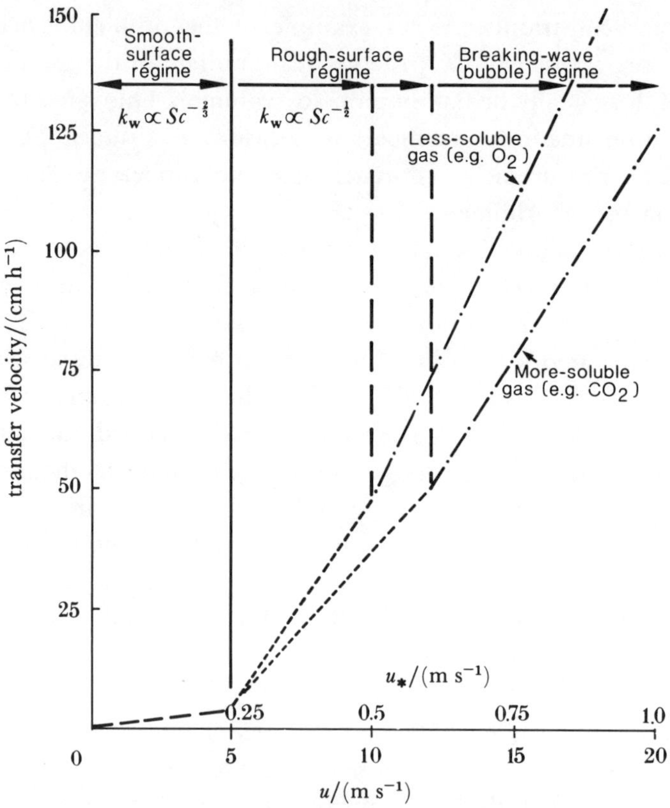

FIGURE 5. Idealized plot of water-phase transfer velocity (k_w) against wind speed (u) and friction velocity (u_*). Based on laboratory results, mainly from the Hamburg wind tunnel (Broecker & Siems 1984). Also shown is the likely dependency of k_w on Schmidt Number (Sc, the ratio of the water viscosity to the molecular diffusivity of the gas in water). (From Liss & Merlivat 1986.)

clearly shown in the wind-tunnel results for O_2 and CO_2 of Broecker & Siems (1984); a theoretical treatment of the effect is given in Merlivat & Memery (1983).

As mentioned earlier, the applicability of such necessarily small-scale laboratory studies to the oceans is open to question. To try to overcome this problem, Liss & Merlivat (1986) have used the lake results with purposefully added SF_6 of Wanninkhof *et al.* 1985 (shown in figure 4) to scale up the general relations illustrated in figure 5 to the field situation. Plainly, Rockland Lake is not the open oceans, but it is the most suitable field data available at present. Using this approach, Liss & Merlivat derive the following equations corresponding to the three régimes shown in figure 5.

$$k_w = 0.17\,u, \text{ for } u \leqslant 3.6, \tag{6}$$

$$k_w = 2.85\,u - 9.65, \text{ for } 3.6 < u \leqslant 13, \tag{7}$$

$$k_w = 5.9\,u - 49.3, \text{ for } u > 13, \tag{8}$$

where k_w is in units of centimetres per hour and the wind speed (u) is in units of metres per second at a height of 10 m.

Equations (6)–(8) are plotted out in figure 6. Also shown are averaged values for k_w as determined by the ^{14}C and ^{222}Rn techniques. The broad agreement between the predictions of (6)–(8) and the field data is reassuring. However, it must be noted that several of the field

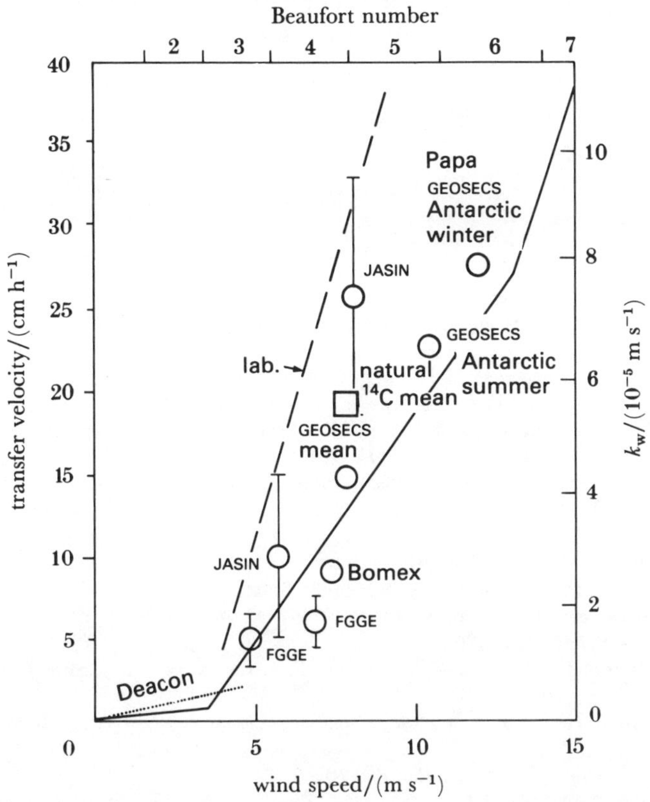

FIGURE 6. Predicted relation between transfer velocity and wind speed according to equations (6)–(8), solid lines. Also shown are oceanic measurements of k_w plotted as a function of wind speed measured at a height of 10 m and the corresponding Beaufort scale number. The dotted line shows predictions based on the Deacon (1977) smooth surface model. The broken line represents the results for intermediate wind speeds from laboratory wind-tunnel studies. All data have been converted to $Sc = 600$, which corresponds to CO_2 at 20 °C, by assuming that $k_w \propto Sc^{-\frac{1}{2}}$ (see caption to figure 5). (After Roether 1986 and Liss & Merlivat 1986.)

results are averages of quite large data sets (for example, from GEOSECS) and in reality show large dispersion in both k_w and wind speed. Also, there is an absence of field results for wind speeds much above 10 m s^{-1}. This, coupled with the fact that the Rockland Lake measurements were limited to wind speeds less than 7 m s^{-1}, means that predictions of k_w from (8) are considerably more uncertain than those from (6) and (7).

4. Conclusions

From this discussion of the use of tracers to estimate k_w for the oceans we can conclude that the globally averaged value of the transfer velocity for a gas like CO_2 is about 20 cm h^{-1}, with a likely error of $\pm 25\,\%$. The only contradictory indications are the estimates from the eddy correlation method, which yields much higher values, but there is considerable uncertainty over the reliability of this technique.

When it comes to estimating the transfer velocity for a particular part of the oceans at a certain time or season then the situation is much less satisfactory. Instantaneous field measurements of k_w are very difficult to make with presently available techniques. Extrapolation from wind-tunnel and lake results has yielded predictive equations for k_w, which is possibly the best approach we have at the moment. The real limitation here is the absence of field data, against which to test and verify the predictions, particularly for high-wind situations. Two possible ways ahead are the further development of micrometeorological techniques and the deployment of purposefully added tracers in coastal and oceanic waters.

References

Broecker, H. C. & Siems, W. 1984 In *Gas transfer at water surfaces* (ed. W. Brutsaert & G. H. Jirka), pp. 229–236. Dordrecht: Reidel.

Broecker, H. C., Petermann, J. & Siems, W. 1978 *J. mar. Res.* **36**, 595–610.

Broecker, W. S. 1965 In *Symposium on diffusion in oceans and fresh water* (ed. T. Ichiye), pp. 116–145. New York: Columbia University.

Broecker, W. S. 1974 *Chemical oceanography*. (214 pages.) New York: Harcourt Brace Jovanovitch.

Broecker, W. S. & Peng, T. H. 1974 *Tellus* **26**, 21–35.

Broecker, W. S., Ledwell, J. R., Takahashi, T., Weiss, R., Merlivat, L., Memery, L., Peng, T. H., Jahne, B. & Munnich, K. O. 1986 *J. geophys. Res.* **91**, 10517–10527.

Buat-Ménard, P. (ed.) 1986 *The role of air–sea exchange in geochemical cycling*. (549 pages.) Dordrecht: Reidel.

Deacon, E. L. 1977 *Tellus* **29**, 363–374.

Hasse, L. 1971 *Bound.-Layer Met.* **1**, 368–379.

Hasse, L. & Liss, P. S. 1980 *Tellus* **32**, 470–481.

Jones, E. P. & Smith, S. D. 1977 *J. geophys. Res.* **82**, 5990–5992.

Lenschow, D. H., Pearson, R. & Stankov, B. B. 1982 *J. geophys. Res.* **87**, 8833–8837.

Liss, P. S. 1971 *Nature, Lond.* **233**, 327–329.

Liss, P. S. 1983 In *Air–sea exchange of gases and particles* (ed. P. S. Liss & W. G. N. Slinn), pp. 241–298. Dordrecht: Reidel.

Liss, P. S. & Merlivat, L. 1986 In *The role of air–sea exchange in geochemical cycling* (ed. P. Buat-Ménard), pp. 113–127. Dordrecht: Reidel.

Merlivat, L. & Memery, L. 1983 *J. geophys. Res.* **88**, 707–724.

Peng, T. H., Broecker, W. S., Mathieu, G. G., Li, Y. H. & Bainbridge, A. E. 1979 *J. geophys. Res.* **84**, 2471–2486.

Redfield, A. C. 1948 *J. mar. Res.* **7**, 347–361.

Roether, W. 1986 In *Dynamic processes in the chemistry of the upper ocean* (ed. J. D. Burton, P. G. Brewer & R. Chesselet), pp. 117–128. New York: Plenum.

Roether, W. & Kromer, B. 1984 In *Gas transfer at water surfaces* (ed. W. Brutsaert & G. H. Jirka), pp. 447–457. Dordrecht: Reidel.

Smethie, W. M., Takahashi, T., Chipman, D. W. & Ledwell, J. R. 1985 *J. geophys. Res.* **90**, 7005–7022.

Smith, S. D. & Jones, E. P. 1985 *J. geophys. Res.* **90**, 869–875.

Smith, S. D. & Jones, E. P. 1986 *J. geophys. Res.* **91**, 10529–10532.

Tsunogai, S. & Tanaka, N. 1980 *Geochem. J.* **14**, 227–234.

Wanninkhof, R., Ledwell, J. R. & Broecker, W. S. 1985 *Science, Wash.* **227**, 1224–1226.

Wesely, M. L. 1986 *J. geophys. Res.* **91**, 10533–10535.

Wesely, M. L., Cook, D. R., Hart, R. L. & Williams, R. M. 1982 *J. geophys. Res.* **87**, 8827–8832.

Discussion

M. Heimann (*Max-Planck-Institut für Meteorologie, Hamburg, F.R.G.*). Is the 10 m wind speed the ideal variable in terms of which to parametrize the gas-transfer velocity? For example, the work of Jähne and co-workers at Heidelberg has shown that the mean square slope of the waves characterizes very well the gas-transfer velocity, opening up interesting possibilities for remote sensing of the latter.

P. S. Liss. Wind speed may not ultimately be the best variable for parametrizing the water-phase transfer velocity. However, it appears to be the best we have for the present. The experiments referred to, that relate gas-transfer velocity to the mean square wave slope, have only been conducted under laboratory conditions; it remains to be seen how well the relation holds at sea. Remote sensing may well become an important technique for air–sea gas-exchange studies, particularly for giving instantaneous wind speed and wave data over a wide area. However, the relations between these parameters and transfer velocity at sea will still have to be established before usable information can be obtained.

W. Roether (*University of Bremen, F.R.G.*). I disagree with Professor Liss's view that the eddy-correlation method for CO_2 has problems because the air–sea transfer resistance is only a small part of the total. I believe the problems to be related to the high atmospheric CO_2 concentrations, the effect of any net interfacial transfer being largely masked. However, there are other gases for which destruction in the atmosphere prevents an atmospheric background building up. Then it is attractive to determine their interfacial flux micrometeorologically; in my view such gases show great promise for obtaining a parametrization of the gas-transfer coefficient.

P. S. Liss. I agree to the extent that the relatively high atmospheric concentration of CO_2 does not help, but there are still fundamental signal:noise ratio problems with approaches involving air measurements. The eddy-correlation method suffers from the difficulties of having to correct for the interfering effects of heat and water-vapour fluxes. To apply the technique to other gases a sensor with a response time of better than 1 s is required; it is not clear that these exist for more than a very few gases. My feeling is that to obtain useful results at sea from either the gradient or eddy-correlation techniques there will have to be considerable improvements in the presently available instrumentation.

The oceanic chemistry of the rare-earth elements

By H. Elderfield

Department of Earth Sciences, University of Cambridge, Madingley Road, Cambridge CB3 0EZ, U.K.

Because of the unique chemical characteristics of the rare-earth elements (REES), they provide information as oceanic tracers of element source and reactivity. The absolute and relative concentrations of the REES in ocean waters reflect their input from rivers, by aeolian transport and from hydrothermal vents; their interaction with the biogeochemical cycle involving removal from surface waters by adsorption and oxidation at particle surfaces (probably with organic coatings) and deeper regeneration; and the effects of advective transport. Within the REE group, Ce anomalies occur in response to oceanic redox conditions; Eu anomalies in response to aeolian and hydrothermal input; and Nd isotopic variations reflect element sources. Both REE concentration patterns and Nd isotopic variations allow the effects of the vertical particle flux and of lateral advection to be separated.

Introduction

One of the most fascinating but difficult areas of marine geochemistry is the interplay of ocean chemistry with biology and physics. Many trace elements added to seawater somehow become involved in the oceans' biogeochemical cycle. They are removed from surface seawater either directly or indirectly as a consequence of biological productivity. Some are regenerated in the deep sea when biogenic tissue is oxidised and their distributions are influenced by deep water circulation. The process of removal of elements from seawater can decouple the associations of elements found in input materials and, similarly, the coupling of elements found within ocean cycles can break down on ultimate burial in sediments.

It is necessary, therefore, to unravel the effects of advection and biogeochemical cycling in defining spatial distributions of the elements in the sea. It is necessary to provide a rigorous chemical treatment of the mechanisms encompassed by the term 'scavenging' (Goldberg 1954; Goldschmidt 1954; Krauskopf 1956; Turekian 1977) for describing the role of particles in controlling the chemical composition of seawater, in the upper water column where biologically formed particles predominate, in the deep oceans near the influence of sediments and at ocean margins where both influences are of importance.

Such work is enhanced by the study of those chemical elements whose properties as tracers of source and/or reactivity make them of particular importance for an understanding of oceanic processes. This paper provides an interpretation of the oceanic chemistry of the rare-earth elements (REES) based upon the recent data on REE concentrations in seawater (Elderfield & Greaves 1981, 1982, 1983; De Baar *et al.* 1983, 1985*a*, *b*, *c*, 1988; Greaves *et al.* 1985; Klinkhammer *et al.* 1983; Palmer 1983) and upon its Nd isotope composition (Piepgras & Wasserburg 1980, 1982, 1983, 1985, 1987; Peipgras *et al.* 1979; Stordal & Wasserburg 1986).

Studies of the concentrations and distributions of the rare-earth elements in seawater offer a unique insight to marine trace-element geochemistry. The REES are affected by nearly all the important processes affecting marine trace-element behaviour (Bender 1982): river, atmospheric and hydrothermal input, adsorption and scavenging onto particles, coprecipitation, ion pairing, redox reactions. Moreover, studies of the REEs make it possible to judge how the behaviour of trace metals is determined by their fundamental chemical properties thus allowing the mechanisms that control their concentrations and distributions to be elucidated. This possibility exists because the fundamental chemical properties of all the REEs are generally rather similar but nevertheless show subtle but significant differences that vary in a predictable manner. It transpires that examining the response of the oceans to these small differences in physicochemical properties provides an extremely useful tool that is much more specific and sensitive than that gained from examination of more poorly related elements.

Fundamental chemical properties

The chemical similarity of different members of the REE group occurs because of their electronic structure. There is a progressive filling of inner 4f electrons with increasing atomic number throughout the series and these electrons are sufficiently shielded by outer electrons not to lead to major differences in chemical reactivity. However, the natural abundances of elements resulting from nucleosynthesis show a rhythmic odd/even variation with atomic number because of the variations in the binding energy of the nucleus as a function of neutron and proton number. Nevertheless, it is easy to correct for this by normalization to a standard. In the example in figure 1, average shale is normalized to meteorite abundances giving a smooth pattern of relative abundance (often, as here, plotted on a logarithmic scale) against atomic number. This 'REE pattern' can be used to characterize different types of materials but,

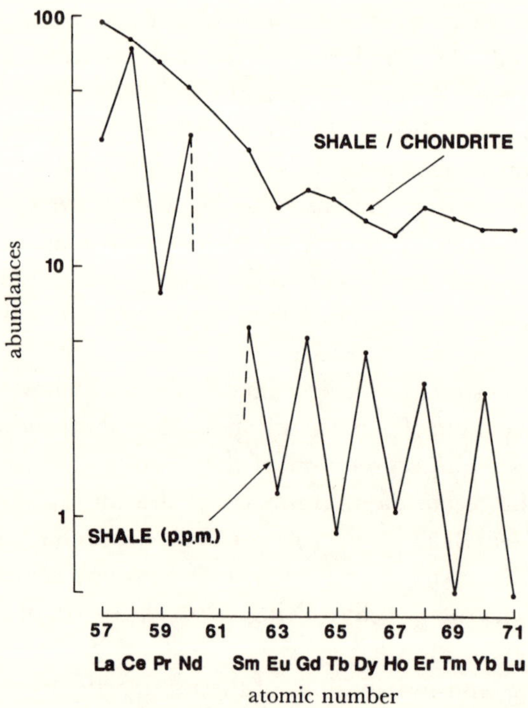

FIGURE 1. Abundances of the REEs in average shale (p.p.m.) and normalized to abundances in chondritic meteorite.

more importantly, it makes it very simple to distinguish the minor mass fractionation between the various REEs relative to a source material and the REEs are unique in this respect.

The mass fractionation of the REEs takes two forms. First, systematic changes in ionic radii (and hence complexation and surface-adsorption constants) occur with increasing atomic number. This may be recognized in the seawater speciation of the REEs (table 1). Rare-earth

TABLE 1. REE SPECIATION IN SEAWATER

(Species as percentage of total dissolved metal ion[a].)

REE	Z	REE^{3+}	$REE(OH)^{2+}$	$REE(F)^{2+}$	$REE(Cl)^{2+}$	$REE(SO_4)^{+}$	$REE(CO_3)^{+}$	$REE(CO_3)_2^{-}$
La	57	6.7	0.9	0.2	3.1	2.8	66.1	20.3
Ce	58	5.1	1.2	0.3	2.9	2.4	65.1	23.0
Pr	59	4.0	1.3	0.2	1.9	2.1	64.4	26.2
Nd	60	3.1	1.3	0.2	1.4	1.7	62.7	29.6
Pm	61	—	—	—	—	—	—	—
Sm	62	1.9	1.1	0.1	0.9	1.2	57.7	37.1
Eu	63	1.6	1.1	0.1	0.9	0.8	54.5	41.1
Gd	64	1.2	0.9	0.1	0.6	0.7	51.1	45.4
Tb	65	1.0	0.7	0.1	0.5	0.6	47.3	49.8
Dy	66	0.8	0.6	0.1	0.4	0.5	43.3	54.3
Ho	67	0.7	0.5	0.1	0.3	0.3	39.3	58.8
Er	68	0.6	0.8	0.1	0.3	0.3	35.1	62.9
Tm	69	0.5	0.9	0.0	0.2	0.3	31.1	67.1
Yb	70	0.4	0.7	0.1	0.2	0.2	27.2	71.2
Lu	71	0.3	1.3	0.1	0.1	0.1	23.5	74.7

[a] For seawater with $[CO_3^{2-}] = 1.39 \times 10^{-4}$ ($TCO_2 = 2.3 \times 10^{-3}$, pH = 7.90); from Cantrell & Byrne (1987) and R. H. Byrne personal communication (1987).

speciation in seawater is dominated by carbonate complexation (Cantrell & Byrne 1987) and the proportion of each REE complexed with seawater carbonate ligands increases with increasing atomic number. Hence, the proportion of the free ion decreases with increasing atomic number, which is important when the ability of the REEs to adsorb onto particle surfaces is considered.

Second, although all the REEs exist in the 3+ oxidation state, cerium and europium also exist in the 4+ and 2+ states, respectively;

$$Ce^{3+} + 2H_2O \rightleftharpoons CeO_2 + 4H^+ + e^-, \tag{1}$$

$$Eu^{3+} + e^- \rightleftharpoons Eu^{2+}. \tag{2}$$

Also, importantly, as shown by equation (1), the higher oxidation state of Ce is present as a solid phase and this is very insoluble in seawater. Therefore, it is possible to separate the role of oxidation in controlling the behaviour of Ce from all other processes affecting it by comparing its concentrations with those of adjacent REEs. The Eu^{3+}/Eu^{2+} redox boundary is below the lower limit of the stability field of water at earth surface conditions but the redox chemistry of europium nevertheless is of significance in the oceans because of the preferential ability of Eu^{2+} to substitute for Ca^{2+} in certain silicate minerals, principally plagioclase, and because of the relevance to the marine geochemical budget of chemical fluxes associated with submarine hydrothermal activity. The formation of acidic, high-temperature fluids at sub-seafloor pressures leads to preferential mobility of Eu^{2+}, relative to the other REEs (Michard *et al.* 1983).

Isotopic- as well as mass-fractionation of some REEs also occurs and although this does not take place within the oceans it is of significance to an understanding of oceanic processes. One of the isotopes of the rare-earth element samarium, ^{147}Sm, is very weakly radioactive, with a half life of 1.06×10^{11} years, and decays to another REE isotope, ^{143}Nd. Because of REE fractionation within the earth, different rocks have evolved with different Sm/Nd ratios and, hence, with different ratios of radiogenic ^{143}Nd to non-radiogenic Nd, e.g. $^{143}Nd/^{144}Nd$. Figure 2 shows the separation of bulk earth into continental crust (lower $^{147}Sm/^{144}Nd$ ratio) and oceanic crust (higher $^{147}Sm/^{144}Nd$ ratio) and the temporal evolution of these geochemical reservoirs with distinctive $^{143}Nd/^{144}Nd$ ratios, expressed in terms of the symbol $\epsilon_{Nd}(0)$ as parts per 10000 (by mass) deviations from the bulk-earth value today. As a result, the isotopic composition of Nd can serve as a 'fingerprint' to identify the sources of the element in seawater. The histogram in figure 3 shows a distinct range in Nd isotope values for sea water ranging from

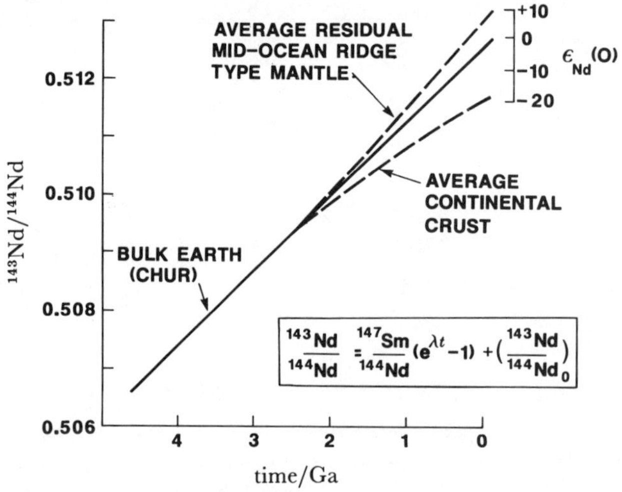

FIGURE 2. Neodymium isotope evolution diagram.

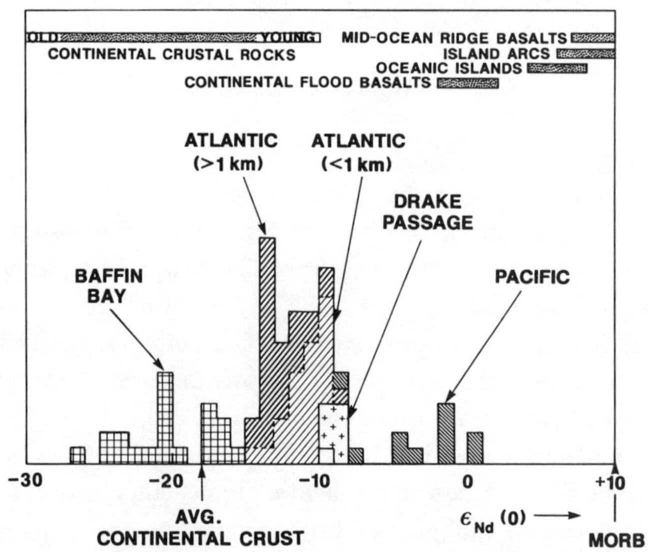

FIGURE 3. Histogram of neodymium isotopic composition of seawater. Data from Piepgras & Wasserburg (1980, 1982, 1983, 1987), Piepgras *et al.* (1979) and Stordal & Wasserburg (1986).

low $\epsilon_{Nd}(0)$ values in the North Atlantic Ocean reflecting drainage from old continental crust through to higher values in the Pacific Ocean. It is clear from figure 3 that neodymium is supplied to seawater predominantly from the continents (i.e. rivers and winds) rather than from oceanic crust (i.e. hydrothermal activity) although there is considerable uncertainty as to the exact proportions because of the unusual $\epsilon_{Nd}(0)$ values reported for hydrothermal fluids (Piepgras & Wasserburg 1985).

There are two further points of interest in the Nd isotope data. First, the differences in $\epsilon_{Nd}(0)$ between the Atlantic and Pacific Oceans indicate a rather short oceanic residence time for neodymium. For example, strontium, which shows similar isotope systematics, has the same isotopic composition everywhere in the oceans because of its long (*ca.* 10^6 years) oceanic residence time. Second, a related point is that the contrast in $\epsilon_{Nd}(0)$ between surface and deep waters in the Atlantic Ocean means that overall there must be significant vertical as well as lateral Nd isotope gradients in the oceans, showing a significant degree of preservation of the isotopic tracer of its sources.

The oceanic distributions of the REEs

Although the available data set is only of moderate size, nevertheless it is now possible to describe the principal features of the vertical and lateral distributions of the REEs in ocean waters, the chief limitation being that results are not yet available for the Indian Ocean.

Examples of depth profiles of REE concentrations in seawater are shown in figure 4. With the

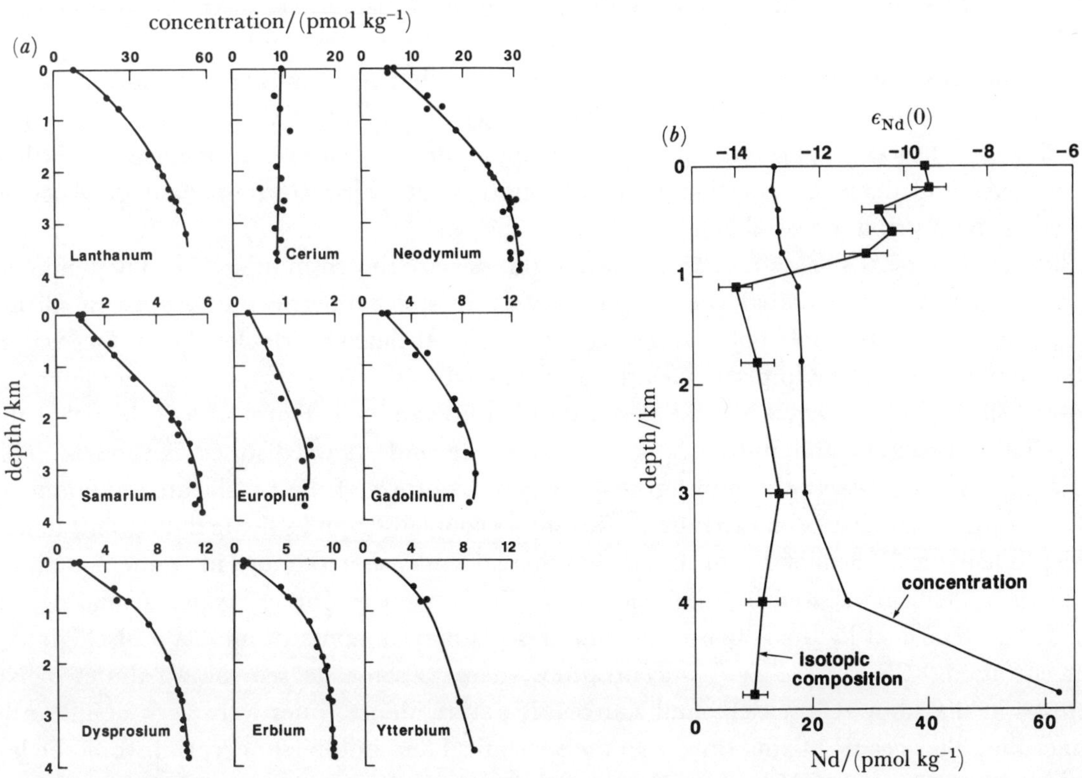

FIGURE 4. Depth profiles of REEs in seawater: (*a*) concentrations of La, Ce, Nd, Sm, Eu, Gd, Dy, Er and Yb (Klinkhammer *et al.* 1983); (*b*) Nd concentration and isotopic composition (modified from Piepgras & Wasserburg 1987).

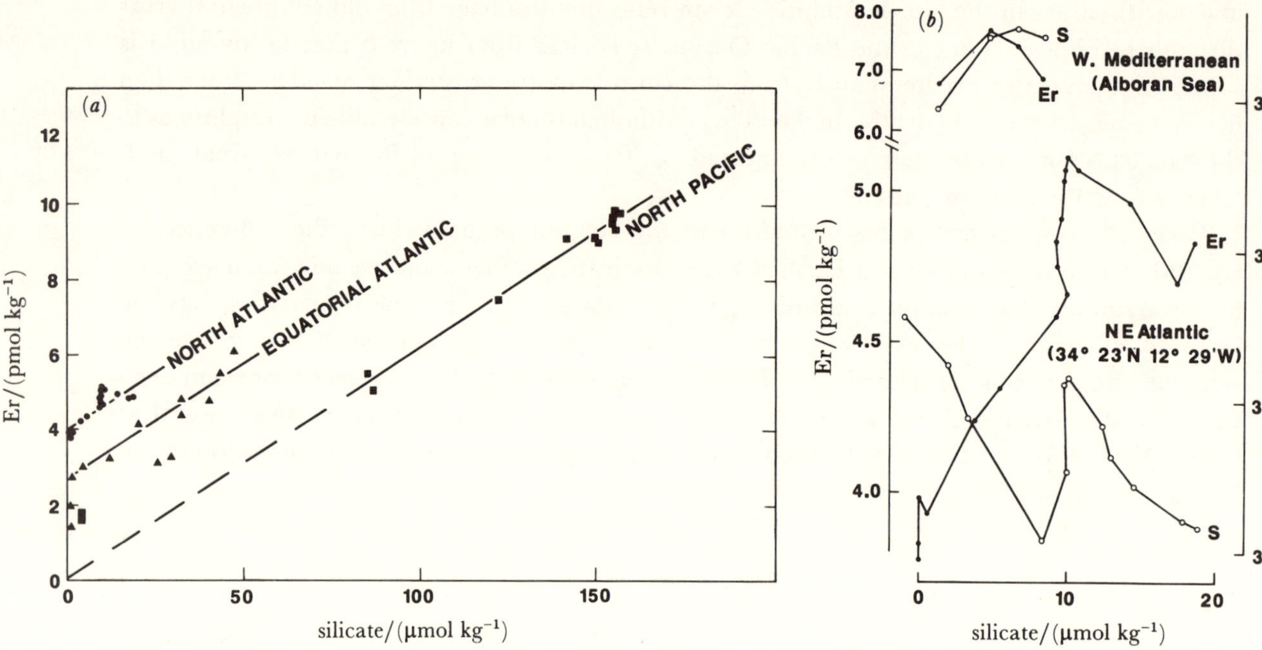

FIGURE 5. Erbium against silicate in seawater. (*a*) Comparison of three stations from N Atlantic, equatorial Atlantic and N Pacific. Data from De Baar *et al.* (1985*a*), Greaves *et al.* (1985) and Palmer (1983). (*b*) Comparison of stations from NE Atlantic and Alboran Sea, also showing salinity (*S*) against silicate. Data from Greaves *et al.* (1985).

exception of Ce, superficially they show the same form as the vertical distribution of silicate, that of depletion in surface waters and enrichment in deep waters. However, examination of results (described below) from different stations shows that the REEs are not coupled tightly with silicate in their oceanic behaviour. Nevertheless, comparison with silicate is useful in defining how the REEs depart from the cycle of assimilation in surface water, followed by settling of particles and deep regeneration that characterizes the vertical component of processes affecting the distribution of silicate.

Plots of Er against silicate for three separate stations are shown in figure 5*a*. They indicate that the REEs are regenerated very approximately in stoichiometric proportions to silicate ($\Delta Er/\Delta Si = 6.2 \times 10^{-8}$ mol/mol) but the degree to which the REEs are depleted in surface waters differs greatly from place to place. Clearly, surface-water REE concentrations are not linked to those of silicate. Moreover, when the larger data set is examined (figure 6) it is clear that the coupling of the REEs and nutrients is, indeed, crude and shows distinct variability. This variability is real as can be seen in figure 5*b*, where a portion of the Er–silicate relation for a NE Atlantic station has been expanded. This shows coupled Er and silicate data points joined in depth sequence. The plot demonstrates that there is a very sharp change in gradient at about 10 μmol kg^{-1} silicate caused by the injection of Mediterranean outflow water at mid-depths that, as is confirmed by the Alboran Sea data also displayed, contains anomalously high REE concentrations. Clearly the REE concentrations share to some degree the attributes of Nd isotopes in that they reflect elemental sources. If REE residence times were very significantly longer than the oceanic mixing time then their chemical and isotopic fingerprint would be less distinct and less useful. However, if REE residence times were extremely short, removal processes would significantly modify the size and pattern of the fingerprint and would also

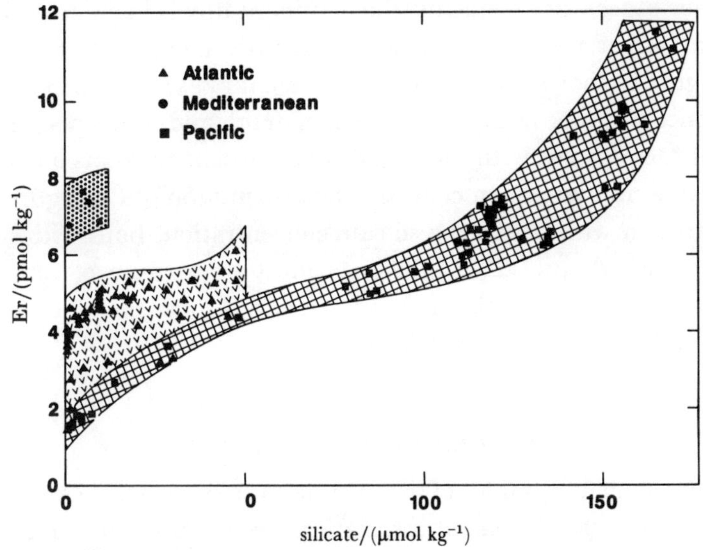

FIGURE 6. Erbium against silicate in seawater for the Atlantic and Pacific Oceans and the Mediterranean Sea. Data from Elderfield & Greaves (1981, 1982), De Baar *et al.* (1985*a, c*), Greaves *et al.* (1985), Klinkhammer *et al.* (1983) and Palmer (1983).

diminish its usefulness. From this example it would seem clear that vertical geochemical cycling does not dominate the processes controlling REE concentrations at depth; the effect of advective transport is readily recognizable from the REE concentrations and patterns.

As shown in figure 4, the depth profile of Ce in seawater is very different from those of the other REEs, Ce shows a gradual decline in concentration with increasing depth unlike the nutrient-like increases of the other REEs. This contrast is also seen in the REE–silicate plot for Ce (figure 7) as compared with those for the other REEs as exemplified by Er (figure 6). As has been seen from the Nd isotope data (figure 3) the principal source of REEs to the oceans is

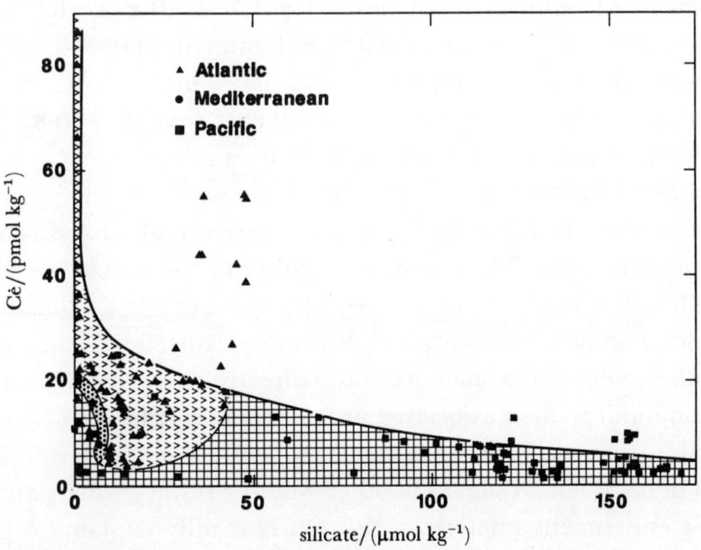

FIGURE 7. Cerium against silicate in seawater for the Atlantic and Pacific Oceans and the Mediterranean Sea. Data as for figure 6. Samples plotting above solid line are from deep waters with probable sediment input.

continental and, as for other trace elements, the dominant flux is likely to be into the Atlantic Ocean because of its larger river and aeolian input rates compared with the Pacific. The distribution of Er (figure 6) shows that the strictly trivalent REEs increase in concentration with increasing silicate concentration, both as a function of depth and in the deep Pacific compared with the deep Atlantic, consistent with their mode of biogeochemical cycling coupled with the flow path of deep-water circulation. In contrast, the distribution of Ce (figure 7) shows that it decreases in concentration with increasing silicate concentration, both with depth and in the deep Pacific against the Atlantic. This difference must be because of removal of Ce from seawater by oxidation and it is clear that oxidative removal must be the dominant process by which the cerium content of seawater is defined.

Mass fractionation of the REEs

Observations of interelement relations of the REEs provide evidence of chemical fractionation within the oceans such that trace-element associations recorded in elemental sources become decoupled. It is obvious from the previous section that the behaviour of Ce differs very significantly from that of the strictly trivalent REEs. The severe mass fractionation of Ce will be treated separately here, as will the behaviour of Eu.

First, bearing in mind that all the REEs have similar fundamental chemical properties, it is useful to compare REEs of very similar character. Figure 8*a* shows the relationship between the oceanic concentrations of Sm ($Z = 62$) and Nd ($Z = 60$), two atomic numbers apart in the REE series. The two elements exhibit no significant mass fractionation. The Sm/Nd ratio is very similar to that of average shale that has been used to typify continental input to the oceans (0.189 mol/mol) and is also similar to ratios measured in river waters (0.218 mol/mol) (Upstill-Goddard *et al.* 1987). Also, the Sm/Nd ratio is near constant in all sea waters and exhibits no interoceanic differences.

Next, figure 8*b* compares Er ($Z = 68$) and Nd ($Z = 60$), eight atomic numbers apart, and shows that there is significant mass fractionation between the heavy and light REEs. All seawaters are enriched in Er relative to Nd (Er/Nd = 0.57–0.10 mol/mol) when compared with shale (0.085 mol/mol) or riverwater (0.088 mol/mol; or 0.093–0.186 mol/mol after correction for fractionation during estuarine removal). Furthermore, Pacific seawaters (0.57–0.20 mol/mol) have higher Er/Nd ratios than Atlantic seawaters (0.33–0.10 mol/mol), i.e. they are even more enriched as compared with shale or riverwater.

Further features of this relation become clear when surface waters are distinguished from intermediate and deep waters. In the Atlantic, many surface samples have a shale-like Er/Nd ratio and high REE concentrations. These samples (figure 8*b*) come from a surface transect in the NE Atlantic from 55° N to the Equator and reflect a signal due to aeolian input of particles to surface waters, dissolution of REEs from which leads to enhanced REE concentrations at the upper portion of depth profiles and a shale-like ratio directly reflecting element input. As will be shown later, aeolian input is often associated with anomalously low Eu concentrations that further characterize this source.

Where this aeolian signal is less strong, as in other Atlantic surface waters and in the Pacific, there is significant Er enrichment relative to Nd that is readily explainable by preferential removal of light REEs onto particles in surface seawater relative to heavy REEs (which show a greater tendency to complex with seawater ligands in solution) with the result that seawater

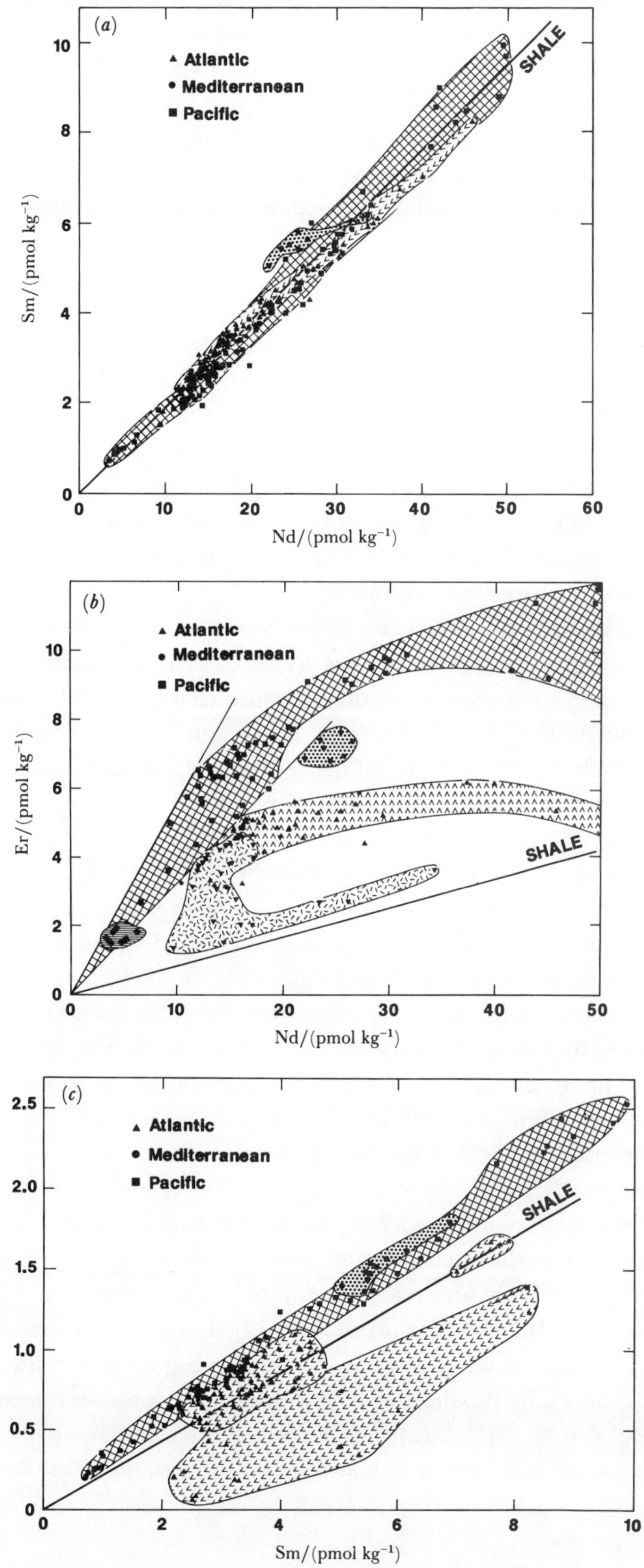

FIGURE 8. REE interelement plots. (*a*) Sm against Nd; (*b*) Er against Nd; (*c*) Eu against Sm. Lines with slopes having REE ratios for average shale are also shown. In (*b*), surface-water samples are distinguished by heavy shading and reversed symbols. In (*c*), surface waters showing evidence of aeolian input (significant negative Eu anomalies and shale-like Er/Nd ratios) are distinguished by heavy shading.

becomes heavy-REE enriched. The degree to which both REEs are depleted reflects the efficiency of this process relative to input fluxes and there is a suggestion in the data that this also may increase mass fractionation: the lower concentrations in Pacific surface waters are associated with slightly higher Er/Nd ratios than for the Atlantic or Mediterranean.

This pattern of heavy-REE enrichment is initially maintained with increasing REE concentration (i.e. depth) but in deep waters in both oceans the trend is reversed: the Er/Nd ratios tend towards shale-like values. One explanation for this is that vertical biogeochemical cycling returns the REEs to solution, i.e. dissolution of REEs from falling particles returns the preferentially removed light REEs to solution leading to decreased Er/Nd ratios. Although this can account for the patterns in the top kilometre or so of the water column there is evidence that this mechanism is not primarily responsible for the deep water ratios. This evidence comes from the persistence of advected REE features (figure 5*b*), as already discussed, and from Nd isotope data (figure 4*b*). If the REE ratios in deep waters reflect only downward transport and regeneration of REEs removed by particles from surface waters, then Nd isotopic compositions should be homogeneous throughout the water column. This is not the case. They appear to be roughly homogeneous only in the top *ca.* 800 m that defines an upper zone of element assimilation and regeneration. Therefore, an additional process or processes must be affecting REE concentrations in the lower zone, at depth.

There are two candidates and it is difficult to evaluate their relative importance at present. One requires very high concentrations of REEs to be carried in sinking surface waters, i.e. preformed REEs. Of course, advective transport is certain to influence the deep-water ratios to some extent, as is shown for the Mediterranean outflow (figure 5*b*), and the lower Er/Nd ratios in the deep Atlantic might be expected to influence deep Pacific ratios in the direction that is observed. The second is that there is significant REE input from the sediments as a result of diagenesis. The extremely low REE concentrations in seawater relative to sediment ($K_{\mathrm{D}\,\mathrm{seawater}}^{\mathrm{sediment}} \approx 10^7$ mol/mol) would seem to favour this; the standing crop of REEs in the 4000 m oceanic water column is equivalent to the amount present in the uppermost 1 mm of ocean sediments. Indeed, very high concentrations of REEs have been measured in pore waters from reducing sediments (Elderfield & Sholkovitz 1987) but totally reliable data have not yet been reported for the deep sea. Superficial examination of the overall curvature of REE against silicate data (see, for example, figure 6) might imply scavenging of REEs, as for Cu, in the deep sea but the above examination of the mass-fractionation effect suggests that this cannot be the dominant process affecting REE behaviour at depth. Indeed, deep scavenging of REEs might be expected to produce heavy-REE enrichment in deep waters, the opposite to what is observed. The effect of mixing seems more important.

This point may be addressed, in principle, by considering Nd concentration against Nd isotope mixing relations in portions of the deep water column where $\theta - S$ and $\epsilon_{\mathrm{Nd}}(0) - S$ (or $\epsilon_{\mathrm{Nd}}(0) - \mathrm{Si}$) plots are linear. This approach is explored in figure 9 from data of Piepgras & Wasserburg (1987) and produces the unfortunate result that the high precision of the isotope data (*ca.* 10 p.p.m. (by mass)) is not enough to distinguish highly non-conservative behaviour from the conservative case. The mixing relations in figure 9 permit an interpretation of either *ca.* 15% Nd removal or *ca.* 15% Nd production, or neither removal nor production! Nevertheless, this approach is a crucial one and should eventually resolve this question.

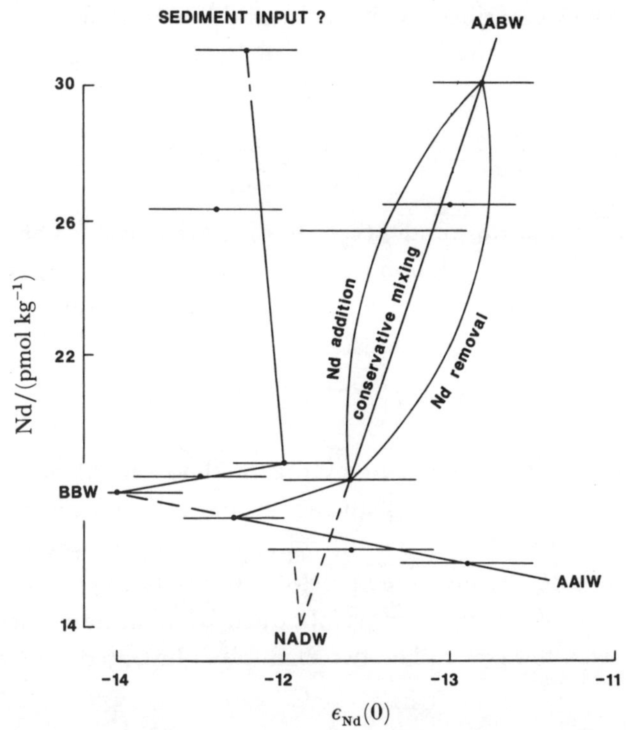

FIGURE 9. Neodymium concentration–isotopic composition diagram for the North Atlantic Ocean based upon data of Piepgras & Wasserburg (1987). The straight lines represent conservative mixing between samples which fall on linear θ–salinity mixing lines. End-member compositions are indicated: NADW, North Atlantic Deep Water; AABW, Antarctic Bottom Water; AAIW, Antarctic Intermediate Water; BBW, Baffin Bay Water; and examples of non-conservative behaviour involving Nd addition and removal are shown for one mixing line.

AN ADSORPTION–OXIDATION SCAVENGING MODEL FOR THE REES

The trivalent REES

The interpretation given above is that REE concentrations in seawater are controlled by adsorptive scavenging by particles, except that cerium behaves anomalously because its principal mode of removal is by oxidation to insoluble Ce^{IV}. This section attempts to treat these removal processes somewhat more rigorously by exploring whether published ideas about particle scavenging as controls on trace-metal concentrations in seawater can explain the mass fractionation of the REEs outlined here.

The approach taken has been to apply a popular scavenging model developed by Schindler (1975) and Balistrieri *et al.* (1981) and used most recently by Whitfield & Turner (1987). The scavenging residence time of the REEs is defined as

$$\tau_{\mathrm{REE}}^{\mathrm{SCAV}} = \frac{[\mathrm{REE}^{\mathrm{III}+}] + [\mathrm{REE} - L] + [\mathrm{REE}]_{\mathrm{ADS}}}{\left(\dfrac{\mathrm{d}[\mathrm{REE}]}{\mathrm{d}t}\right)_{\mathrm{SCAV}}}, \tag{3}$$

which describes the competition between adsorption of the free metal ion onto particled surfaces (ADS) and complexation with ligands in solution (L), which make less of the metal available for adsorption (see Balistrieri *et al.* (1981) for details and more general application).

Because REE speciation in seawater is dominated by carbonate complexation (table 1) the competition referred to can be described by the ratios of the conditional REE stability constants for seawater (β_1 and β_2) to the formation constant (K_{Me}) for a REE–particle surface complex (see also Cantrell & Byrne 1987)

$$\tau_{REE}^{SCAV} = \tau_P + A\left(\frac{\beta_1}{K_{Me}} + \frac{\beta_2}{K_{Me}}[L]\right), \tag{4}$$

where A is a constant and τ_P is the residence time of the particles. This relation has been simplified further by establishing the relation between β and K_{Me} by using the term 'm', which is designated the 'binding constant': large values of m reflect a strong binding capacity of particles for REEs,

$$\tau_{REE}^{SCAV} = \tau_P + A\left(\beta_1^{1-m} + \frac{\beta_2}{\beta_1^m}[CO_3^{2-}]\right). \tag{5}$$

The REE scavenging residence times obtained by this method should be proportional to the ratios of the concentrations of each REE in seawater to that in input material.

Greater details of this approach and results will be given elsewhere and only a brief application is given here. Figure 10 shows the model-derived pattern of relative REE scavenging residence times as a function of atomic number obtained by plotting values of the β term in parentheses in (5) normalized to $La = 1$. The two examples illustrated were obtained by using

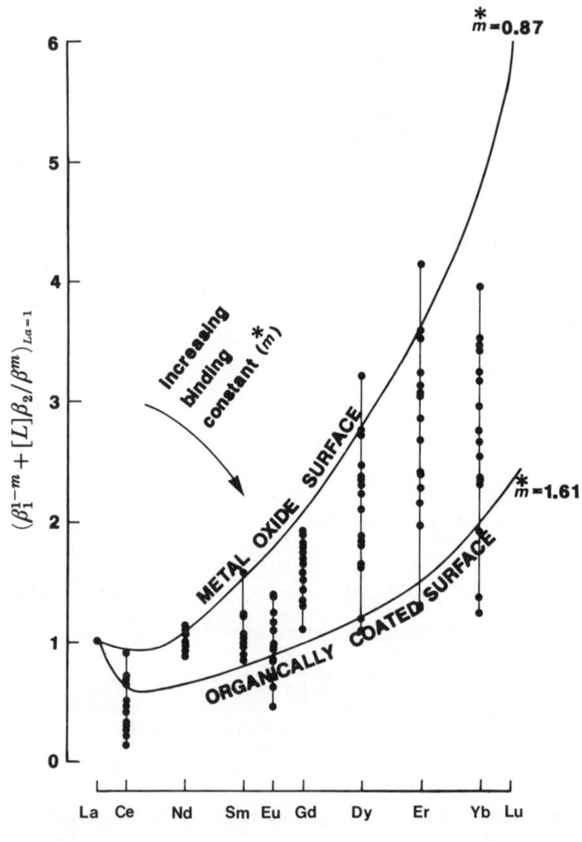

FIGURE 10. Adsorption scavenging model for the REEs. Plot of β term (normalized to $La = 1$) in (5) of text against REE atomic number showing model-derived patterns for metal oxide surface and organically coated surface together with patterns for REEs in northeast Atlantic seawaters (Elderfield & Greaves 1982) normalized to average shale and $La = 1$.

values of binding constant equivalent to those used in the Balistrieri model: a lower value for an inorganic surface and a higher value for an organically coated surface.

The model makes a number of predictions. First, the heavy REEs have longer scavenging residence times than the light REEs. Second, the contrast in residence times depends on the strength of the surface adsorption complex. Organically coated surfaces have stronger binding characteristics than oxide surfaces and therefore allow less mass fractionation in solution. Third, it is interesting that the model predicts that the REE with the lowest residence time is not the lightest REE (La).

Figure 10 also compares the model with one example of REE concentrations from a station in the northeast Atlantic Ocean, normalized to average shale. Clearly, Ce fits the model prediction poorly but the other REEs fit the model with some degree of success mostly falling between the two curves and La does not have the lowest normalized abundance, in agreement with the model prediction. Perhaps a better alternative to normalizing the seawater data to shale is to normalize to average river input. Although there is yet no generally agreed REE pattern for average riverwater it now seems clear that rivers are to some degree heavy-REE enriched relative to shale (Upstill-Goddard *et al.* 1987). Therefore, agreement between river normalized data and model prediction requires adsorption by surfaces with strong binding capacity. Adsorption by inorganic oxide surfaces would predict much more REE fractionation than is observed in nature.

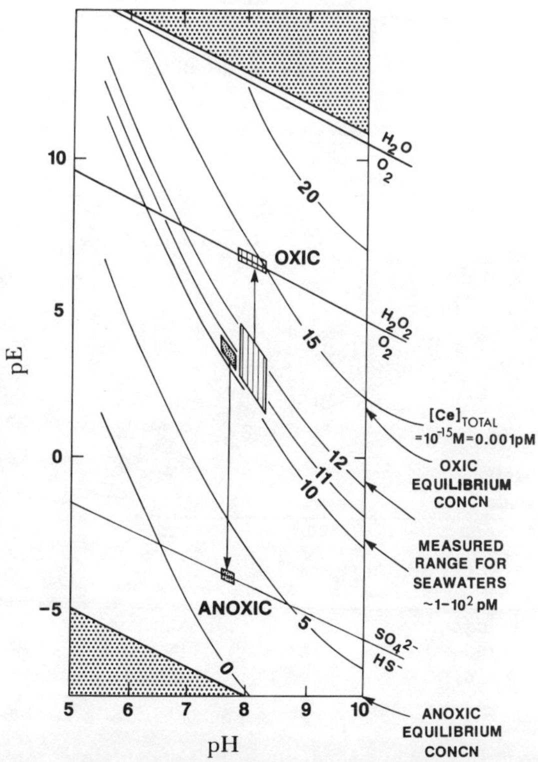

FIGURE 11. A pE–pH diagram showing the predicted isopleths of total dissolved Ce^{III} concentrations (as pCe_{TOTAL}) in equilibrium with solid CeO_2. Also shown are upper and lower stability boundaries for water, pE–pH relations for H_2O/O_2 and H_2O_2/O_2 in equilibrium with atmospheric oxygen and for SO_4^{2-}/HS^-. The shaded areas show the actual and predicted Ce concentrations for oxic and anoxic seawaters. Modified from De Baar *et al.* (1988). (In this paper, $pE = -\lg a_{e^-}$, where a is activity.)

The behaviour of cerium

The adsorptive scavenging model outlined above describes the behaviour of Ce poorly. Cerium concentrations are much lower than predicted (figure 10). Nevertheless pE–pH relations (figure 11) show that Ce^{3+} is supersaturated in seawater and equilibrium concentrations are even lower than those observed. Therefore, the scavenging model that is an equilibrium model cannot apply. Ce is under kinetic control and is removed from seawater by oxidation.

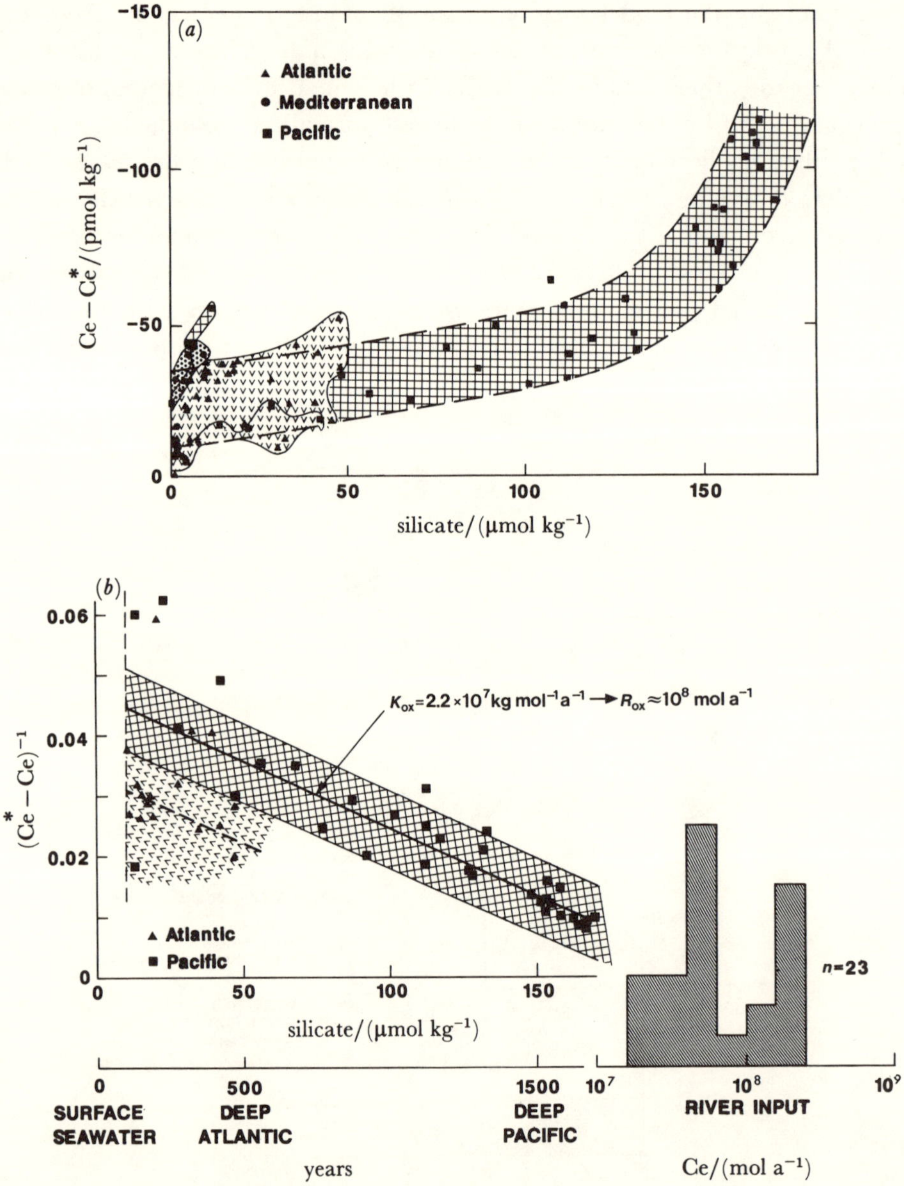

FIGURE 12. Ce removal from seawater by oxidation. (*a*) Plot of oxidative Ce removal against silicate. $\overset{*}{Ce}$ is the Ce concentration expected if Ce behaves like the trivalent REES and obtained by linear interpolation between La and Nd. (*b*) Plot of reciprocal of Ce removal term against silicate. Also shown is crude timescale for deep-water flow leading to quoted second-order oxidation rate constant (K_{ox}) and removal rate (R_{ox}). The histogram shows the estimated river input flux of Ce based upon Ce concentrations in 23 rivers (Upstill-Goddard *et al.* 1987) and the total annual river discharge rate. Estuarine removal would decrease the flux to *ca.* 60 % of these values.

The role of oxidation can be defined by subtracting the observed Ce concentrations in seawater from the concentrations that would be expected if Ce behaved like its REE neighbours. This treatment shows increasing Ce removal from seawater in the direction of deep-water flow (figure 12*a*). Furthermore, by placing a crude timescale upon this pattern it is possible to estimate the rate of Ce removal. The pattern best fits removal by second-order kinetics (implying that Ce oxidation is autocatalytic, like that of Mn in seawater) and shows (figure 12*b*) that Ce is removed from seawater by oxidation at approximately the same rate as the rate

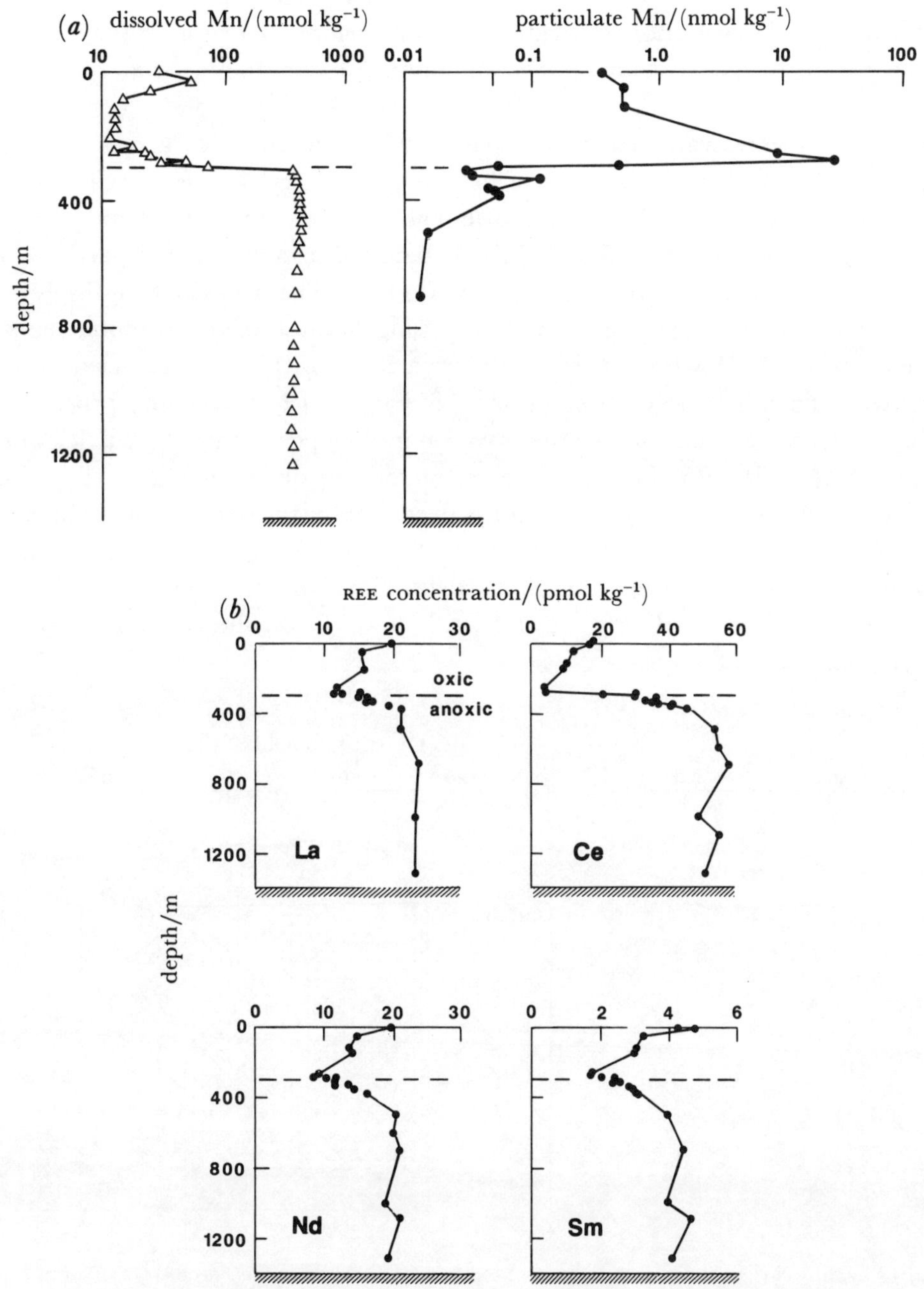

FIGURE 13. Concentration profiles in the Cariaco Trench. (*a*) Dissolved and particulate Mn. (*b*) Examples of REE concentrations. Data from De Baar *et al.* (1988).

of Ce input to the oceans from rivers (*ca.* 10^8 mol a^{-1}). Because the negative Ce normally in the normalized REE pattern for seawater is roughly an order of magnitude in scale, Ce must be removed from seawater by oxidation about an order of magnitude more rapidly than by the other processes affecting REE removal. Therefore, the oxidation-removal rate term must be the dominant one for cerium.

Evidence from anoxic basins

The concentrations of the REEs in ocean waters, and their variabilities, are consistent, therefore, with adsorption and autocatalytic oxidation by particles and both the equilibrium scavenging model and the oxidation-rate calculation for cerium offer support for this concept. Some key supporting evidence comes from an unusual environment, that of the anoxic basin.

Figure 13 shows the seawater concentrations of Mn and the REEs across the O_2–H_2S boundary in the Cariaco Trench. Because of its redox chemistry, Mn is recycled across the boundary, being reduced and dissolved in the lower anoxic water and oxidised and precipitated just above the oxic–anoxic interface to produce a standing crop of reactive particles (figure 13*a*). The REEs are incorporated into this cycle (figure 13*b*). The behaviour of dissolved Ce is identical to that of dissolved Mn; this is not surprising because of their similar oxidation–reduction characteristics. However, the strictly trivalent REEs also are recycled across the O_2–H_2S interface and this must be because of a similar particle-scavenging process to that described for the open ocean but, unlike the situation in the open ocean, the particles bearing the scavenged REEs settle into the lower waters of the anoxic basin where they dissolve. Therefore, one would predict that the REEs in the deeper anoxic waters should (unlike deeper oxic ocean waters) by heavy-REE depleted because of the preferential enrichment of the light REEs, which are transported downwards in particles and dissolve. This is exactly what is observed (figure 14*a*); the Er/Nd ratio is significantly lower below the O_2–H_2S interface than in the overlying oxic seawater.

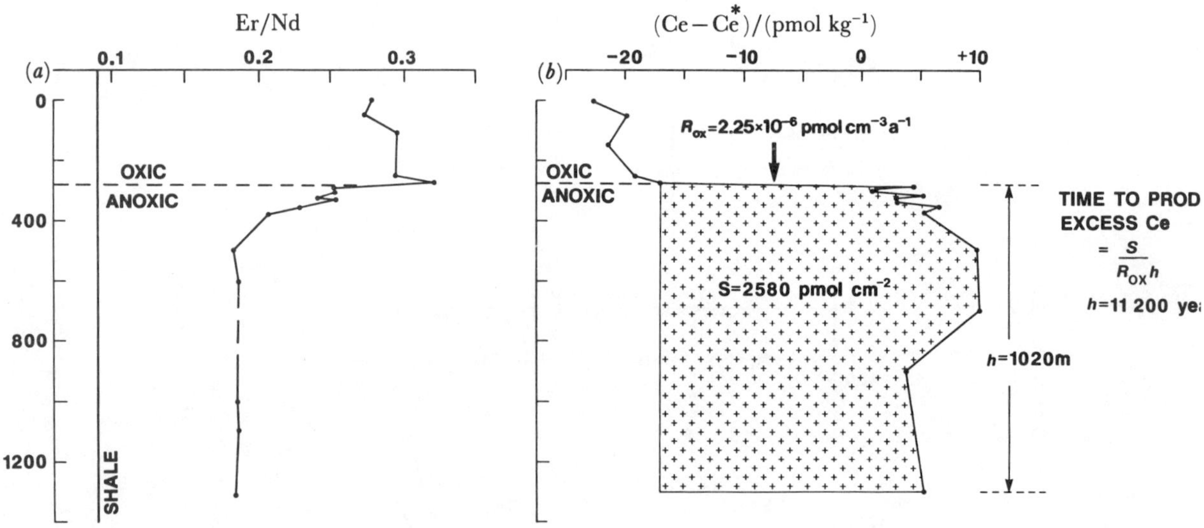

FIGURE 14. (*a*) Er against Nd in the Cariaco Trench. Modified from De Baar *et al.* (1988). (*b*) Plot of excess Ce against depth in the Cariaco Trench (symbols as for figure 12). R_{ox} is based on K_{ox} given in figure 12; S is the standing crop of Ce below the O_2/H_2S boundary.

Furthermore, the Ce data supports the description of the open ocean process too. Examination of the cerium concentrations for the anoxic waters on the pE–pH diagram (figure 11) shows that whereas oxic seawater is supersaturated with Ce, anoxic seawater is considerably undersaturated. Therefore, the concentration of Ce attained in the deep waters of the Cariaco Trench should record the progressive enrichment of reduced Ce in the anoxic seawater. Figure 14*b* shows that the water column standing crop of excess Ce would have taken 11200 years to accumulate at the Ce oxidation rate obtained previously (figure 12*b*) and assuming instantaneous reduction of Ce^{IV} within the anoxic water column. Observations of water-column stability of the Cariaco Trench suggest that it could overturn within less than 200 years (Scranton *et al.* 1987). However, the bottom sediment record reveals that the Trench has been anoxic for the past 11000 years (Heezen *et al.* 1959). The more rapid flushing time would require a 50 times faster production rate of reduced Ce, presumably through dissolution of detrital Ce and/or enhanced uptake of Ce from oxic seawater in association with the recycling across the O_2–H_2S interface.

Europium anomalies in seawater

Figure 8*c* shows the relation between the concentrations in oceanic waters of Eu and Sm, elements of adjacent atomic numbers. Figure 8*a* shows, as was discussed earlier, that Nd and Sm, two atomic numbers apart, behave extremely coherently with no significant mass fractionation.

What then is the reason for the range of Eu/Sm ratios? Some surface waters have anomalously low Eu/Sm ratios because they record the pattern of aeolian input that in the Atlantic appears to contain a negative Eu anomaly and these plot below the shale-ratio line in figure 8*c*. However, many sea waters have high Eu/Sm ratios, especially those in the Pacific Ocean that are higher than the generally shale-like values for Atlantic samples not affected by aeolian input. A river-input ratio significantly different from shale is one possible explanation although the similarity between the shale ratio (0.212 mol/mol) and the non-aeolian Atlantic values (0.21 mol/mol) and the currently available river-water data (0.229 mol/mol) would argue against this. Another explanation is that the concept of comparing REE behaviour in terms of linear mass fractionation is too simplistic, as the scavenging model indeed implies.

Nevertheless there remains a likelihood that the anomalously high Eu/Sm ratios in seawater are not a consequence of the river-input ratio or of mass fractionation with the oceans. If so, it is inescapable that the Eu/Sm ratio records hydrothermal input to seawater and there is evidence to support this explanation. Figure 15*a* shows the seawater-normalized REE pattern of a typical high-temperature vent fluid. This contains more than 3000 times more Eu than normal seawater but only 100 times more Sm. In addition, this high ratio is preserved during mixing with ambient seawater as illustrated in figure 15*b*. The samples shown contain less than 0.15% of hydrothermal end member yet show very elevated Eu/Sm ratios that extrapolate to high end-member ratios such as those recorded in vent fluids. If this interpretation is supported by future work, the Eu/Sm ratio could prove to be useful as a hydrothermal tracer because it is evident from figure 15*b* that this mass ratio behaves somewhat as an isotope ratio and is not affected by removal processes. It may help in addressing the importance of ridge-crest hydrothermal processes to the geochemical mass balance of the oceans.

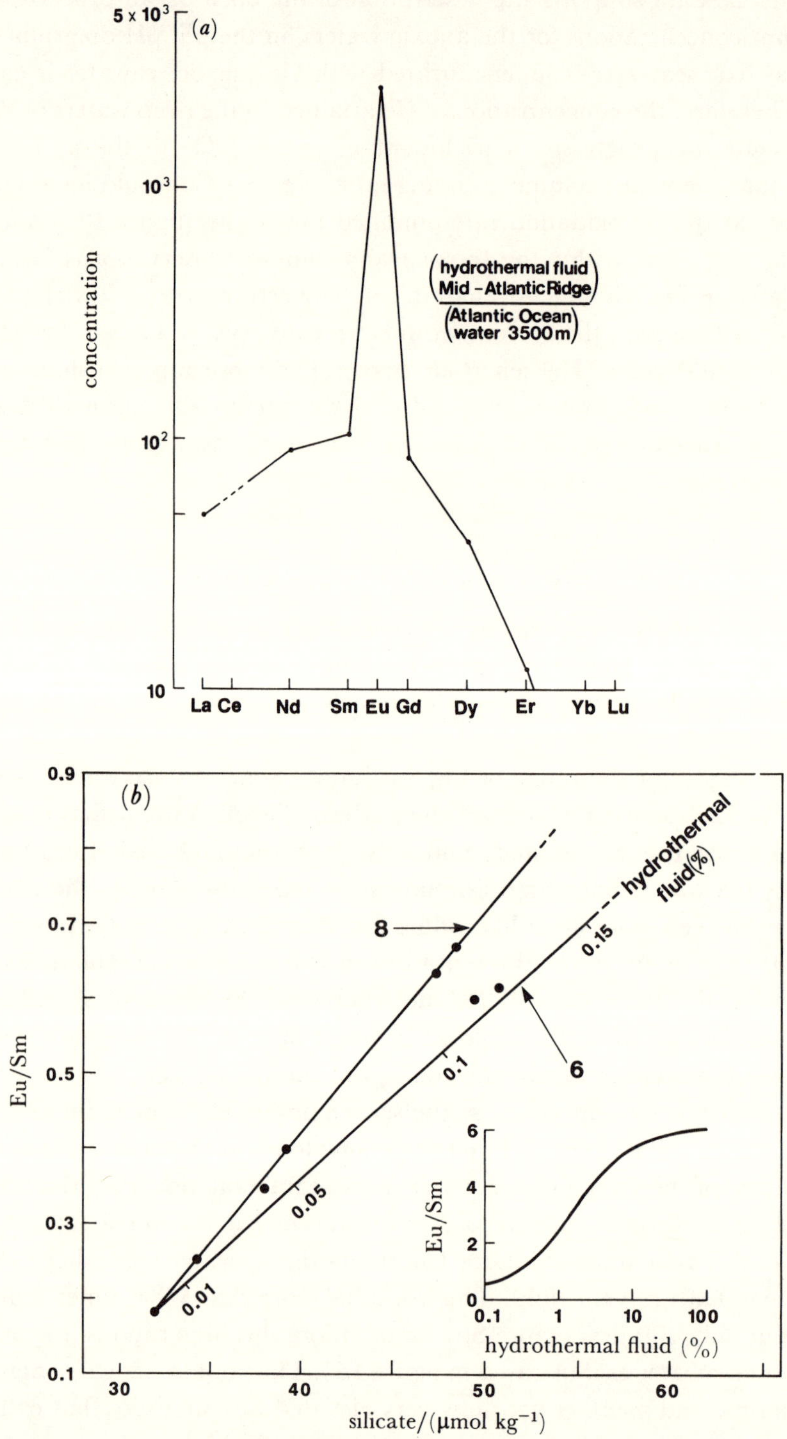

FIGURE 15. (*a*) REE concentrations in hydrothermal fluid from Mid-Atlantic Ridge normalized to concentrations in Atlantic ocean water. (*b*) Eu/Sm ratio against silicate for samples of hydrothermal plume in Mid-Atlantic Ridge median valley. Data from Elderfield *et al.* (1987). The values indicated by the arrows are for $(Eu/Sm)_{HF}$.

Concluding comment

Work on the REE chemistry of seawater has occupied only a very recent period in the history of rare-earth element research. The first concentration measurements were reported almost 25 years ago (Goldberg *et al.* 1963) but it is only in the last 5 years that there has been a concerted effort to try and understand what controls their oceanic chemistry. Nevertheless, this year is a most appropriate one in which to present this review because 1987 is the 200th anniversary of the discovery of the rare earths by Lieutenant Carl Axel Arrhenius who found a black mineral in a quarry in the village of Ytterby near Stockholm in 1787. However, it took a further 120 years' work to complete the separation of the individual REE, a process that, as shown above, the oceans perform with consumate ease.

Finally, it is worth reporting a further historical event that occurred 100 years ago, almost exactly at the half-way point of REE research. On the Friday evening of 18 February 1887, William Crookes, F.R.S., delivered a lecture at the Royal Institution that was published one week later in the Chemical News (Crookes 1887). The lecture concerned the genesis of the elements and Crookes was particularly intrigued by the metals of the rare earths, which he discussed as part of a category 'of elements associated with other elements more or less closely allied to themselves', no doubt impressed by their chemical coherence. Earlier in his article, Crookes writes 'These elements perplex us in our researches, baffle us in our speculations, and haunt us in our very dreams. They stretch like an unknown sea before us – mocking, mystifying and murmuring strange revelations and possibilities'. This seems a most apt summary of the state of understanding of marine trace-element chemistry from which we are now emerging. It is hoped that this review will be of use in describing some of the 'strange revelations and possibilities' obtained from studying rare-earth elements in the sea.

I acknowledge the contributions of the past and present members of the Marine Geochemistry Group at Cambridge who have worked with me on oceanic REE chemistry, and the Natural Environment Research Council for supporting this work. I am grateful to Carma Schilpp for tracing the quotation by Sir William Crookes. This is Cambridge Earth Sciences Series Contribution no. 1010.

References

Balistrieri, L., Brewer, P. G. & Murray, J. W. 1981 Scavenging residence times of trace metals and surface chemistry of sinking particles in the deep ocean. *Deep Sea Res.* **28**A, 101–121.

Bender, M. L. 1982 Trace elements and ocean chemistry. *Nature, Lond.* **296**, 203, 204.

Cantrell, K. J. & Byrne, R. H. 1987 Rare earth element complexation by carbonate and oxalate ions. *Geochim. cosmochim. Acta* **51**, 597–605.

Crookes, W. 1887 Genesis of the elements. *Chem. News, Lond.* **55**, 83–99.

De Baar, H. J. W., Bacon, M. P. & Brewer, P. G. 1983 Rare-earth distributions with a positive Ce anomaly in the Western North Atlantic Ocean. *Nature, Lond.* **301**, 324–327.

De Baar, H. J. W., Bacon, M. P., Brewer, P. G. & Bruland, K. W. 1985*a* Rare earth elements in the Atlantic and Pacific Oceans. *Geochim. cosmochim. Acta* **49**, 1943–1959.

De Baar, H. J. W., Brewer, P. G. & Bacon, M. P. 1985*b* Anomalies in rare earth distributions in Seawater: Gd and Tb. *Geochim. cosmochim. Acta* **49**, 1961–1969.

De Baar, H. J. W., Elderfield, H. & Bruland, K. W. 1985*c* Rare earth elements at VERTEX IV in the Central North Pacific glyre. *Trans. Am. geophys. Un.* **66**, 1291.

De Baar, H. J. W., German, C. R., Elderfield, H. & Van Gaans, P. 1988 Rare earth element distributions in anoxic waters of the Cariaco Trench. *Geochim. cosmochim. Acta.* (In the press.)

Elderfield, H. & Greaves, M. J. 1981 Negative cerium anomalies in the rare earth element patterns of oceanic ferromanganese nodules. *Earth planet. Sci. Lett.* **55**, 163–170.

Elderfield, H. & Greaves, M. J. 1982 The rare earth elements in seawater. *Nature, Lond.* **296**, 214–219.
Elderfield, H. & Greaves, M. J. 1983 Determination of the rare earth elements in seawater. In *Trace metals in seawater* (ed. C. S. Wong, E. Boyle, K. W. Bruland, J. D. Burton & E. D. Goldberg), pp. 427–445. New York & London: Plenum Press.
Elderfield, H. & Sholkovitz, E. R. 1987 Rare earth elements in the pore waters of reducing nearshore sediments. *Earth planet. Sci. Lett.* **82**, 280–288.
Elderfield, H., Greaves, M. J., Klinkhammer, G. P. & Edmond, J. M. 1987 Rare earth elements in hydrothermal fluids and plumes: influence on geochemical mass balance. Abstract, IUGG Symposium, Vancouver 1987.
Goldberg, E. D. 1954 Marine geochemistry 1. Chemical scavengers of the sea. *J. Geol.* **62**, 49–265.
Goldberg, E. D., Koide, M., Schmitt, R. A. & Smith, R. H. 1963 Rare-earth distribution in the marine environment. *J. geophys. Res.* **68**, 4209–4217.
Goldschmidt, V. M. 1954 *Geochemistry* (730 pages.) Oxford: Clarendon Press.
Greaves, M., Klinkhammer, G., Elderfield, H., Lambert, C. & Risler, N. 1985 Rare earth geochemistry in the Alboran Sea. *Terra cogn.* **5**, 190.
Heezen, B. C., Menzies, R. J., Broecker, W. S. & Ewing, M. 1959 Stagnation of the Cariaco Trench. In *Preprints International Oceanographic Congress* (ed. M. Sears), pp. 99–102. Washington D.C.: Am Assoc. Adv. Science.
Klinkhammer, G., Elderfield, H. & Hudson, A. 1983 Rare earth elements in sea water near hydrothermal vents. *Nature, Lond.* **297**, 185–188.
Krauskopf, K. B. 1956 Factors controlling the concentration of thirteen rare metals in sea water. *Geochim. cosmochim. Acta* **9**, 1–32.
Michard, A., Albarede, F., Michard, G., Minster, G. F. & Charlou, J. L. 1983 Rare-earth elements and uranium in high temperature solutions from East Pacific Rise hydrothermal vent field (13° N). *Nature, Lond.* **303**, 795–797.
Palmer, M. 1983 Rare earth elements and Nd and Sr isotopes in the Atlantic Ocean. Ph.D. thesis, University of Leeds. (284 pages.)
Piepgras, D. J. & Wasserburg, G. J. 1980 Neodymium isotopic variations in sea water. *Earth planet. Sci. Lett.* **50**, 128–138.
Piepgras, D. J. & Wasserburg, G. J. 1982 Isotopic composition of neodymium in waters from the Drake Passage. *Science, Wash.* **217**, 207–214.
Piepgras, D. J. & Wasserburg, G. J. 1983 Influence of the Mediterranean outflow on the isotopic composition of neodymium in waters of the North Atlantic. *J. geophys. Res.* **88**, 5997–6006.
Piepgras, D. J. & Wasserburg, G. J. 1985 Strontium and neodymium isotopes in hot springs on the East Pacific Rise and Guaymas Basin. *Earth planet. Sci. Lett.* **72**, 341–356.
Piepgras, D. J. & Wasserburg, G. J. 1987 Rare Earth element transport in the western North Atlantic inferred from Nd isotopic observations. *Geochim. cosmochim. Acta* **51**, 1257–1271.
Piepgras, D. J., Wasserburg, G. J. & Dasch, E. J. 1979 The isotopic composition of Nd in different ocean masses. *Earth planet. Sci. Lett.* **45**, 223–236.
Schindler, P. W. 1975 Removal of trace metals from the oceans: a zero order model. *Thalassia jugosl.* **11**, 101–111.
Scranton, M. I., Sayles, F. L., Bacon, M. P. & Brewer, P. G. 1987 Temporal changes in the hydrography and chemistry of the Cariaco Trench. *Deep Sea Res.* **34**, 945–963.
Stordal, M. C. & Wasserburg, G. J. 1986 Neodymium isotopic study of Baffin Bay water: Sources of REE from very old terranes. *Earth planet. Sci. Lett.* **77**, 259–272.
Turekian, K. K. 1977 The fate of metals in the oceans. *Geochim. cosmochim. Acta* **41**, 1139–1144.
Upstill-Goddard, R., Elderfield, H. & Sholkovitz, E. R. 1987 Rare earth elements in rivers and estuaries. (In preparation.)
Whitfield, M. & Turner, D. R. 1987 The role of particles in regulating the composition of sea water. In *Aquatic surface chemistry* (ed. W. Stumm), pp. 457–493. New York: Wiley.

Discussion

M. WHITFIELD (*Marine Biological Association, Plymouth, U.K.*). Most stability constants for REE complexation in seawater are obtained by analogy rather than by experiment. Therefore, arguments that depend on the magnitude of carbonate complexation rather than the relative trends must be treated with caution. Is Dr Elderfield sure that the model he presents can distinguish between 'inorganic' and 'organic' interactions?

H. ELDERFIELD. Yes. The model I presented used experimentally determined data by Cantrell & Byrne (1987) rather than the earlier data of Turner *et al.* (1981), which were obtained by analogy.

Reference

Turner, D. R., Whitfield, M. & Dickson, A. G. 1981 The equilibrium speciation of dissolved components in fresh water and sea water at 25 °C and 1 atm pressure. *Geochim. cosmochim. Acta* **45**, 855–881.

M. WHITFIELD. Dr Elderfield's data show a similarity between the REE chemistry of Mediterranean and Pacific waters. How does he explain this?

H. ELDERFIELD. They are similar in the sense that both Mediterranean and Pacific surface waters show similar REE mass fractionation as shown by their Er/Nd ratios. This is a consequence of the removal of REEs from surface seawater by adsorption at particle surfaces, which is a ubiquitous process. However, surface-water concentrations in the Mediterranean are much higher than in the Pacific. This has been recognized for other trace metals (see, for example, Boyle *et al.* 1985) and has been explained by the anti-estuarine circulation of the Mediterranean that permits less vertical recycling of nutrients and hence less removal of metals from surface waters relative to deep waters.

Reference

Boyle, E. A., Chapnick, S. D., Bai, X. X. & Spivack, A. 1985 Trace metal enrichments in the Mediterranean Sea. *Earth planet. Sci. Lett.* **74**, 405–419.

J. D. BURTON (*Department of Oceanography, University of Southampton, U.K.*). The possible importance of transport of rare earths in colloidal forms in rivers was mentioned. Would Dr Elderfield like to comment on the importance of destabilization and aggregation of colloids in estuaries in relation to the fluxes from continental weathering processes?

H. ELDERFIELD. This is a process that we have been considering recently (Hoyle *et al.* 1984; Upstill-Goddard *et al.* 1987). It is striking that the variations of REE concentrations in six estuaries now examined invariably are similar to those in iron. The highest REE concentrations we have measured in rivers are from rivers containing iron-rich colloidal materials and it seems very likely that the REEs are associated with such phases.

Reference

Hoyle, J., Elderfield, H., Gledhill, A. & Greaves, M. 1984 The behaviour of the rare earth elements during mixing of river and sea waters. *Geochim. cosmochim. Acta* **48**, 143–149.

M. P. BACON (*Woods Hole Oceanographic Institution, Woods Hole, U.S.A.*). Dr Elderfield has tended to use REE ratios in average shale for comparison with the seawater data. To what extent does he consider that shale typifies the pattern of REE input to the oceans from the continents?

H. ELDERFIELD. Although I tended to display shale REE ratios for comparative purposes, I mentioned that the conclusions reached also apply were REE ratios for riverwaters used. Recently we have measured REE concentrations in 23 rivers (Upstill-Goddard *et al.* 1987) and have found that their patterns do not always resemble that of average shale. After correction for REE fractionation during estuarine removal the pattern of river input is heavy-REE enriched but not to the degree found in ocean waters. Therefore, *some* of the REE fractionation implied to have occurred in seawater by normalizing patterns to average shale may be an artefact of the normalization procedure but it cannot explain the major effects such as differences between different ocean basins. There are four problems in the choice of an average riverwater pattern for normalization. There is a variability in pattern between different rivers. Temporal variations in pattern occur in individual rivers. The effect of estuarine removal on the REE input pattern has not yet been evaluated thoroughly. Other sources of REE to seawater are important locally, principally aeolian input.

P. S. LISS (*School of Environmental Sciences, University of East Anglia, U.K.*). In view of the similar redox chemistry of cerium and manganese, is there any evidence either from the field or the laboratory for photochemical reduction of Ce^{IV} to Ce^{III} in the presence of dissolved organic matter, as appears to occur for manganese ($Mn^{IV} \rightarrow Mn^{II}$)?

H. ELDERFIELD. This is an interesting suggestion, in particular because the free-energy changes for Ce^{IV} and Mn^{IV} reduction are very similar. Surface-water enrichments of Ce have been observed at some stations but these are invariably associated with enrichments of other REES too, and may be attributed to aeolian input. However, a positive Ce anomaly in western North Atlantic surface waters has been recorded (De Baar *et al.* 1983) and although it was attributed to advective input from reducing marginal sediments, photochemical reduction is an equally plausible explanation.

Trace metals as tracers in the ocean

By J. D. Burton and P. J. Statham

Department of Oceanography, University of Southampton, Southampton SO9 5NH, U.K.

The trace metals that have the greatest potential value as tracers for circulation and mixing in the ocean are markedly non-conservative, with short oceanic residence times. Their distributions strongly reflect the influence of sources either at boundaries or within the water column itself. The signals arising from these inputs can act as indicators for sources and transport pathways. These characteristics are well exemplified by dissolved manganese, which shows variations in concentration within the ocean that reflect the advection and mixing of waters that have acquired enhanced concentrations as a result of three principal processes. First, injection at the sea floor in hydrothermal fluids leads to anomalies in deep-water concentrations that are detectable over about 1000 km in parts of the Pacific Ocean. At the Mid-Atlantic Ridge, the advection of hydrothermal inputs is largely restricted by the topography of the axial rift valley, but the signals have proved valuable in exploration for hydrothermal sites. Secondly, pronounced maxima are associated with the highly oxygen-deficient waters of the oxygen minimum in parts of the eastern North Pacific Ocean. They are accounted for by *in situ* remobilization from particulate forms of manganese and by lateral advection from sources at the ocean margins. Interpretation of the data for the manganese maximum layer with two-dimensional box models has illustrated the potential to estimate lateral advective velocities on the basis of fluxes of manganese. Thirdly, the occurrence of maxima in the surface mixed layer of the ocean, which is particularly marked in areas where aeolian inputs appear to be an important source for dissolved manganese, suggests its potential use as a tracer for mixing and transport in the upper ocean. In all three cases, a more detailed and quantitative knowledge of the geochemical processes that determine the observed features will be needed if the potential for tracer applications is to be fully realized.

Dissolved aluminium varies in concentration in deep-ocean waters by about two orders of magnitude, suggesting that it may be useful as a tracer for the movement and mixing of water masses. Systematic differences in concentration occur in water masses in the northwest Atlantic Ocean, but the origins of the differences remain to be clarified. Beryllium may also serve to identify water masses in the Atlantic Ocean. The few data available for cobalt suggest that this metal also may find applications as a tracer.

Introduction

Trace metals would probably have received little attention in any discussion of tracers in the sea before the late 1970s. The major advances that began a few years earlier, in clean sampling and in the handling and analysis of samples under clean conditions, have led to a much improved knowledge of the distributions of trace metals (Burton & Statham 1982; Bruland 1983). Some aspects of the processes of input and removal and of the mechanisms of internal cycling within the ocean have now begun to be investigated more fully.

These studies have established that a number of trace metals behave conservatively in the ocean, or show only minor divergences from such behaviour. Examples include the alkali metals, lithium and caesium. Molybdenum, which forms a relatively stable oxy-anion, appears to behave essentially conservatively in most marine environments. Methylated germanium species have recently provided an unexpected addition to the group of conservative constituents (Lewis *et al.* 1984). The conservative behaviour of uranium leads to important applications because it gives rise to uniform generation of radioactive daughter products in the uranium and actino-uranium series and these nuclides provide valuable means of estimating geochemical parameters such as scavenging rates.

A second group of trace metals contains those that show more or less strong positive correlations with micronutrients such as phosphate and dissolved silicon, and accordingly tend to increase in concentration along the direction of the main advective flow of deep water through the ocean basins. These correlations reflect involvement of the metals with the major vertical flux of particulates from the euphotic zone into deeper waters and their release during the decomposition of organic material or the dissolution of skeletal carrier phases, such as opal and calcium carbonate. Cadmium, zinc, and the inorganic forms of germanium show such behaviour to a very marked degree.

Of greater potential interest as tracers for transport and mixing are those metals that have distributions that strongly reflect the presence of sources, either at boundaries or within the water column itself, leading to distinctive signatures that can act as indicators of sources and transport pathways. These metals are markedly non-conservative, with short oceanic residence times, so that the concentrations in waters distant from sources become depleted in concentration relative to those close to inputs.

The difficulty in using these constituents quantitatively as tracers for physical processes can be summarized in terms of the one-dimensional advection–diffusion–scavenging model of Craig (1974) for trace elements in the deep sea. This model shows the distribution of a dissolved constituent to be dependent upon the ratio of the advective velocity to the scavenging rate constant; scavenging is assumed here to be a first-order reaction. In geochemical interpretations with this model (see also, for example, Weiss 1977) the advective velocity is constrained through other measurements to estimate the scavenging rate constant. To make the inverse calculation would require quantitative information on the scavenging rate that is at present essentially lacking. Nevertheless, the qualitative value of certain trace metals as tracers for inputs and as labels on water masses is apparent. It is the purpose of this paper to review, in this context, the geochemical behaviour of two principal examples, namely manganese and aluminium, and to consider data that indicate their potential for more quantitative applications; brief comments are included on two other metals, beryllium and cobalt, which may find some uses as tracers.

Manganese

The dominant factor influencing the aquatic geochemistry of manganese is the change of oxidation state of the element between reducing and oxidizing environments, coupled with the marked difference between the oxidation states in terms of their forms and reactivity.

For the reaction:

$$MnO_2(s) + 4H^+ + 2e^- \rightleftharpoons Mn^{2+} + 2H_2O, \tag{1}$$

it may be calculated (Stumm & Morgan 1981) that in seawater of pH 8.1, at 25 °C, the pE† at which the activities of the oxidized and reduced species are equal is 5.3. Seawater containing free oxygen at the concentrations that characterize most oceanic waters has a considerably

† In this paper, $pE = -\lg a_{e^-}$, where a is activity.

higher pE and on thermodynamic grounds manganese should be present as Mn^{IV}, in which oxidation state the element has a very low solubility. Under reducing conditions, Mn^{II} is the stable oxidation state, and the element is then much more soluble and geochemically mobile.

The oxidation of Mn^{II} in the environment produces a solid phase in which the average oxidation number of the metal is less than (IV) (see, for example, Grill 1982). It has been argued that the solid phase in equilibrium with Mn^{II} in seawater may be Mn_3O_4 (hausmannite) (Klinkhammer & Bender 1980) or γ-MnOOH (Grill 1982). The initially formed phase becomes more highly oxidized following deposition. The oxidation rate of Mn^{II} is sufficiently slow for the reduced form to persist in natural waters for significant periods. In oceanic deep waters the scavenging rate, which appears to reflect both removal to particles by oxidation at the particle surface and adsorption, particularly onto calcium carbonate (Martin & Knauer 1983), may be of the order of 0.01 a^{-1} (Landing & Bruland 1980). A similar value of 0.02 a^{-1} was derived by Weiss (1977) for the scavenging rate constant, assumed to be first order, in a plume of hydrothermal manganese above the Galapagos Spreading Centre. Where the scavenging rate is low relative to the rate of input or release of Mn^{II}, enhanced concentrations of the metal are observed; these can provide distinctive signatures for certain water bodies.

In studies of manganese in ocean waters a number of operationally defined parameters have been used in addition to dissolved and particulate fractions resolved by conventional filtration methods. These other parameters, total dissolved manganese and reactive manganese, are nearly equivalent to dissolved manganese in most of the situations discussed here and in such cases are referred to in the text as dissolved manganese without further qualification.

The distribution of dissolved manganese is shown in figure 1 for a vertical oceanic profile

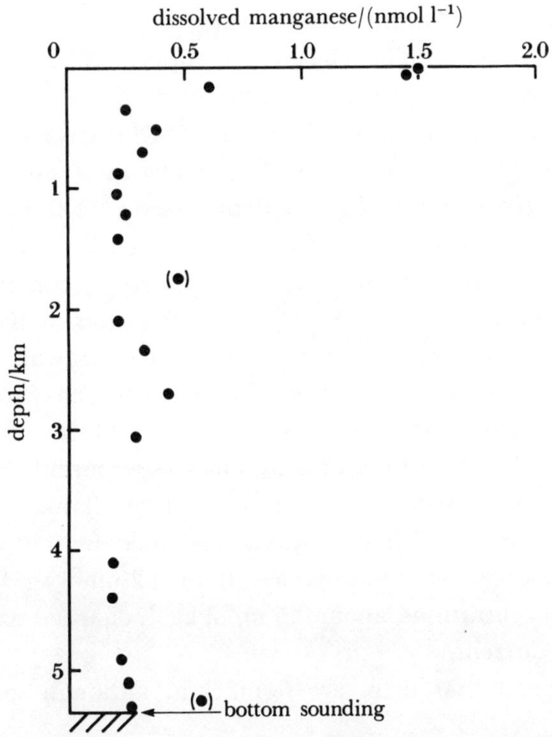

FIGURE 1. Vertical profile of concentration of dissolved manganese in the eastern North Atlantic Ocean (19° 22′ N, 12° 51′ W). Points in parentheses are considered analytically suspect. (From Statham *et al.* (1985).)

typical of areas where there are no significant inputs to intermediate and deep waters. The characteristic features of such areas are a maximum or several maxima in the mixed layer, sometimes occurring at the surface, a marked decrease within the upper few hundred metres, leading in some profiles to a minimum around the top of the thermocline, and relatively uniform and low values in the deep waters. This general pattern has been observed in many regions (Bender *et al.* 1977; Klinkhammer & Bender 1980; Martin & Knauer 1980; Landing & Bruland 1980, 1987; Bruland & Franks 1983; Burton *et al.* 1983; Yeats & Bewers 1985; Statham *et al.* 1985; Statham & Burton 1986).

The concentrations in deep waters are towards the lower end of the range of analytical methods and often close to the detection limits. Thus, there are uncertainties about the comparability of results reported by different workers for different oceanic regions. Nevertheless, some data have been reported indicative of regional variations in deep-water concentrations. Yeats & Bewers (1985) report regional differences in concentration of dissolved manganese in deep waters of the northwestern Atlantic Ocean. For example, concentrations in waters deeper than 2000 m at about 50° N averaged 0.56 nmol l^{-1}, whereas at 32° N concentrations averaging about 0.3 nmol l^{-1} were found at 1100–1450 m, which were the greatest depths sampled there. These workers suggest that differences between these two regions reflect the increased effects of scavenging during the southward transport of deep water that originally has a relatively high concentration of manganese as a result of its recent formation by sinking from the surface. Bruland & Franks (1983), however, report a concentration of 0.56 nmol kg^{-1} at 2962 m in the Sargasso Sea at 34° N, and Jickells (1986) has reported similar values for this region.† For the northeastern Atlantic Ocean, concentrations averaging about 0.25 nmol l^{-1} have been reported for waters below about 1000 m (Statham *et al.* 1985; Statham & Burton 1986); the results from these studies suggest no significant differences between deep Mediterranean and northeastern Atlantic Ocean waters.

Consideration of data for the Pacific Ocean is complicated not only by analytical uncertainties but also by the effects of hydrothermal inputs on the deep-water concentrations at some locations, as discussed below. Landing & Bruland (1980) gave a value of approximately 0.2 nmol kg^{-1} for deep waters in the northeastern Pacific Ocean; at an open-ocean station their results are in the range of 0.10–0.15 nmol kg^{-1} for depths below 3000 m. Further measurements on profiles in the central and eastern Pacific Ocean between 37° N and 20° S (Landing & Bruland 1987) have yielded notably smooth and correspondingly convincing profiles in which some features in deeper water are apparent. At the VERTEX II station, in the eastern tropical North Pacific Ocean, and at a central oceanic station southwest of Hawaii, concentrations in the depth range corresponding to the Pacific Deep Water averaged about 0.2–0.3 nmol kg^{-1}, values similar to those reported for the northeastern Atlantic Ocean. At the remaining open-ocean station, southwest of Tahiti, a broad maximum was apparent within the Pacific Deep Water that was provisionally attributed to advection from flank sediments of islands or seamounts of the region. For all offshore stations examined in this work, the lowest concentrations of dissolved manganese, typically less than 0.2 nmol kg^{-1}, occurred in bottom waters. Higher average concentrations, about 0.5 nmol kg^{-1}, characterized deep waters at two stations in the California Current.

Overall, these data suggest that there are significant, although small, variations in the

† We have chosen to quote measured concentrations in the units reported by the workers concerned.

concentrations of dissolved manganese in deep waters within the major ocean basins. Although in some cases variability may reflect the differences in age between waters that have sunk very recently from the surface and waters that are older in relation to this process, it is clear that variations of comparable magnitude can arise in other ways. Some differences may reflect local sources, as in the example of the possible influence of seamount or island flank sediments, quoted above. Accurate measurements on sections carefully chosen in relation to the patterns of deep-water flow are needed to evaluate and interpret the variations. It is clear that any systematic difference along the direction of the major global deep water circulation is small, although the lowest concentrations found in Pacific Ocean waters deeper than about 3000 m are less than any reported so far for the Atlantic Ocean. This is probably attributable to the rapid scavenging of manganese that leads to depletion even in recently formed deep waters. It may be that over much of the deep ocean the concentrations of dissolved manganese are close to a background value determined by a quasi-equilibrium between manganese in solution and the solid phase intially formed by its oxidation, as suggested by Klinkhammer & Bender (1980).

Much more clearly defined variations in concentrations of dissolved manganese are associated with the advection and mixing of waters that have received localized inputs of mobilized Mn^{II}. These phenomena are associated with three principal processes that mobilize manganese.

(1) Injection at the sea bed in hydrothermal fluids.

(2) Release from either bottom sediments or suspended particulate material under suboxic or anoxic conditions.

(3) Input to surface ocean waters, which at least in central areas appears to arise mainly by release from particulate material introduced by aeolian transport.

Hydrothermal inputs of manganese

The penetration of seawater through oceanic crustal material at ocean-ridge spreading centres can lead to high-temperature water–basalt interactions, producing hydrothermal fluids that are returned to the ocean through vents at the sea floor. The chemical processes that occur have been extensively described by, among others, Edmond *et al.* (1979, 1982), Von Damm *et al.* (1985*a*, *b*) and Bowers *et al.* (1985). Concentrations of manganese in the high-temperature hydrothermal fluids have been reported to range from 0.13 to 1.5 mmol kg^{-1} (Von Damm *et al.* 1985*a*, *b*; Klinkhammer & Hudson 1986), about a million times higher than those in ocean-bottom waters.

The hot buoyant water emanating from vents rises to some few hundred metres above the sea bed before attaining density stability and spreading laterally, large amounts of water from deeper in the water column becoming entrained in the process (Lupton *et al.* 1985). These processes were observed to produce large anomalies in the concentrations of dissolved and particulate manganese in the waters above the Galapagos Spreading Centre (Klinkhammer *et al.* 1977; Weiss 1977; Bolger *et al.* 1978). Corresponding features have been observed in other areas of hydrothermal activity in the Pacific Ocean, namely on the East Pacific Rise at 21° N (Lupton *et al.* 1980), over the Juan de Fuca Ridge (Jones *et al.* 1981), and in the Guaymas Basin (Campbell & Gieskes 1984). Lateral spreading leads to anomalies in the concentrations of dissolved manganese in deep waters over extensive areas, the concentrations reflecting the dilution of the water enriched in manganese and the scavenging of the metal. Klinkhammer *et al.* (1977) and Weiss (1977) estimated from a one-dimensional dispersion model that the horizontal propagation distance for the inputs in the Galapagos Spreading Centre region was of the order of 1000 km. This has been confirmed by observation of anomalies probably attributable to distant hydrothermal sources, in the Guatemala Basin (Klinkhammer 1980) and the detection, in the data set of Klinkhammer & Bender (1980), of a slight anomaly at a station (GEOSECS 317) about 2000 km west of the ridge axis (Klinkhammer & Hudson 1986).

A detailed examination of the dispersal patterns in the region of the East Pacific Rise between about 10° S and 40° S has been made by Klinkhammer & Hudson (1986). A profile for a station on the ridge axis, with a pronounced maximum above the sea floor, is shown in figure 2*a*. In this region the flow of deep water at depths corresponding to those of plume injection is dominantly westward but at 30° S the flow is eastward. An approximately north–south section east of the East Pacific Rise, at 91°–94° W, shows a core of manganese-rich water centred on about 30° S at a level corresponding to that of plume injection at the ridge axis (figure 3). West of the Rise, at about 12°–19° S, the westward flow of deep water is reflected in anomalies, generally with several maxima, arising from the effects of several plumes arising from vents at different depths.

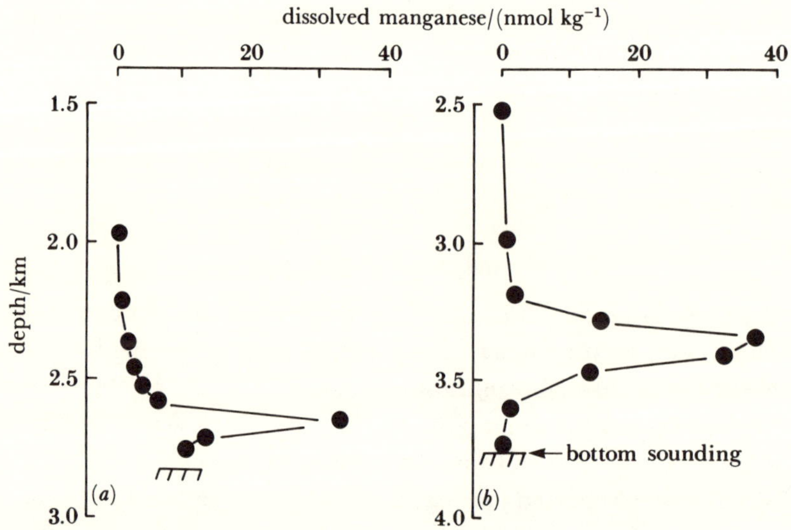

FIGURE 2. Vertical profiles, showing anomalies in concentration of dissolved manganese associated with advection from hydrothermal sources, at (*a*) the East Pacific Rise at 20° S (Klinkhammer & Hudson 1986) and (*b*) the Mid-Atlantic Ridge at 26° N (Klinkhammer *et al.* 1986). See text for explanation of use of the term dissolved manganese.

Studies on the Pacific Ocean have demonstrated that anomalies in the concentration of manganese, like those in the conservative tracer helium-3 (see, for example, Weiss (1977), Lupton *et al.* (1977, 1980) and Lupton & Craig (1981)), can provide information about the patterns of deep-water circulation in regions influenced by hydrothermal inputs. They have also shown the potential for the use of manganese as a geochemical indicator in exploring for hydrothermal vents. This approach is particularly valuable because measurements of manganese can be carried out fairly readily on board ship.

The value of measurements of manganese in detecting hydrothermal inputs has been effectively realized in recent work in the Atlantic Ocean. Evidence of hydrothermal activity on the Mid-Atlantic Ridge, where spreading rates are considerably lower than those on the East-

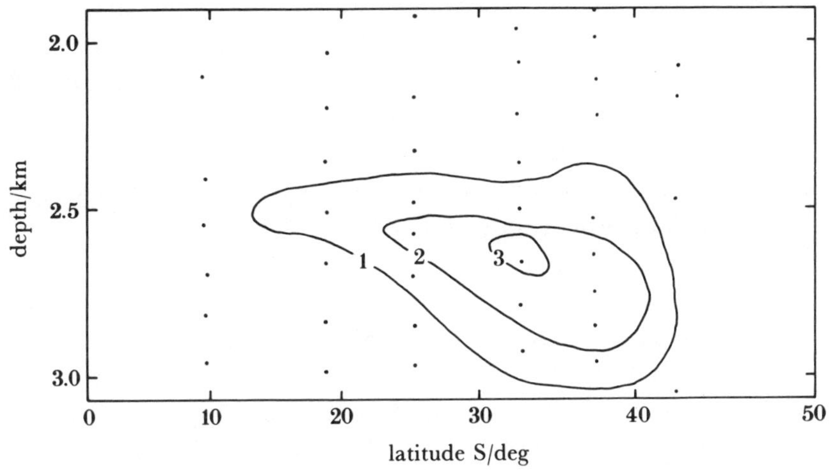

FIGURE 3. Concentrations of total dissolvable manganese (nanomoles per kilogram) along a section in the Galapagos Basin at about 93° W (Klinkhammer & Hudson 1986).

Pacific Rise, has been obtained from findings of metalliferous deposits and sediments with anomalously high concentrations of trace metals such as iron and manganese, from observations of suspended particulate material enriched in manganese and iron, and from anomalies of temperature and helium-3 in the water column, as well as from magnetic measurements on crustal material. These discoveries, which have been summarized by Rona (1980) and Rona *et al.* (1984), have focused on the Trans-Atlantic Geotraverse (TAG) hydrothermal field at 26° N, on the east wall of the rift valley. Rona *et al.* (1984) concluded that the observed features were consistent with alternating periods of low-temperature and high-temperature venting.

The first observations of anomalies in manganese were those of Klinkhammer *et al.* (1985) at stations above the rift valley of the Ridge, extending from the TAG area at 26° N to the Vema Fracture Zone (11° N). These observations showed anomalies of dissolved manganese up to about 2 nmol kg^{-1}, in maxima above the sea bed. The amounts of acid-soluble particulate manganese present varied considerably between stations, probably reflecting differences in the degree of scavenging of dissolved manganese on to particulate phases and thus in the age of the manganese relative to its time of injection, i.e. differences in distance from the vent. From such differences, together with variations in the depth of the maxima, and the extent and shape of the anomalous section of the profile, it was concluded that there were at least five sources within this section of the rift valley.

An anomaly in concentration of manganese was also found in the median valley of the Mid-Atlantic Ridge at 43° N (Hydes *et al.* 1986). In the section of the ridge between 42°–47° N, there occurs a series of major segments of the valley, each of which contains, below 3000 m, water of uniform potential temperature (Saunders & Francis 1985). The properties of the water that enters the valley are determined by the shallowest depth at which exchange with an exterior source can occur, in effect a sill depth analogous to that of a fjord. The observations on manganese were made in the valley designated by Saunders & Francis (1985) as valley A with a sill depth of 2450 m. The water within the valley was found to be very well mixed, as described by Saunders & Francis (1985) for valleys to the north. Concentrations of manganese, like those of copper, nickel and cadmium, were uniform below 2400 m but unlike these other elements, manganese was significantly enriched relative to typical Atlantic Ocean deep waters, with a concentration of about 1.7 nmol l^{-1}. The anomaly in dissolved manganese was attributed to a hydrothermal source. It was concluded from the shape of the profile above the sill depth, and a consideration of horizontal and vertical dispersion rates, that other hydrothermal inputs occurred along the ridge and that a major component of deep-water flow was along the axis of the ridge, which is consistent with other observations (Keller *et al.* 1975).

Further measurements of manganese were made in the TAG area as part of an investigation (Rona *et al.* 1986) that led to the discovery of active vents on the wall of the rift valley, with a range of associated phenomena, including active and inactive chimneys, and vent biota. The data for manganese have been given by Klinkhammer *et al.* (1986). They show the existence of advective plumes with concentrations of total reactive manganese exceeding 30 nmol kg^{-1} in maxima in the water column, both the concentrations, and the shapes of the maxima (figure 2*b*), resembling those observed on the East Pacific Rise. The dominant contrast between the Atlantic and Pacific Oceans is that in the former there is an effective topographic restriction on the dispersal of the manganese introduced from vents. Thus, observations of anomalous concentrations of manganese, associated with hydrothermal inputs, have so far been restricted to the rift valley. Klinkhammer *et al.* (1985) point out the possibility that leakage of manganese could occur through fracture zones. Labelling of deep water in this way could potentially provide a useful tracer for deep-water movements but the present indications are that any signals would be small.

Inputs resulting from depletion of oxygen in the water column or in sediments at the ocean margin

In regions of the Pacific Ocean where the oxygen concentration at depths near the oxygen minimum falls below about 100 μmol l^{-1}, a significant increase in the concentration of dissolved manganese was reported by Klinkhammer & Bender (1980). This phenomenon has been extensively studied only in the eastern Pacific Ocean where substantial anomalies in dissolved manganese are generally associated with the suboxic conditions that occur in the oxygen minima over extensive areas (Murray *et al.* 1983; Martin & Knauer 1984, 1985; Martin *et al.* 1985; Jones & Murray 1985). A few examples of small anomalies associated with much less intense oxygen minima have been reported for the Atlantic Ocean (Statham & Burton 1986).

The occurrence of enhanced concentrations of dissolved manganese at intermediate depths in parts of the North Pacific Ocean is illustrated by the results for VERTEX stations on a section approximately normal to the western coastline of the U.S.A., between 28° N and 36° N (Martin *et al.* 1985). The variations with distance offshore are shown in figure 4; the enhanced concentrations in the surface waters are discussed subsequently. The section shows a core of water with enhanced concentrations of dissolved manganese (0.5–1.4 nmol l^{-1}) centred on about 600–700 m, with decreasing concentrations in the offshore direction. Similar features were described by Jones & Murray (1985) for a section at about 47° N, with concentrations of dissolved manganese in the oxygen minimum decreasing from 4.7 nmol kg^{-1} close to the continental slope to about 0.7 nmol kg^{-1} offshore.

Two principal processes may contribute to the increase in dissolved manganese in the dissolved oxygen minimum: the diffusion and advection of manganese from the continental slope region and the release of manganese from particles sinking from the surface waters. Such release is probably primarily caused by the dissolution of an oxidized phase under the conditions of lowered pE, by the decomposition of organic material, or by a combination of

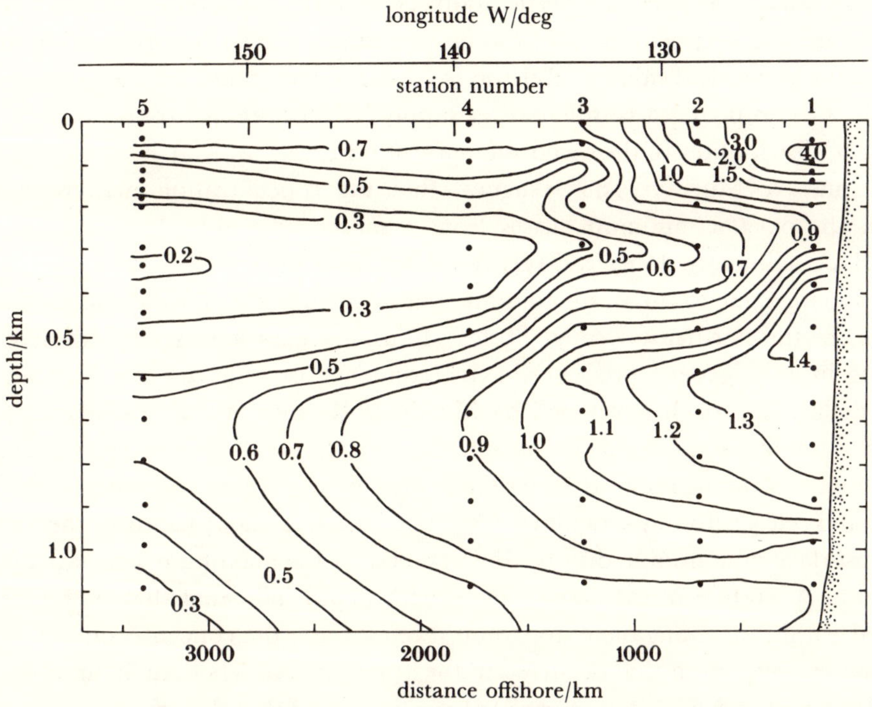

FIGURE 4. Concentrations of dissolved manganese (nanomoles per kilogram) along a section (VERTEX V) from Monterey Bay to the VERTEX IV station north of Hawaii (Martin *et al.* 1985).

these processes. The occurrence of such processes is indicated by the fact that particulate material in the oxygen minimum is markedly depleted in leachable manganese (Martin & Knauer 1984).

The clearest evidence concerning the relative importance of these processes has come from interpretations by box models of measurements made during VERTEX cruises. Martin & Knauer (1984) determined manganese in the material collected by sediment traps at various depths at the VERTEX station at 18° N, 108° W, and the concentrations of dissolved and particulate manganese in the water column. The data were interpreted by using a box model with eight compartments representing sections of the water column in the vertical, with depth limits defined by the depths of deployment of the sediment traps. Vertical fluxes of particulate manganese were determined from the measurements on material collected in the sediment traps; these values excluded the refractory fraction of particulate manganese, that held in mineral lattice structures, which forms only a small and rather uniform component of the total particulate manganese throughout the water column. Vertical fluxes of dissolved manganese were calculated by using estimated values of the eddy diffusivity coefficient and the vertical advection velocity. Mass-balance calculations with the box model indicated that in the depth range containing the core of the maximum in dissolved manganese (200–400 m), with oxygen concentrations below 10 μmol kg^{-1}, the net release from particles was inadequate to sustain the observed concentrations of dissolved manganese and it was necessary to invoke a balancing flux due to horizontal diffusion and advection, amounting to about twice that due to net release from particles.

Calculations made by Landing & Bruland (1987) for the same station led to a very similar conclusion. These workers applied the vertical advection–diffusion model of Craig (1974) to a region of linear mixing between 150 and 750 m, to estimate the loss of dissolved manganese from the maximum by vertical mixing. Release from particulate material accounted for only about half of the loss, implying a compensating supply by offshore horizontal advection.

The patterns of distribution of dissolved manganese observed by Martin *et al.* (1985) (see figure 4) and Jones & Murray (1985) suggest that the concentration maximum that is associated with the oxygen minimum arises from a source of dissolved manganese at the interface with slope sediments. Jones & Murray (1985) have discussed the processes that can supply dissolved manganese to the water column. They conclude that for the continental slope off Washington, in the northeast United States, diffusive fluxes across the sediment–water interface, arising from diagenetic processes in the sediments, provide the major input.

The box-modelling approach developed by Martin & Knauer (1984) was also applied to data from four of the five stations on the section shown in figure 4 (Martin *et al.* 1985). The results for the fluxes of manganese at different depths in the water column indicated that scavenging by, and release from, particulates were not usually a major factor in the formation of either the subsurface minimum in dissolved manganese or the maximum associated with the oxygen minimum. At station 5, increasing fluxes with depth indicated that manganese was being scavenged within the manganese maximum even though the concentration of dissolved oxygen at the oxygen minimum that occurred in the same box was less than 30 $\mu mol\ kg^{-1}$. Use of the box-modelling approach, as summarized above, indicated that the subsurface manganese minimum must be maintained by horizontal advection shoreward of water containing low concentrations of dissolved manganese, whereas maintenance of the maximum requires the offshore advection of water with a high concentration of dissolved manganese. The approach also allowed calculation of the net horizontal advective velocities necessary to balance the net vertical fluxes of dissolved and particulate manganese. With the data for stations 1–4 offshore velocities in the manganese maximum layer were estimated as about 0.4 $cm\ s^{-1}$, whereas estimates of onshore velocities in the manganese minimum layer ranged from 0.9 $cm\ s^{-1}$ at station 4 to 4.4 $cm\ s^{-1}$ at station 1.

Similar calculations were made by Martin & Knauer (1985) for the whole section shown in figure 4, including station 5. A plot of dissolved manganese against σ_t for station 5 and a location close to station 1 (figure 5) shows a clear similarity in shape between the depths of 500 and 1100 m, which interval includes the oxygen minimum layer. Assuming that station 1 is representative of the marginal regions that supply manganese to the gyre circulation, the net advective velocity over the 3150 km dividing the stations was estimated as 0.9 $cm\ s^{-1}$, within the depth intervals considered.

These calculations are essentially illustrative. They involve a number of simplifying assumptions, one of the most important being implied by the use of two-dimensional models. The importance of longshore horizontal mixing in the eastern tropical North Atlantic Ocean is emphasized by Landing & Bruland (1987). Sampling grids extending alongshore as well as offshore will be needed to realize the very considerable potential for the use of manganese as a tracer for large-scale circulation processes in this area.

Manganese in the surface ocean

A surface or near-surface maximum in dissolved manganese is a ubiquitous feature in the ocean, although the concentrations found vary considerably. The results shown in figure 4

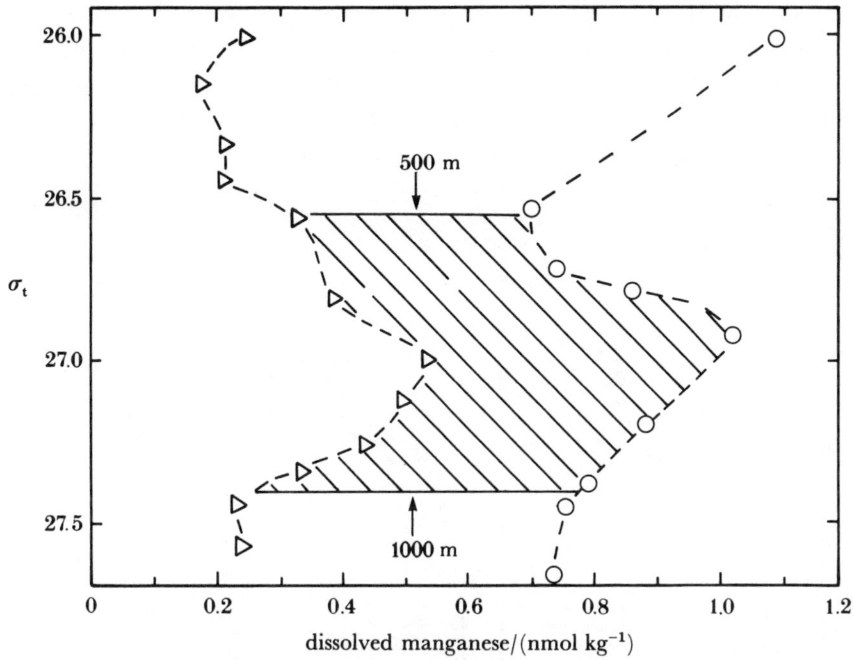

FIGURE 5. Relation between concentration of dissolved manganese and σ_t at station 5 (△) and a station close to station 1 (○) in figure 4. The hatched area was used in the box-model calculations summarized in the text. (From Martin & Knauer (1985).)

suggest the existence of a source of manganese at the ocean margin supplying the element to surface waters. Such a source was invoked by Landing & Bruland (1980) to explain the marked increase in concentrations between the California Current and the adjacent coastal waters. Similar features to those shown in figure 4 were found by Jones & Murray (1985) off the Washington coast. The features in both cases are consistent with inputs from shelf waters where the metal can be enhanced by river inputs and diffusion from, and resuspension of, reducing sediments. Upwelling of deeper water, which is rich in manganese in these environments, is another possible mechanism of supply (Jones & Murray 1985).

Transport of manganese from shelf waters may influence areas at considerable distance from the sources. Thus, Landing & Bruland (1987) found evidence for transport of manganese from the eastern boundary region of the North Pacific Ocean by the Equatorial Current. There is substantial evidence, however, for a role of aeolian input as a source of manganese for surface waters in many regions. Klinkhammer & Bender (1980) considered this source to be important at the stations they studied in the Pacific Ocean, a view that is supported by an increase in the concentrations of dissolved manganese in the surface mixed layer between the western boundary of the California Current and central gyre waters. Data for the shelf break off northwestern Europe (Kremling 1985) do not indicate any major advection of dissolved manganese to the adjacent ocean, and the distribution of dissolved manganese over extensive areas of the northeastern Atlantic Ocean is consistent with the influence of aeolian inputs (Kremling 1985; Statham & Burton 1986), high concentrations occurring in the latitudes most affected by the Saharan dust plume.

In areas where inputs of aeolian dust imprint signatures of enhanced concentrations of dissolved manganese on surface waters there may be considerable potential for using these signatures to study mixing and transport in the upper ocean. In figure 6, vertical profiles for dissolved manganese are shown for a station in the eastern North Atlantic Ocean, occupied on two separate occasions. Data for salinity are also given and these show the pronounced deepening of the mixed layer at the time of the winter occupation. With the deepening of the mixed layer, the dissolved manganese maximum is displaced to about 150 m. The complex

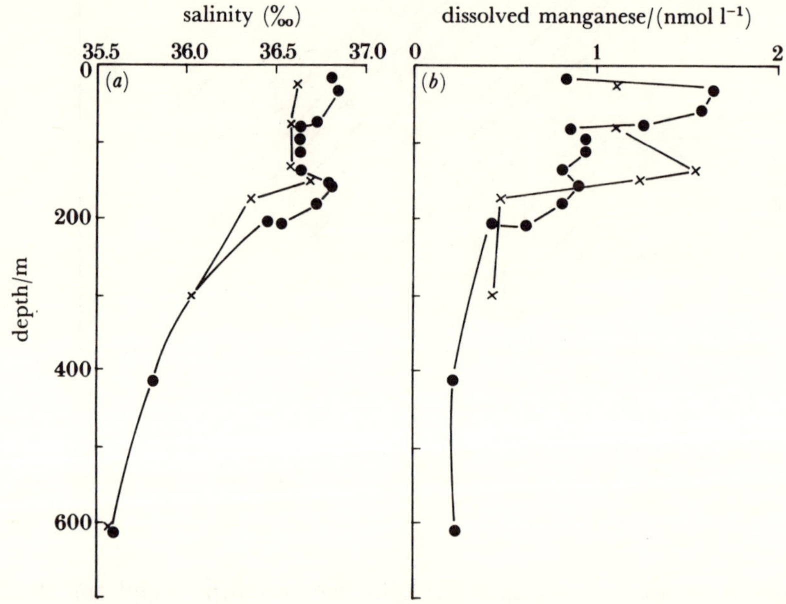

FIGURE 6. Temporal variations in the near-surface distribution of (*a*) salinity and (*b*) dissolved manganese at a station in the eastern Atlantic Ocean (31° 28′ N, 24° 30′ W). Crosses indicate data for February 1982 and circles those for June 1982. (From Statham & Burton (1986).)

distribution of dissolved manganese shown under the conditions of shallower stratification prevailing on re-occupation, with subsidiary maxima below 100 m, may reflect the decay by scavenging of signals that had been produced at these depths by the downward mixing of water from the surface, where high concentrations of dissolved manganese had been produced by *in situ* processes.

Although cycling involving organisms may play a part in the chemistry of manganese in the surface waters, the most probable hypothesis concerning the formation of near-surface maxima is that the manganese is released from particles by desorption and dissolution under the conditions prevailing in the surface ocean. Laboratory experiments with an artificially prepared manganese oxide phase have demonstrated a release of manganese to solution, the rate being enhanced by light and organic matter (Sunda *et al.* 1983). The mechanism of photoreduction of manganese oxide phases is not understood. One possibility is that it occurs through the presence of hydrogen peroxide, the pE for the O_2–H_2O_2 couple being lower than that for the O_2–H_2O couple (Breck 1974). The presence of significant concentrations of hydrogen peroxide in precipitation and in surface seawater has been established (see, for example, Cooper & Zika (1983) and Zika *et al.* (1982, 1985)). Its distribution, with surface maxima and a rapid decrease in concentration with depth, parallels somewhat that of dissolved manganese. Irrespective of the mode of formation of Mn^{II} its relatively slow oxidation is a likely factor contributing to its often complex distributions in the mixed layer. It seems probable that these reflect, in addition to the effects of mixing and convection, the balance between release of Mn^{II} from particles, by desorption and by dissolution under the direct and indirect influences of photochemical reactions, and scavenging onto particle surfaces. Unravelling and quantifying these processes are prerequisites for exploitation of the potential of manganese as a tracer in the upper ocean.

Aluminium

The measurements of aluminium discussed here have been made on both filtered and unfiltered samples. As pointed out by Orians & Bruland (1986), analyses on unfiltered samples that have been acidified to low pH can for some environments lead to substantial differences between the results thus obtained and those for filtered samples. In most studies where unfiltered samples have been analysed, however, storage under acidic conditions has not been employed and contributions of particulate fractions to the reactive aluminium have been small. For the purposes of this discussion it therefore seems permissible to regard the aluminium measured on either filtered or unfiltered samples as being dissolved.

Aluminium is a particle-reactive element with a short residence time in oceanic waters; for example, the residence time of dissolved aluminium in deep Pacific Ocean waters has been estimated as 50–150 years (Orians & Bruland 1986). Concentrations are markedly lower in the older deep waters of the Pacific Ocean than in deep Atlantic waters. The latter typically contain some 10–30 nmol kg^{-1}, whereas concentrations in the deep and bottom waters of the Panama Basin are 1–6 nmol kg^{-1} (Measures *et al.* 1984) and the deep waters of the central gyres of the Pacific Ocean contain 0.5–2.0 nmol kg^{-1} (Orians & Bruland 1985, 1986).

With the low concentrations in Pacific Ocean waters, effects of inputs at the atmospheric and benthic boundaries are more readily seen; they are strikingly apparent in the data of Orians & Bruland (1986) for the VERTEX IV station, north of Hawaii (figure 7). The enhanced

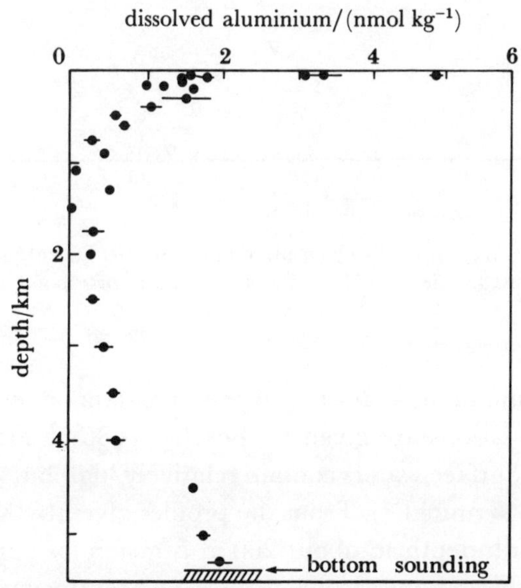

FIGURE 7. Vertical profile of concentration of dissolved aluminium in the central North Pacific gyre (28° 15′ N, 155° 07′ W). (From Orians & Bruland (1986).)

concentrations in deep waters may be attributable to dissolution in the benthic boundary layer or to diffusion from sediments. The fact that the surface water concentrations increase from the California Current region to the central gyre indicates that aeolian input is a major source for aluminium in the latter waters. The importance of this source in the Atlantic Ocean is suggested by the data of Hydes (1983) for dissolved aluminium in surface waters of the eastern Atlantic Ocean, showing an association of higher concentrations with the region most influenced by the Saharan dust plume.

As observed by Orians & Bruland (1986) the range of the concentration of dissolved aluminium in the deep waters of various ocean regions is greater than that shown by any other metal that has been investigated. There is a difference by two orders of magnitude between the concentrations in the deep waters of the North Pacific gyre and those reported by Caschetto & Wollast (1979) for the Mediterranean Sea. Such differences suggest that aluminium should be a useful element in the study of the movement and mixing of water masses. At present, the scope for the interpretation of measurements of the element in terms of circulation models is restricted by limited knowledge of the geochemical factors that initially determine and subsequently modify the signatures on water masses.

Measures *et al.* (1986) determined dissolved aluminium in samples from four profiles in the Sargasso Sea. Their data for subsurface waters are shown in figure 8 in the form of a plot of

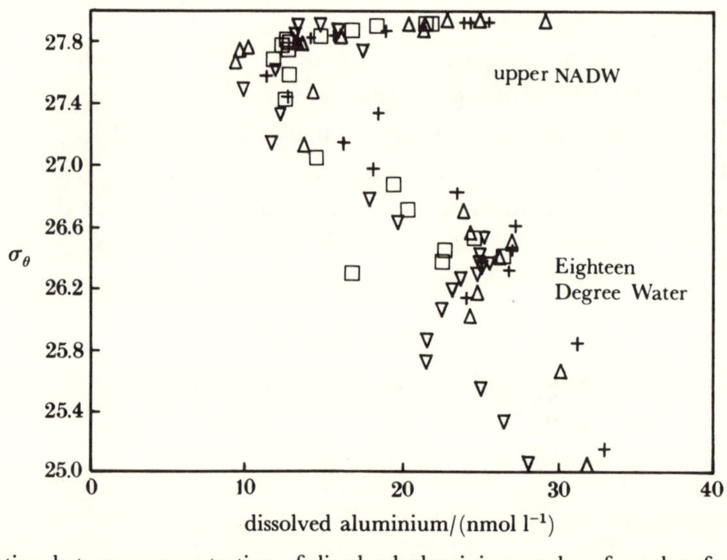

FIGURE 8. Relation between concentration of dissolved aluminium and σ_θ for subsurface waters in the western Atlantic Ocean between about 32°–38° N. The different symbols correspond to different stations. (From Measures *et al.* (1986).)

concentration of dissolved aluminium against σ_θ. Additional data on temporal variations in the upper waters at one of the stations are given by Jickells (1986). Certain general trends are apparent from these studies. Surface waters contain relatively high but variable concentrations of dissolved aluminium (18–45 nmol l^{-1}). From the profiles given by Measures *et al.* (1986) a decrease in concentration with depth to about 200 m is generally apparent. The time series data of Jickells (1986) show this to be a seasonal feature. Concentrations tend to increase in the Eighteen Degree Water, centred on about 300 m, and then decrease to a broad minimum, with concentrations of about 10 nmol l^{-1}, centred on about 1400 m. Below this zone concentrations increase with depth, typically to about 25 nmol l^{-1}. Features that are consistent with these were reported earlier for the northwestern Atlantic Ocean by Hydes (1979) and Moore & Millward (1984).

In this region the North Atlantic Deep Water (NADW) with an upper boundary at about 1200 m can be considered as consisting of Upper NADW to about 2000 m, overlying the Iceland–Scotland overflow and Denmark Strait overflow waters that have their origins in the upper waters north of Iceland. The Upper NADW has contributions of Labrador Sea Water and Mediterranean Outflow Water, the latter being weakly represented so far west. Measures *et al.* (1986) have suggested that the increased concentrations of dissolved aluminium in the Lower NADW reflect rapid transport of aluminium-rich waters from the northwestern European shelf region into the Norwegian Sea and the relatively short transit time between the formation of the deep water and its appearance in the Sargasso Sea. Data for a station south of Iceland at about 60° N (Ólafsson 1983) show deep-water concentrations of dissolved aluminium very similar to those in the Sargasso Sea. The rapidity of transfer of material from northern regions of the Greenland Sea to deep water south of Denmark Strait is shown by measurements of caesium -137, which is introduced anthropogenically from British shelf waters (Livingston *et al.* 1985). These suggest that the transit time may be only a few years.

The Labrador Sea Water, which is a major component of the Upper NADW, also has a short transit time between formation and appearance in the Sargasso Sea. Measurements of chlorofluoromethanes in the Atlantic Ocean (Weiss *et al.* 1985) have demonstrated the rapid transport of these substances along a western boundary undercurrent of Upper NADW. Interpretation of the results using a mixing model suggests a mean velocity of about 450 km a^{-1}. On this basis the transit time to the Sargasso Sea for the Labrador Sea Water is of the order of 10 years. The low concentration of dissolved aluminium in the Upper NADW is attributed by Measures *et al.* (1986) to the fact that this water has a major source in Arctic surface water transported in the East Greenland Current; surface waters in the Arctic Ocean have been shown by Moore (1981, 1983) to be low in dissolved aluminium (less than 5 nmol l^{-1}).

The hypothesis of Measures *et al.* (1986) raises questions regarding the role of shelf waters as a source of aluminium for the Norwegian and Greenland Seas. Measurements of dissolved aluminium in northwestern European shelf waters have shown high concentrations, averaging over 150 nmol kg^{-1}, in the Western Approaches, English Channel and southern North Sea in early Spring 1982 (Kremling 1985). Values averaging only 28 nmol kg^{-1} have been reported, however, by Kremling & Hydes (1988) for the Celtic Sea and English Channel in November 1985, and concentrations lower by a factor of about three were found under summer conditions. Oceanic surface waters close to the shelf-break in the section examined by Kremling (1985) were not enriched in dissolved aluminium relative to surface Atlantic Ocean waters generally, although there was no clear depletion in the boundary current region as has been observed for the California Current (Orians & Bruland 1986) and the Gulf Stream (Moore & Millward 1984; Measures *et al.* 1984). A role for the shelf waters in supplying high concentrations of aluminium to the Norwegian Sea has thus not been established.

A further uncertainty arises from the absence, in the findings of Moore (1981), of a signal of increased concentration of aluminium in Arctic Sea waters originating in the Atlantic Ocean. Much more detailed studies will be needed to clarify the ways in which the distribution of aluminium in the North Atlantic Ocean and Arctic Sea arises by the interaction of source functions, scavenging, and the formation and circulation of water masses. Concomitant studies of aluminium and conservative and time-dependent tracers are particularly required.

Other metals

Beryllium

Recent measurements suggest that beryllium may provide a label for water masses in the Atlantic Ocean, although relatively few data are so far available. Measures & Edmond (1982, 1983) have shown that this metal has a mean oceanic residence time of the order of a thousand years, longer than those of aluminium and manganese, and that the concentrations in deep water of the Pacific Ocean are slightly higher than those in the Atlantic Ocean. Different water masses in the Atlantic Ocean appear to carry somewhat different signatures of dissolved beryllium (Measures *et al.* 1984), presumably reflecting differences in supply in the source regions. The NADW has a high concentration relative to water masses of Antarctic origin. In the western basin, at 8° S the concentration in the NADW is 31 pmol kg^{-1}, compared with 19 pmol kg^{-1} in the Antarctic Intermediate Water and 26 pmol kg^{-1} in the Antarctic Bottom Water. These differences are not large but the potential of the metal as a label for waters of different origins may be worth investigating further in regions where deep waters are formed.

Cobalt

Information on cobalt in the ocean is very limited, because it has been difficult to achieve analytical detection limits that are adequate in relation to the concentrations of the element in seawater, which are usually below 0.1 nmol l^{-1}. The distribution of this metal is thus poorly known, but there are preliminary indications that it may be of potential interest as a tracer. Results for the northeastern Pacific Ocean (Knauer *et al.* 1982; Martin 1985) suggest overall that concentrations in surface waters decrease from nearshore (60–120 pmol kg^{-1}) to offshore (*ca.* 20 pmol kg^{-1}) waters. In the former, the concentrations are highest in surface waters, and thus the distributions resemble those of manganese, but offshore they increase with depth. Similar deep-water concentrations of less than 20–30 pmol kg^{-1} have been reported for both nearshore and offshore stations (Knauer *et al.* 1982; Martin 1985). The offshore data also show a mid-depth maximum for cobalt associated with the oxygen-minimum layer; as with manganese, this is contributed to by lateral transport as well as by release from particles sinking in the water column. It thus appears that cobalt may resemble manganese in that the concentrations in oceanic surface water nearshore are enhanced by marginal inputs, probably dominated by fluvial sources (Knauer *et al.* 1982), and that mobilization occurs in intense oxygen minima. Enhancement in open oceanic surface waters is not, however, apparent in the few profiles examined in the Pacific Ocean.

Measurements made on northeastern Pacific Ocean waters by Bruland & Landing, summarized by Bruland (1983), generally tend to confirm these features. They found, however, a shallow maximum, with a concentration of about 100 pmol kg^{-1}, in a profile in the California Current, and deep-water concentrations as low as about 10 pmol kg^{-1} in an open ocean profile.

Surface maxima have been reported for the Atlantic Ocean (Jickells 1986) and also for the Mediterranean Sea, where concentrations higher than those in the Pacific Ocean, by a factor of about two, have been found (Huyng Ngoc & Whitehead 1986). Jickells (1986) gives two profiles for the Sargasso Sea that indicate a rapid decrease in concentration below the surface waters, containing 18–44 pmol l^{-1}, to values of less than 15–20 pmol l^{-1} from 200–600 m, followed by a slow increase in concentration with depth to values comparable with or higher than those in surface waters. The concentrations were too close to the analytical detection limit to sustain a detailed interpretation but it was suggested that the increase in the concentrations in the deep waters of this area might reflect the advection of waters that had recently sunk from the surface at higher latitudes, carrying initially enhanced concentrations of the metal, as has been suggested for manganese and aluminium.

Concluding comments

The foregoing sections by no means exhaust the list of metals with potential tracer applications. The distribution of lead, which has a low scavenging residence time, is strongly influenced by anthropogenic aeolian inputs (Schaule & Patterson 1981, 1983; Flegal & Patterson 1983) making it a potentially useful time-dependent tracer. Investigations have been limited because of the great difficulties of avoiding contamination, particularly in sampling. Measurements by Boyle *et al.* (1986) of stable lead and ^{210}Pb in the mixed layer of the water column near Bermuda have shown the response of surface-water concentration to the decreasing anthropogenic input of stable lead and also a considerable seasonal variability in the concentrations of both constituents. The supply of stable lead and ^{210}Pb to the thermocline was shown to be mainly by lateral transfer along isopycnals. Iron and tin may be useful elements from the standpoint of tracer applications but contamination is a major problem and so far the data are very restricted. The use of rare-earth elements is discussed by Elderfield (this symposium). The present account has concentrated on non-conservative metals. Metals that show conservative or quasi-conservative behaviour may, however, be very useful in following the mixing history of water bodies that become labelled with distinctively high concentrations as a result of localized inputs.

References

Bender, M. L., Klinkhammer, G. P. & Spencer, D. W. 1977 *Deep Sea Res.* **24**, 799–812.

Bolger, G. W., Betzer, P. R. & Gordeev, V. V. 1978 *Deep Sea Res.* **25**, 721–733.

Bowers, T. S., Von Damm, K. L. & Edmond, J. M. 1985 *Geochim. cosmochim. Acta* **49**, 2239–2252.

Boyle, E. A., Chapnick, S. D., Shen, G. T. & Bacon, M. P. 1986 *J. geophys. Res.* **91**, 8573–8593.

Breck, W. G. 1974 In *The sea*, vol. 5 (marine chemistry) (ed. E. D. Goldberg), pp. 153–179. New York: Wiley-Interscience.

Bruland, K. W. 1983 In *Chemical oceanography*, vol. 8 (ed. J. P. Riley & R. Chester), pp. 157–220. London: Academic Press.

Bruland, K. W. & Franks, R. P. 1983 In *Trace metals in sea water* (ed. C. S. Wong, E. Boyle, K. W. Bruland, J. D. Burton & E. D. Goldberg), pp. 395–414. New York: Plenum Press.

Burton, J. D. & Statham, P. J. 1982 In *Environmental chemistry*, vol. 2, pp. 234–265. London: Royal Society of Chemistry.

Burton, J. D., Maher, W. A. & Statham, P. J. 1983 In *Trace metals in sea water* (ed. C. S. Wong, E. Boyle, K. W. Bruland, J. D. Burton & E. D. Goldberg), pp. 415–426. New York: Plenum Press.

Campbell, A. C. & Gieskes, J. M. 1984 *Earth planet. Sci. Lett.* **68**, 57–72.

Caschetto, S. & Wollast, R. 1979 *Mar. Chem.* **7**, 141–155.

Cooper, W. J. & Zika, R. G. 1983 *Science, Wash.* **220**, 711–712.

Craig, H. 1974 *Earth planet. Sci. Lett.* **23**, 149–159.

Edmond, J. M., Measures, C. I., McDuff, R. E., Chan, L. H., Collier, R., Grant, B., Gordon, L. I. & Corliss, J. B. 1979 *Earth planet. Sci. Lett.* **46**, 1–18.

Edmond, J. M., Von Damm, K. L., McDuff, R. E. & Measures, C. I. 1982 *Nature, Lond.* **297**, 187–191.

Flegal, A. R. & Patterson, C. C. 1983 *Earth planet. Sci. Lett.* **64**, 19–32.

Grill, E. V. 1982 *Geochim. cosmochim. Acta* **46**, 2435–2446.

Hydes, D. J. 1979 *Science, Wash.* **205**, 1260–1262.

Hydes, D. J. 1983 *Geochim. cosmochim. Acta* **47**, 967–973.

Hydes, D. J., Statham, P. J. & Burton, J. D. 1986 *Sci. total Environ.* **49**, 133–145.

Huyng Ngoc, L. & Whitehead, N. E. 1986 *Oceanologica Acta* **9**, 433–438.
Jickells, T. D. 1986 Ph.D. thesis, University of Southampton.
Jones, C. J. & Murray, J. W. 1985 *Limnol. Oceanogr.* **30**, 81–92.
Jones, C. J., Johnson, H. P. & Delaney, J. R. 1981 *Geophys. Res. Lett.* **8**, 873–876.
Keller, G. H., Anderson, S. H. & Lavelle, J. W. 1975 *Can. J. Earth Sci.* **12**, 703–710.
Klinkhammer, G. P. 1980 *Chem. Geol.* **29**, 211–226.
Klinkhammer, G. P. & Bender, M. L. 1980 *Earth planet. Sci. Lett.* **46**, 361–384.
Klinkhammer, G. & Hudson, A. 1986 *Earth planet. Sci. Lett.* **79**, 241–249.
Klinkhammer, G. P., Bender, M. & Weiss, R. F. 1977 *Nature, Lond.* **269**, 319–320.
Klinkhammer, G. P., Rona, P., Greaves, M. & Elderfield, H. 1985 *Nature, Lond.* **314**, 727–731.
Klinkhammer, G. Elderfield, H., Greaves, M., Rona, P. & Nelsen, T. 1986 *Earth planet. Sci. Lett.* **80**, 230–240.
Knauer, G. A., Martin, J. H. & Gordon, R. M. 1982 *Nature, Lond.* **297**, 49–51.
Kremling, K. 1985 *Deep Sea Res.* **82**, 531–555.
Kremling, K. & Hydes, D. J. 1988 *Continental Shelf Res.* **8**, 89–106.
Landing, W. M. & Bruland, K. W. 1980 *Earth planet. Sci. Lett.* **49**, 45–56.
Landing, W. M. & Bruland, K. W. 1987 *Geochim. cosmochim. Acta* **51**, 29–43.
Lewis, B. L., Froelich, P. N. & Andreae, M. O. 1984 *Nature, Lond.* **313**, 303–305.
Livingston, H. D., Swift, J. H. & Ostlund, H. G. 1985 *J. geophys. Res.* **90**, 6971–6982.
Lupton, J. E. & Craig, H. 1981 *Science, Wash.* **214**, 13–18.
Lupton, J. E., Weiss, R. F. & Craig, H. 1977 *Nature, Lond.* **267**, 603–604.
Lupton, J. E., Klinkhammer, G. P., Normark, W. R., Hayman, R., Macdonald, K. C., Weiss, R. F. & Craig, H. 1980 *Earth planet. Sci. Lett.* **50**, 115–127.
Lupton, J. E., Delaney, J. R., Johnson, H. P & Tivey, M. K. 1985 *Nature, Lond.* **316**, 621–623.
Martin, J. H. 1985 *Trans. Am. geophys. Un.* **66**, 1291.
Martin, J. H. & Knauer, G. A. 1980 *Earth planet. Sci. Lett.* **51**, 266–274.
Martin, J. H. & Knauer, G. A. 1983 *Deep Sea Res.* **30**, 411–425.
Martin, J. H. & Knauer, G. A. 1984 *Earth planet. Sci. Lett.* **67**, 35–47.
Martin, J. H. & Knauer, G. A. 1985 *Nature, Lond.* **314**, 524–526.
Martin, J. H., Knauer, G. A. & Broenkow, W. W. 1985 *Deep Sea Res.* **32**, 1405–1427.
Measures, C. I. & Edmond, J. M. 1982 *Nature, Lond.* **297**, 51–53.
Measures, C. I. & Edmond, J. M. 1983 *Earth planet. Sci. Lett.* **66**, 101–110.
Measures, C. I., Grant, B., Khadem, M., Lee, D. S. & Edmond, J. M. 1984 *Earth planet. Sci. Lett.* **71**, 1–12.
Measures, C. I., Edmond, J. M. & Jickells, T. D. 1986 *Geochim. cosmochim. Acta* **50**, 1423–1429.
Moore, R. M. 1981 *Geochim. cosmochim. Acta* **45**, 2475–2482.
Moore, R. M. 1983 In *Trace metals in sea water* (ed. C. S. Wong, E. Boyle, K. W. Bruland, J. D. Burton & E. D. Goldberg), pp. 131–142. New York: Plenum Press.
Moore, R. M. & Millward, G. E. 1984 *Geochim. cosmochim. Acta* **48**, 235–242.
Murray, J. W., Spell, B. & Paul, B. 1983 In *Trace metals in sea water* (ed. C. S. Wong, E. Boyle, K. W. Bruland, J. D. Burton & E. D. Goldberg), pp. 643–669. New York: Plenum Press.
Ólafsson, J. 1983 In *Trace metals in sea water* (ed. C. S. Wong, E. Boyle, K. W. Bruland, J. D. Burton & E. D. Goldberg), pp. 475–485. New York: Plenum Press.
Orians, K. J. & Bruland, K. W. 1985 *Nature, Lond.* **316**, 427–429.
Orians, K. J. & Bruland, K. W. 1986 *Earth planet. Sci. Lett.* **78**, 397–410.
Rona, P. A. 1980 *J. geol. Soc. Lond.* **137**, 385–402.
Rona, P. A., Thompson, G., Mottl, M. J., Karson, J. A., Jenkins, W. J., Graham, D., Mallette, M., Von Damm, K. & Edmond, J. M. 1984 *J. geophys. Res.* **89**, 11365–11377.
Rona, P. A., Klinkhammer, G., Nelsen, T. A., Trefry, J. H. & Elderfield, H. 1986 *Nature, Lond.* **321**, 33–37.
Saunders, P. M. & Francis, T. J. G. 1985 *Progr. Oceanogr.* **14**, 527–536.
Schaule, B. K. & Patterson, C. C. 1981 *Earth planet. Sci. Lett.* **54**, 97–116.
Schaule, B. K. & Patterson, C. C. 1983 In *Trace metals in sea water* (ed. C. S. Wong, E. Boyle, K. W. Bruland, J. D. Burton & E. D. Goldberg), pp. 487–504. New York: Plenum Press.
Statham, P. J. & Burton, J. D. 1986 *Earth planet. Sci. Lett.* **79**, 55–65.
Statham, P. J., Burton, J. D. & Hydes, D. J. 1985 *Nature, Lond.* **313**, 565–567.
Stumm, W. & Morgan, J. J. 1981 *Aquatic chemistry*, 2nd edn (780 pages). New York: Wiley.
Sunda, W. G., Huntsman, S. A. & Harvey, G. R. 1983 *Nature, Lond.* **301**, 234–236.
Von Damm, K. L., Edmond, J. M., Grant, B., Measures, C. I., Walden, B. & Weiss, R. F. 1985*a* *Geochim. cosmochim. Acta* **49**, 2197–2220.
Von Damm, K. L., Edmond, J. M., Measures, C. I. & Grant, B. 1985*b* *Geochim. cosmochim. Acta* **49**, 2221–2237.
Weiss, R. F. 1977 *Earth planet. Sci. Lett.* **37**, 257–262.
Weiss, R. F., Bullister, J. L., Gammon, R. H. & Warner, M. J. 1985 *Nature, Lond.* **314**, 608–610.
Yeats, P. A. & Bewers, J. M. 1985 *Mar. Chem.* **17**, 255–263.
Zika, R. G., Saltzman, E. S., Chameides, W. L. & Davis, D. D. 1982 *J. geophys. Res.* **87**, 5015–5017.
Zika R. G., Saltzman, E. S. & Cooper, W. J. 1985 *Mar. Chem.* **17**, 265–275.

Discussion

H. Elderfield (*Department of Earth Sciences, University of Cambridge, U.K.*). It is interesting that the shape of the seawater Mn profile from the Mid-Atlantic Ridge at 43° N, which Dr Burton ascribes to hydrothermal activity, is unlike the shapes of all the hydrothermal manganese plumes previously observed. Geophysical models of plumes from magma chambers imply that the top of the buoyant hydrothermal plumes in seawaters can act as a lid so that the median valley can be thought of as a 'leaky box' where, elsewhere in the Atlantic, there is significant lateral leakage at fracture zones. This suggests two possible explanations for the profile at 43° N. (1) There is a very large hydrothermal flux and (2) the ridge topography is such that there is nowhere for the manganese to escape. Has Dr Burton any comments?

J. D. Burton. The shape of the profile is certainly different. The profile given by Klinkhammer *et al.* (1985) for their station 8LI (16° 47.7′ N, 46° 26.6 W′) resembles it more closely than do any others reported, but that profile still shows a significant advective maximum within the depth range of elevated concentrations. There is no evidence concerning the possible magnitude of a hydrothermal flux in the region of 43° N. The anomaly of dissolved manganese in excess of background concentrations in deep water is relatively small (*ca.* 1.5 nmol l^{-1}) but if the profile is typical of the whole valley the total amount of dissolved manganese attributable to hydrothermal input would be of the order of 10^5 moles. The fact that the valley contains an isolated and well-mixed body of water is clear from the vertical and horizontal uniformity of potential temperatures observed by Saunders & Francis (1985); the vertical uniformity was confirmed at the time the measurements of manganese were made. If the source of manganese is within the valley then it is possible either that the flux from an active vent or vents is small enough for the vertical concentration gradient at this station to be rendered undetectable by mixing or that venting is episodic with sufficient time having elapsed since input occurred to have allowed the injected material to become well mixed.

Tracers of chemical scavenging in the ocean: boundary effects and large-scale chemical fractionation

By M. P. Bacon

Woods Hole Oceanographic Institution, Woods Hole, Massachusetts 02543, *U.S.A.*

Several nuclides in the natural radioactive decay series display a strong non-conservative behaviour in the oceanic water column because of their removal to the sediments by uptake at the solid–solution interface. This removal process is commonly referred to as scavenging, and it is believed to be important in governing the behaviour of many heavy metals and other trace substances in the ocean. Measurements of radioactive disequilibrium in seawater yield estimates of the rate at which scavenging occurs. Results show that removal times by scavenging are comparable to within-basin lateral mixing times. It follows that the distribution of the scavenged tracers should be sensitive to the spatial distribution of their sinks, and this has been demonstrated by observation, most notably in the case of ^{210}Pb. The distribution of ^{210}Pb suggests a strong control by intensified uptake at boundaries, especially at the ocean margins. This conclusion has been confirmed by sediment traps and by measurements of ^{210}Pb accumulations in the sediment column. A similar phenomenon of intensified scavenging at ocean margins has been observed for ^{231}Pa. Studies with ^{230}Th, on the other hand, show that its delivery to the sediments is mainly caused by the local flux of sinking particles. Thus two transport pathways for removal by scavenging are envisioned, one with a strong horizontal component due to the intensified uptake at the margins and the other with a strong vertical component due to the particle flux. Important large-scale chemical fractionations, best illustrated by comparative studies of ^{230}Th and ^{231}Pa, are produced by differential partitioning of elements between these two pathways. Model calculations suggest that horizontal mixing rate and particle flux are important variables controlling the partitioning of reactive chemical substances between pelagic sediments of the interior ocean and hemipelagic sediments of the ocean margins. Changes in the extent of this partitioning through time may be recorded as changes in the ^{230}Th/^{231}Pa ratio through the sediment column.

1. Introduction

Geochemists have long recognized the importance of adsorption at solid surfaces in the long-term control of trace-metal concentrations in the ocean (Goldschmidt 1954; Goldberg 1954; Krauskopf 1956; Turekian 1977). The removal of chemical substances from the oceanic water column by uptake on the surfaces of particles and subsequent delivery to the sediments is commonly referred to as chemical scavenging. Certain nuclides belonging to the natural radioactive decay series can serve as tracers that allow the estimation of scavenging rates and the delineation of transport pathways. This information is important to our understanding of how the oceans assimilate materials added to them. An important outcome of recent tracer studies is the conclusion that chemical scavenging is intensified at the ocean margins and that there occur significant net fluxes of some tracers from the ocean interior into the sediments accumulating at the margins. In this paper I summarize the evidence supporting that conclusion and examine some of its consequences.

2. Radioactive disequilibrium in the ocean

The ability to infer rates of oceanic processes, such as chemical scavenging, from simple concentration measurements in seawater derives from the existence of radioactive disequilibrium within a decay series. Figure 1 illustrates the radioactive disequilibria that occur within the ^{238}U series in deep-ocean water. It can be seen that the daughter nuclides ^{230}Th and ^{210}Pb are markedly depleted relative to their parent nuclides in the decay series. Additional examples of radioactive daughter depletions are found in the ^{232}Th and ^{235}U series. The depletions are caused by the continual scavenging of the daughter nuclides, and it can easily be shown that the rate of removal J_d of a daughter nuclide is proportional to the difference between parent activity A_p and daughter activity A_d:

$$J_d = \lambda_d(A_p - A_d), \qquad (1)$$

where λ_d is the radioactive decay constant of the daughter.

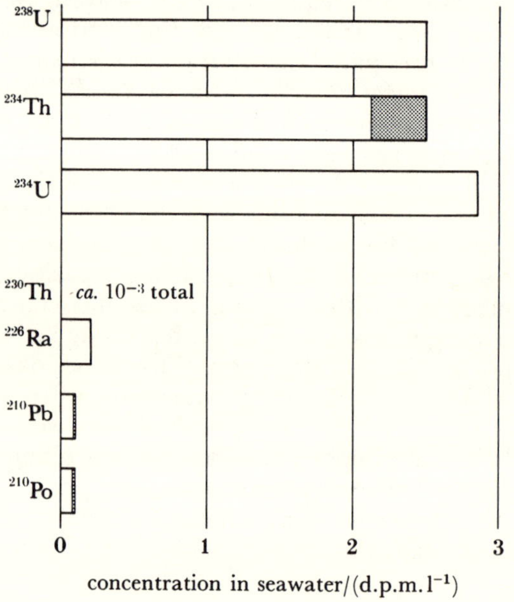

Figure 1. Concentrations of the longer-lived members of the ^{238}U decay series in seawater arranged in descending order within the series. The values chosen are representative of deep ocean water. Somewhat different relations are found in the surface waters. Shaded areas represent the suspended particulate fraction. Deficiencies of ^{230}Th and ^{210}Pb relative to their parent nuclides result from scavenging. A ^{234}Th deficiency is not found because of the short half-life, but a significant uptake by the suspended particles is evident. (1 d.p.m. (disintegration per minute) = 60 Bq.)

The first study of radioactive disequilibrium to estimate rate of removal by scavenging was done by Moore & Sackett (1964), who, on the basis of ^{230}Th–^{234}U and ^{231}Pa–^{235}U measurements, showed that the residence time of Th and Pa with respect to scavenging in the deep sea is less than 100 years. The discovery by Craig *et al.* (1973) of a ^{210}Pb–^{226}Ra disequilibrium in the deep sea showed the residence time of Pb to be approximately 50 years, much shorter than had previously been believed. Numerous subsequent investigations have confirmed these results and developed the use of these radionuclides as chemical tracers in the ocean (see reviews by Cochran 1982; Bacon 1984).

3. Intensified scavenging at ocean margins

The measurements of radioactive disequilibrium discussed above can yield exact and unambiguous determinations of removal rates, but the mechanisms and sites of removal that are deduced from them can depend on differing interpretations. In earlier studies it was usually assumed implicitly that the removal from a given column of water was accomplished entirely by the vertical flux of particles through that column. More recently, however, several lines of evidence have converged to show that chemical scavenging occurs more rapidly at the ocean margins than in the ocean interior and that, for some chemical species, much of the removal from the interior is caused by lateral transport to sites of intensified scavenging at the margins.

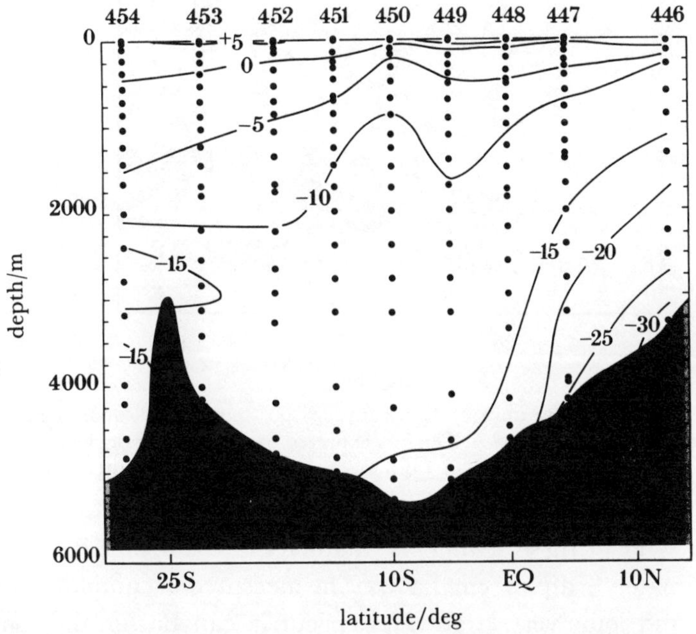

Figure 2. Vertical section showing the $^{210}Pb-^{226}Ra$ activity difference (disintegrations per minute per 100 kg) in the central Indian Ocean. From Cochran *et al.* (1983).

Figure 2 is a vertical section of the activity difference $^{210}Pb-^{226}Ra$ measured in the central Indian Ocean. Positive values (i.e. excess ^{210}Pb) are found at the ocean surface because of ^{210}Pb supply from the atmosphere. At depth are found the systematic ^{210}Pb deficiencies due to scavenging. The apparent rate of scavenging, as measured by the activity difference (equation (1)), increases towards the seafloor and is greatest at the northern end of the section at the station closest to land. As a result there is a horizontal gradient of decreasing ^{210}Pb concentration towards the ocean margin in the north, and it is expected that a net flux of ^{210}Pb from the ocean interior into the margin sediments must occur. Other examples of this phenomenon have been published by Bacon *et al.* (1976), Nozaki *et al.* (1980) and Spencer *et al.* (1981).

A net transport of ^{210}Pb to the margin should support an accumulation of ^{210}Pb in the margin sediments that exceeds the local rate of supply. There are no data yet available to test this hypothesis in the northern Indian Ocean, but a good example from the west coast of the U.S.A. has been given by Carpenter *et al.* (1981) and is shown here in figure 3. The expected

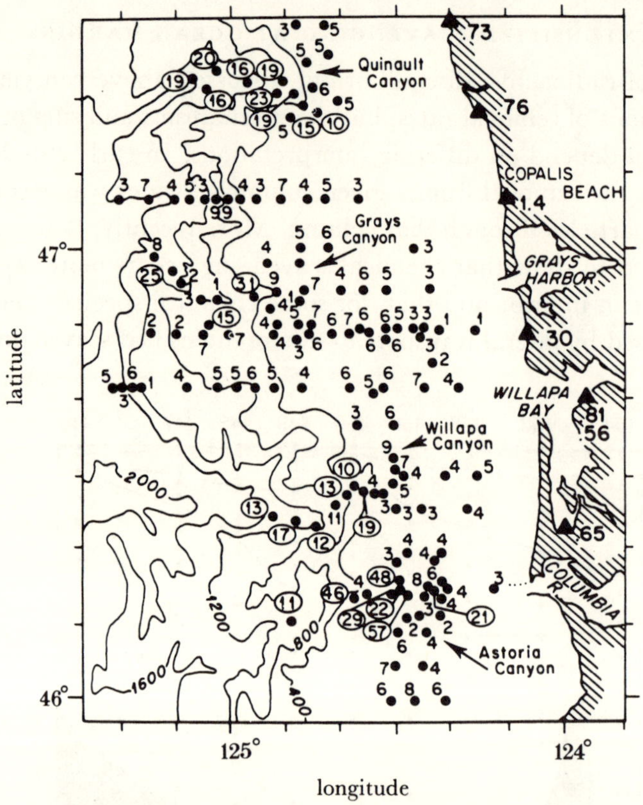

FIGURE 3. Accumulation rates of ^{210}Pb (in units of disintegrations per minute per square centimetre per year) in sediments off the coast of Washington, U.S.A. The flux expected from atmospheric deposition is estimated from soil profiles on land (triangles) and is less than 1 d.p.m. cm^{-2} a^{-1}. From Carpenter *et al.* (1981).

accumulation rate of ^{210}Pb in these sediments, mainly the ^{210}Pb that is deposited from the atmosphere, is approximately 1 d.p.m. cm^{-2} a^{-1}.† The measured accumulations usually exceed this value severalfold, and some very large values occur in canyons on the continental slope. It can be shown that the contribution of ^{210}Pb by rivers is small, and one is forced to conclude that most of the ^{210}Pb that is accumulating in these sediments is derived from offshore. More recently Buesseler *et al.* (1985/86) have compiled data from the east coast of the U.S.A. that show smaller accumulations of ^{210}Pb than are observed on the west coast. It remains to be determined which location is more representative of ocean margins in general.

Recent years have seen an impressive development in the use of deep-sea sediment traps to measure the vertical transport of materials by the flux of particles. Sediment traps provide the opportunity to compare direct measurements of radionuclide fluxes with predictions from radioactive disequilibrium measurements in the overlying water. Such a comparison for three radionuclides is shown in table 1. In the case of ^{230}Th, more than 70% of the expected flux is found in the trap. In contrast, much smaller fractions of the expected amounts of ^{231}Pa and ^{210}Pb are found. The result for ^{210}Pb is consistent with the observations discussed above, which indicate that much of the ^{210}Pb escapes uptake by the vertical flux of particles and is lost from the ocean interior by lateral transport and deposition in sediments accumulating at the margins. It appears that a similar percentage of the ^{231}Pa may be lost in the same way.

† 1 d.p.m. (disintegration per minute) = 60 Bq.

TABLE 1. RADIONUCLIDE FLUXES AT 3200 m, SARGASSO SEA NEAR BERMUDA

(Fluxes are given in units of disintegrations per minute per square metre per year. Measured fluxes are one-year averaged values from Bacon *et al.* (1985).)

nuclide	half-life/years	known scavenging rate[a]	measured flux	measured/expected
^{230}Th	75200	84	60	0.71
^{231}Pa	32500	7.8	2.8	0.36
^{210}Pb	22.3	11900	3500	0.29

[a] Expected fluxes based on radioactive disequilibrium in overlying water column. The value for ^{210}Pb includes the atmospheric input.

4. ^{230}Th/^{231}Pa FRACTIONATION

(a) *Historical review*

The long-lived radionuclides ^{230}Th and ^{231}Pa have an extended history of use in the estimation of deep-sea sediment accumulation rates (Ku 1976; Cochran 1982). They are produced in seawater from the decay of isotopes of U, which is a conservative element, and thus have a source that is uniformly dispersed throughout the oceans. Because of their efficient scavenging from the water column, ^{230}Th and ^{231}Pa accumulate in the sediments. The reduction of their activities downcore due to radioactive decay gives a measure of sediment accumulation rate, if the accumulation rate is assumed to be constant with time.

In the 1960s it was independently suggested by Sackett (1960) and by Rosholt *et al.* (1961) that measurements of the ^{230}Th/^{231}Pa ratio in sediments should allow an absolute age determination and not require the assumption of constant accumulation rate. Because of the different half-lives of the two nuclides, the ratio increases with age at a known rate, and the ratio at zero age should be equal to the production-rate ratio, which is rigidly fixed at a value of 11 (activity ratio) by the ^{234}U/^{235}U ratio in seawater. Actual measurements in surface sediments, however, gave results that departed markedly from this value. In pelagic sediments the ^{230}Th/^{231}Pa ratio at the surface was usually found to exceed the production-rate ratio, often by a factor of 2 or 3 (Ku 1966). The later work on ^{210}Pb and the finding of its intensified scavenging at the ocean margins led to the suggestion that ^{231}Pa might be similarly affected (Bacon *et al.* 1976). The work of Anderson *et al.* (1983*a*, *b*) confirmed this by showing low ^{231}Pa in the vertical flux of particles collected by sediment traps at mid-ocean sites (see also table 1) and by showing significant horizontal ^{231}Pa concentration gradients indicating a net transport to the margins. DeMaster (1981), Shimmield *et al.* (1986) and Yang *et al.* (1986) have shown that surface sediments around the margins of the Pacific Ocean usually contain ^{230}Th/^{231}Pa ratios that are substantially lower than the production-rate ratio, in contrast with the abyssal sediments underlying the interior regions.

In figure 4 is reproduced the map of ^{230}Th/^{231}Pa ratios in surface sediments from the Pacific Ocean that was published by Yang *et al.* (1986). It is clear from this map that the redistribution processes described above have a large, ocean-wide impact. Relative to ^{230}Th, ^{231}Pa is systemically depleted in sediments from the centre of the ocean and enriched in sediments at the margins. This large-scale chemical fractionation comes about because of the differential partitioning of the two elements between a vertical flux due to uptake by the downward flux of particles and a lateral flux driven by intensified uptake at the margins. The differential partitioning is caused by differences in reactivity between the two elements. Figure 5 summarizes results of ^{230}Th/^{231}Pa ratio measurements in deep Pacific waters, which show that the

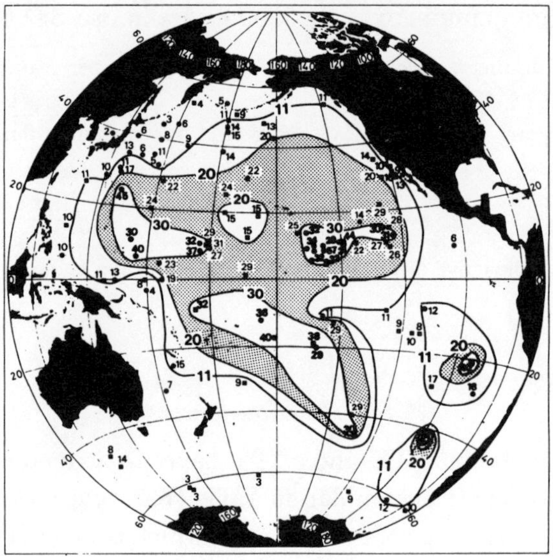

FIGURE 4. Activity ratios $^{230}Th/^{231}Pa$ in Pacific Ocean surface sediments. The production-rate ratio is 11. From Yang *et al.* (1986).

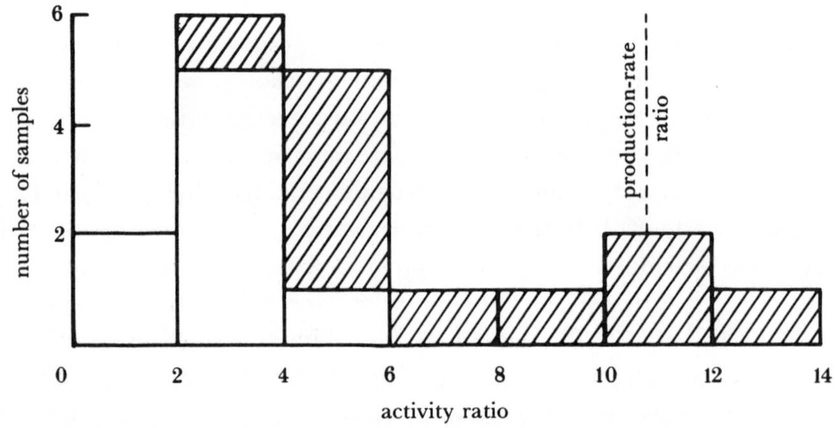

FIGURE 5. Summary of $^{230}Th/^{231}Pa$ ratios measured in deep water from the Pacific Ocean. Samples less than 1000 m below the surface and less than 1000 m above the bottom were not included. Open areas, interior Pacific (Nozaki & Nakanishi 1985); shaded areas, Guatemala/Panama basins (Anderson *et al.* 1983*b*).

ratio is usually less than half the production-rate ratio. This implies a longer oceanic residence time for Pa than for Th, allowing a larger fraction of the Pa to be transported to sites of intensified removal at the ocean margins. In what follows I attempt to formalize and quantify the description of the Th/Pa fractionation process with a simple analytical model.

(*b*) *Pacific Ocean model*

Figure 6 is an idealized representation of the Pacific Ocean in which an interior box I and a margin box M are defined by the 4000 m isobath. From the data of Kossinna (1921; cited in Sverdrup *et al.* 1942, p. 21), 20% of the volume of the ocean is assigned to M and 65% to I. A third box, B, contains waters below 4000 m, which make up 15% of the total volume. Horizontal fluid exchange between I and M occurs at the rate v_{ex}, but vertical exchange

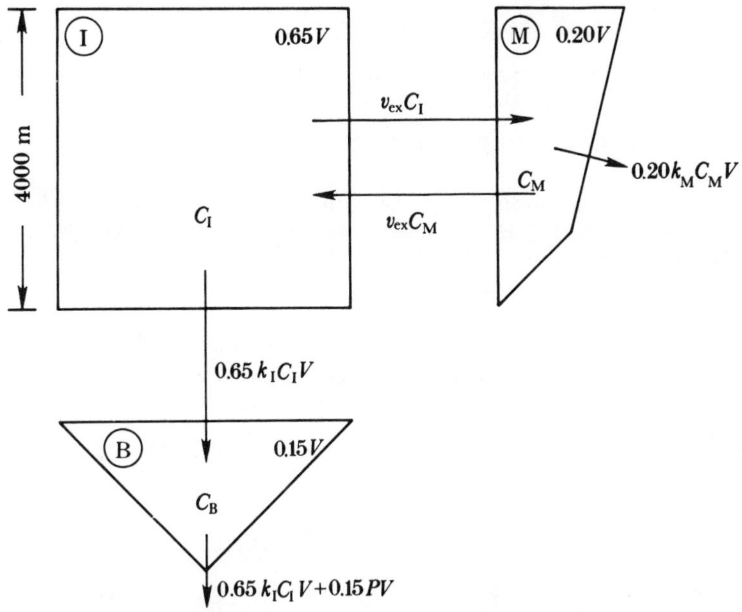

FIGURE 6. Pacific Ocean box model. See text for explanation.

between I and B is ignored. Each box has average tracer concentrations, C, and scavenging rates are assumed to be proportional to tracer concentration and given by rate coefficients k. Because of the long half-lives of ^{230}Th and ^{231}Pa, radioactive decay is negligible, and the sum of the deposited fluxes out of M and B ($0.20\ k_M C_M V + 0.65\, k_I C_I V + 0.15\, PV$) must balance the total *in situ* tracer production PV.

Steady-state material balance is assumed to hold for each of the tracers in I and M as follows.

$$0.65\, P^{Th} V = 0.65\, k_I^{Th} C_I^{Th} V + v_{ex}(C_I^{Th} - C_M^{Th}), \tag{2}$$

$$0.65\, P^{Pa} V = 0.65\, k_I^{Pa} C_I^{Pa} V + v_{ex}(C_I^{Pa} - C_M^{Pa}), \tag{3}$$

$$0.20\, P^{Th} V + v_{ex}(C_I^{Th} - C_M^{Th}) = 0.20\, k_M^{Th} C_M^{Th} V, \tag{4}$$

$$0.20\, P^{Pa} V + v_{ex}(C_I^{Pa} - C_M^{Pa}) = 0.20\, k_M^{Pa} C_M^{Pa} V. \tag{5}$$

The concentrations C are all measurable quantities. It is also possible to obtain a measure of the deposited flux ratio R from measurements of ^{230}Th/^{231}Pa ratios in sediment traps or in surface sediments:

$$R_I = k_I^{Th} C_I^{Th} / k_I^{Pa} C_I^{Pa}, \tag{6}$$

$$R_M = k_M^{Th} C_M^{Th} / k_M^{Pa} C_M^{Pa}. \tag{7}$$

(Strictly speaking, the ratio found in surface sediments is not an exact measure of R_I as defined here, because of the additional deposition of tracer produced in the bottom box B. Surface-sediment ratios can also be modified by bioturbation. In the calculations that follow sediment-trap data are used to define R_I and R_M. Note that a knowledge of the sediment trapping efficiency is not required.) Table 2 summarizes the data that are available and the values that were chosen for making computations. It is obvious that the data base is very small, and quantitative conclusions based on it must be regarded as tentative and subject to revision as more data become available.

TABLE 2. DATA USED IN MODEL CALCULATIONS

(The range and number of samples refer to those samples that were collected in mid-water, more than 1000 m below the surface and more than 1000 m above the bottom.)

quantity	reported range	number of samples	selected value	reference
C_I^{Th} (d.p.m. m^{-3})	0.8–1.8	8	1.2	a
C_I^{Pa} (d.p.m. m^{-3})	0.41–0.69	9	0.53	a
C_M^{Th} (d.p.m. m^{-3})	0.21–0.62	12	0.35	b
C_M^{Pa} (d.p.m. m^{-3})	0.044–0.113	11	0.058	b
R_I	21–35	11	26	c, d
R_M	5–8	5	7	b

[a] Nozaki & Nakanishi (1985); central, western Pacific.
[b] Anderson *et al.* (1983*b*); Guatemala and Panama basins.
[c] Anderson *et al.* (1983*a*); central Pacific, Sargasso Sea, equatorial Atlantic.
[d] Bacon *et al.* (1985); Sargasso Sea.

TABLE 3. MODEL PARAMETERS

(P^{Th} and P^{Pa} are constants based on the U isotope content of seawater. Other parameters were calculated as described in the text.)

production rates/(d.p.m. m^{-3} a^{-1})
$P^{Th} = 0.0261$
$P^{Pa} = 0.00242$

rate constants/a^{-1}	reciprocals/years
$v_{ex}/V = 0.0014$	710
$k_I^{Th} = 0.020$	50
$k_M^{Th} = 0.092$	11
$k_I^{Pa} = 0.0026$	380
$k_M^{Pa} = 0.10$	10

(*c*) *Model results*

The system described by equations (2–7) is overdetermined. The procedure that was followed was to use (2), (3) and (6) for the interior box to solve for v_{ex}/V by eliminating k_I^{Th} and k_I^{Pa}. The value obtained was 0.0021 a^{-1}. The procedure was then repeated with (4), (5) and (7) for the margin box and yielded the value 0.00075 a^{-1}. The discrepancy between these two estimates of v_{ex}/V could be eliminated by adjusting the starting concentrations in table 2. However, for purposes of the illustrative calculations shown below, I adopted the simpler procedure of taking the average value 0.0014 a^{-1}, and the scavenging rate coefficients k were then determined uniquely from (2–5). The values obtained are listed in table 3. They show that, within the interior ocean, Th is more efficiently scavenged than is Pa ($k_I^{Th} > k_I^{Pa}$). As pointed out by Anderson *et al.* (1983*a*), this is consistent with suspended-particle data showing that Th is bound more strongly by the particles than is Pa (i.e. a larger distribution coefficient K_d for Th than for Pa). At the ocean margin, scavenging of both nuclides is, as expected, more rapid than it is in the interior ($k_M > k_I$), and the contrast between Th and Pa has now virtually disappeared ($k_M^{Th} \approx k_M^{Pa}$). This result was also found by Anderson *et al.* (1983*b*), who attributed it to an abundance of manganese-oxide surfaces, with similar distribution coefficients for both Th and Pa, in suspended particles in the Guatemala and Panama basins. Whether or not these qualitative differences in particle-surface chemistry between ocean interior and ocean margin are truly representative of the whole ocean, or whether the Guatemala and Panama basins are a unique environment because of their very high biological productivity, is an important unanswered question (Yang *et al.* 1986).

Having computed values for the model parameters from the existing data, we can now examine how sensitive the observed chemical fractionation between the two elements might be to oceanographic change. This is of interest for understanding how the chemistry of the present ocean is controlled, and it is also of interest for examining whether past changes in oceanographic conditions might have been recorded as variations in the $^{230}Th/^{231}Pa$ ratio in the sediment column. Horizontal mixing rate must be an important variable controlling the degree of fractionation, and I have performed a series of computations to illustrate this. The scavenging rate coefficients k were all held constant at the values listed in table 3, and v_{ex}/V was allowed to vary. Figure 7 shows how the deposited fluxes of each tracer out of each box respond, and figure 8 shows explicitly how the ultimate fate of each tracer is controlled by the mixing rate. At very low mixing rates an exact balance between *in situ* production and the deposited flux is approached, and in figure 7 the curves for both the interior and margin boxes converge to the production-rate line. As mixing rate increases, there is increasing loss of tracer from the interior to the margin, and the curves diverge. The shaded region on each plot includes the selected value of v_{ex}/V ($0.0014\ a^{-1}$) and threefold variations about it (*ca.* ± 0.5 log unit). It is intended to represent a reasonable range of oceanographic conditions. Within this range the degree of ^{231}Pa transfer from interior to margin is quite sensitive to the mixing rate, and large transfers can occur. Over the same range, ^{230}Th deposition is less sensitive to change, and one would expect more spatial uniformity in its flux to the underlying sediments.

Figure 9 shows the effect of horizontal mixing rate on the $^{230}Th/^{231}Pa$ ratio in the deposited flux. It is this ratio that is preserved in deep-sea sediments and potentially could give a measure

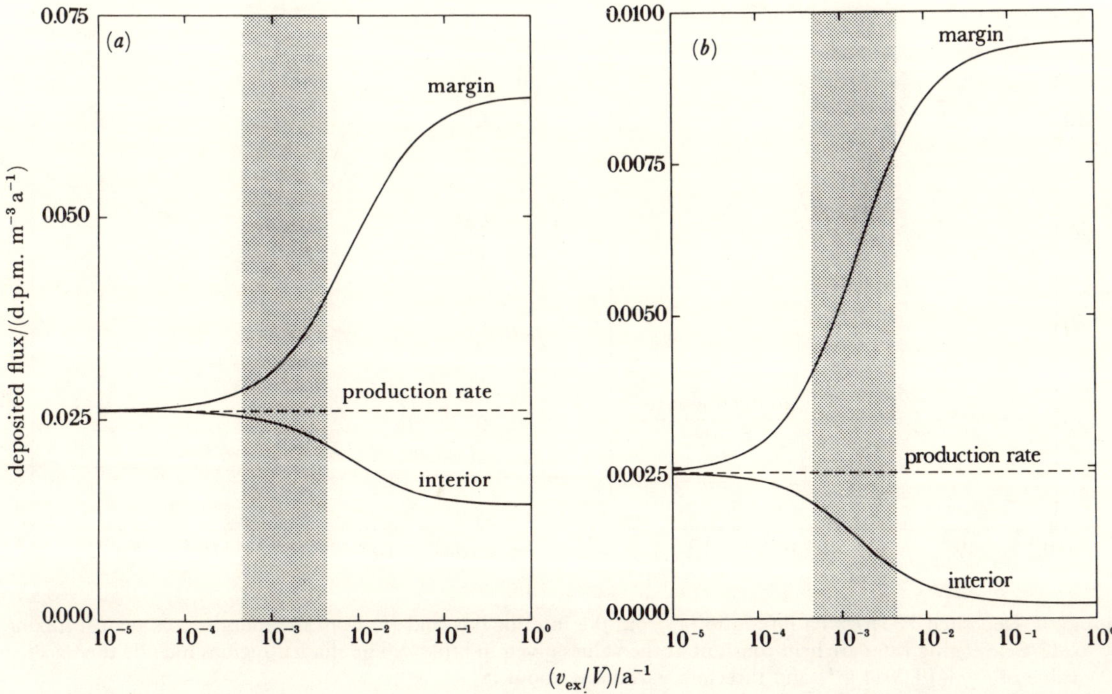

FIGURE 7. Deposited flux of (*a*) Th and (*b*) Pa against mixing rate. Scavenging rates are held constant at the values given in table 3. The shaded regions include the selected value of v_{ex}/V ($0.0014\ a^{-1}$) and threefold variations about it.

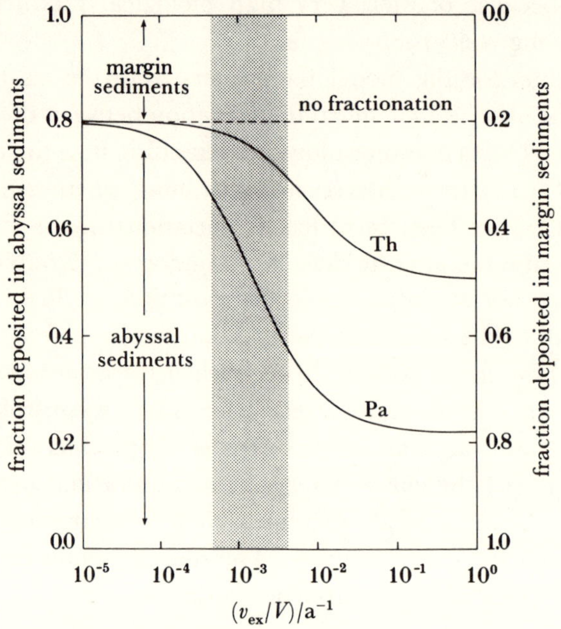

FIGURE 8. Distribution of tracer inventory between abyssal and margin sediments against mixing rate. The inventories include the amounts produced in the bottom box (0.15 *PV*). Scavenging rates are held constant at the values given in table 3. The shaded region includes the selected value of v_{ex}/V (0.0014 a^{-1}) and threefold variations about it.

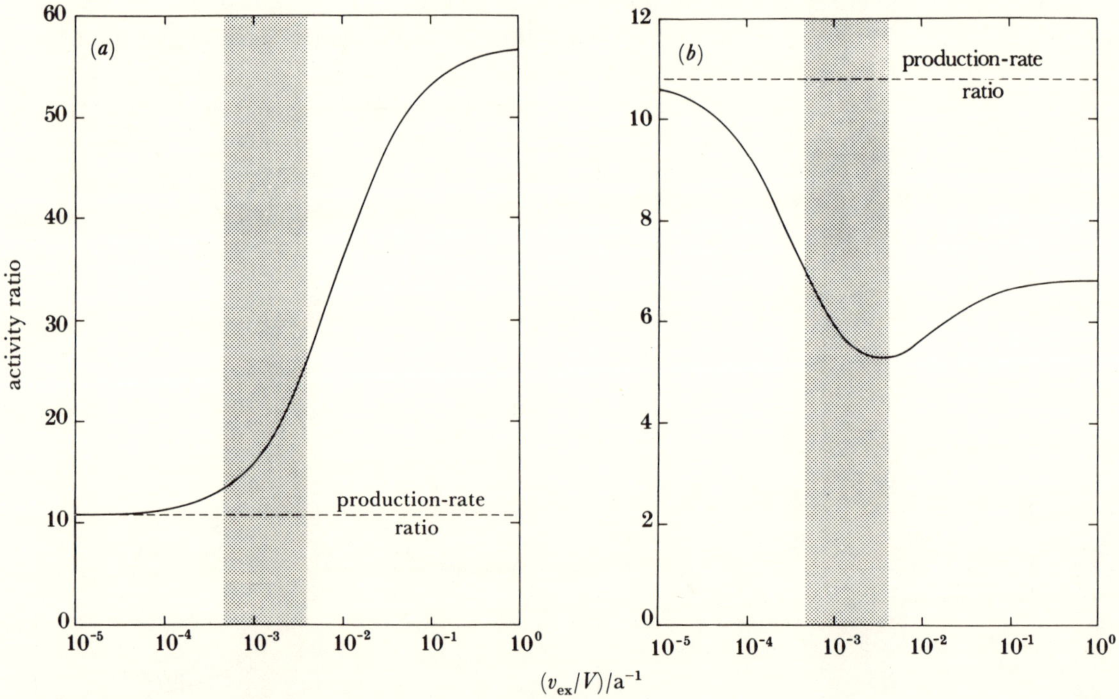

FIGURE 9. Deposited $^{230}Th/^{231}Pa$ flux ratio (*a*) from the interior box and (*b*) from the margin box against mixing rate. Scavenging rates are held constant at the values given in table 3. The shaded regions include the selected value of v_{ex}/V (0.0014 a^{-1}) and threefold variations about it.

of past environmental change. In the interior (figure 9*a*) the ratio is very sensitive to mixing rate, and it is not surprising that the ratio method for dating deep-sea sediments has not proven, in most cases, to be superior to methods based on individual radionuclides (Ku 1976). Irregular (non-monotonic) downcore variations in $^{230}Th/^{231}Pa$ were found by Bacon & Rosholt (1982) in a core from the Bermuda Rise, indicating past variations in the deposited flux ratio. At the ocean margin (figure 9*b*) there is a minimum in the curve, and the flux ratio is less sensitive to the exchange rate (note the difference in scale of the vertical axis between figures 9*a* and *b*).

It follows from the existence of a strong boundary scavenging effect that the oceanic residence time of a scavenged element should be dependent on the rate of large-scale mixing. The residence time of a tracer in a multi-compartment system is given by

$$\tau = \sum_{i} C_i V_i / PV, \tag{8}$$

where C_i is the tracer concentration in each compartment, V_i is the volume of each compartment, P is the uniform rate of tracer production, and V is the total volume of the system. In the present case, (8) becomes

$$\tau = (0.20\, C_M + 0.65\, C_I + 0.15\, P/k_I)/P. \tag{9}$$

As shown in figure 10, a significant dependence on horizontal mixing rate is indeed found for Pa. The dependence is much weaker for the more reactive element Th. (In equation (9) it is assumed that the scavenging rate coefficient in the bottom box B is the same as in the interior box I.)

The other important variable that controls the degree of fractionation between Th and Pa is the scavenging rate. It would be possible to examine the effect of varying each of the four

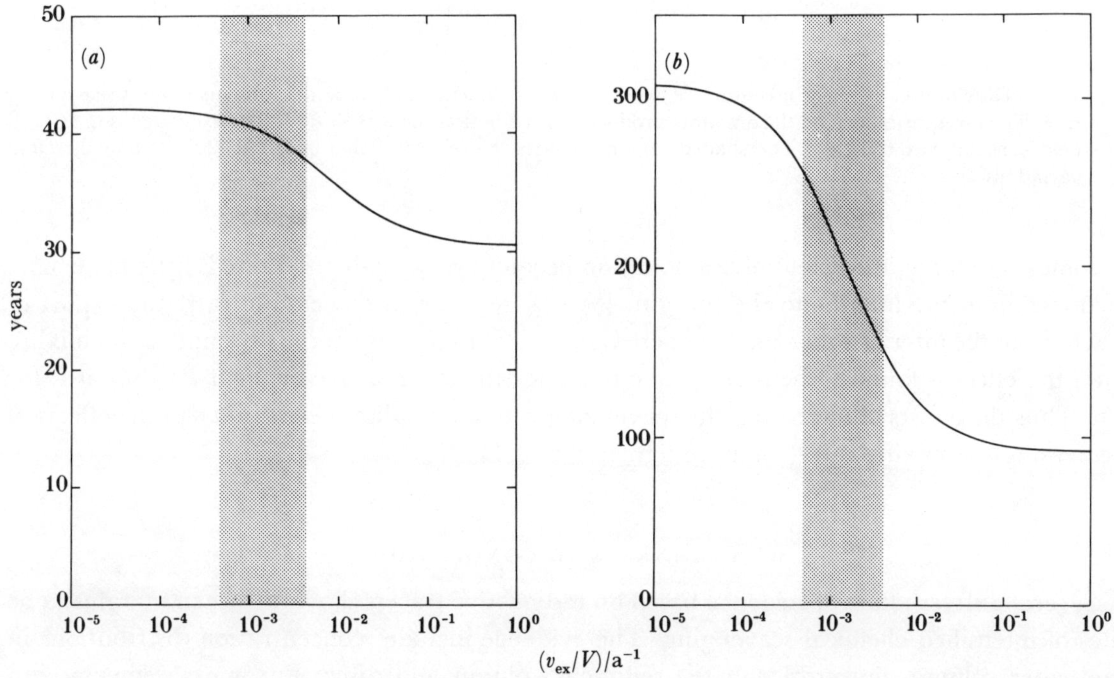

FIGURE 10. Oceanic residence time of (*a*) Th and (*b*) Pa against mixing rate. Scavenging rates are held constant at the values given in table 3. The shaded regions include the selected value of v_{ex}/V (0.0014 a^{-1}) and threefold variations about it.

k values independently, but because of space and time limitations I chose for present purposes to examine just the effect of varying the total scavenging rate while holding constant the relative proportions of the individual k values. In other words, I maintained the proportions $k_I^{Th}:k_M^{Th}:k_I^{Pa}:k_M^{Pa} = 1:4.6:0.13:5$, as they are in table 3, while simultaneously varying all of the k values. This would represent the effect of ocean-wide variations in scavenging rate due to changes in biological productivity or sedimentation rate but without considering changes in their spatial distribution. The mixing rate v_{ex}/V was held constant at $0.0014\ a^{-1}$. I use just one figure to illustrate the effects of varying the scavenging rate. Figure 11 shows that as the scavenging rate

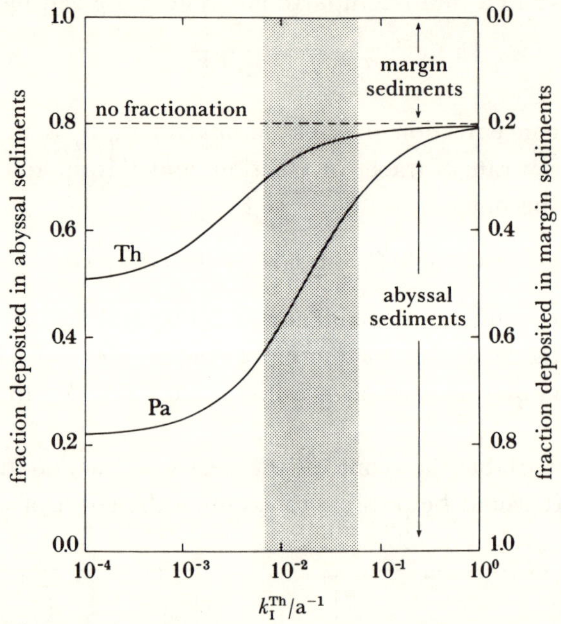

FIGURE 11. Distribution of tracer inventory between abyssal sediments and margin sediments against scavenging rate. The inventories include the amounts produced in the bottom box (0.15 PV). The exchange rate v_{ex}/V is held constant at $0.0014\ a^{-1}$. The shaded region includes the selected value of k_I^{Th} ($0.020\ a^{-1}$) and threefold variations about it.

becomes very large, the extent of fractionation becomes very small and there is little net export of tracer from the interior to the margin. With a decrease in scavenging rate, the export of tracer from the interior to the margin increases, and, as in the case of the dependence on mixing rate, the effect is found to be more pronounced for the less reactive element Pa than it is for Th. Thus the effects of increasing the scavenging rate are qualitatively similar to the effects of decreasing the mixing rate (compare figures 8 and 11).

5. SUMMARY AND CONCLUSIONS

Several different lines of evidence based on radioactive tracers show the ocean margins to be sites of intensified chemical scavenging. The evidence includes concentration distributions in the water column, inventories in the sediment column and direct flux measurements with sediment traps. Because of the strong boundary effect, the oceanic residence time of a scavenged element is dependent on oceanic mixing rate.

The observed fractionation of ^{230}Th and ^{231}Pa in the oceans is explained by a differential partitioning of the two elements between a vertical flux via sinking particles and a horizontal flux to ocean-margin sinks. Two conditions bring this partitioning about: (1) a difference in reactivity towards particle surfaces between the two elements and (2) the existence of different regions of the ocean, characterized by different scavenging rates for reactive elements, that are coupled by horizontal exchange. Under these two conditions, there is a natural tendency for the less reactive element to accumulate preferentially in the region where the scavenging rate is higher (see also Broecker & Peng 1982, pp. 208–210).

Model calculations suggest that horizontal mixing rate is an important variable controlling the fractionation between ^{230}Th and ^{231}Pa. It may seem surprising to conclude that tracers displaying such strongly non-conservative behaviour could play a useful role in the study of ocean mixing rates. The simple calculations presented here, however, suggest that once the concentrations of the tracers in the water column and their ratio in the deposited flux are fixed by measurement, the allowable range of mixing rates is quite narrow. The potential for using the tracers in this way should be explored further with more sophisticated models.

Results of the model calculations indicate that the $^{230}Th/^{231}Pa$ ratio in the deposited flux should be sensitive to oceanographic change. Systematic downcore variations should be sought in deep-sea sediments to ascertain whether systematic glacial-to-interglacial variations occurred.

Additional examples of large-scale chemical fractionation should be sought among elements known to be controlled by scavenging. Possible candidates might be the cosmogenic radionuclides ^{26}Al (possibly a highly reactive element like Th) and ^{10}Be (possibly a less reactive element like Pa). A second possibility would be a comparison between tetravalent Ce and the trivalent rare-earth elements, the tetravalent Ce being more reactive.

Numerous discussions with Dr R. F. Anderson over the past several years have contributed to the ideas expressed in this paper. My research in this area has been generously supported by the U.S. National Science Foundation (grant OCE-8417910) and the U.S. Department of Energy (contract DE-AC02-76EV03566). This paper is contribution no. 6643 from the Woods Hole Oceanographic Institution.

References

Anderson, R. F., Bacon, M. P. & Brewer, P. G. 1983*a* *Earth planet. Sci. Lett.* **62**, 7–23.
Anderson, R. F., Bacon, M. P. & Brewer, P. G. 1983*b* *Earth planet. Sci. Lett.* **66**, 73–90.
Bacon, M. P. 1984 In *Global ocean flux study*, pp. 180–205. Washington, D.C.: National Academy Press.
Bacon, M. P., Huh, C.-A., Fleer, A. P. & Deuser, W. G. 1985 *Deep Sea Res.* **32**, 273–286.
Bacon, M. P. & Rosholt, J. N. 1982 *Geochim. cosmochim. Acta* **46**, 651–666.
Bacon, M. P., Spencer, D. W. & Brewer, P. G. 1976 *Earth planet. Sci. Lett.* **32**, 277–296.
Broecker, W. S. & Peng, T.-H. 1982 *Tracers in the sea.* Palisades, New York: Lamont-Doherty Geological Observatory.
Buesseler, K. O., Livingston, H. D. & Sholkovitz, E. R. 1985/86 *Earth planet. Sci. Lett.* **76**, 10–22.
Carpenter, R., Bennett, J. T. & Peterson, M. L. 1981 *Geochim. cosmochim. Acta* **45**, 1155–1172.
Cochran, J. K. 1982 In *Uranium series disequilibrium: applications to environmental problems* (ed. M. Ivanovich & R. S. Harmon), pp. 384–430. Oxford: Clarendon Press.
Cochran, J. K., Bacon, M. P., Krishnaswami, S. & Turekian, K. K. 1983 *Earth planet. Sci. Lett.* **65**, 433–452.
Craig, H., Krishnaswami, S. & Somayajulu, B. L. K. 1973 *Earth planet. Sci. Lett.* **17**, 295–305.
DeMaster, D. J. 1981 *Geochim. cosmochim. Acta* **45**, 1715–1732.
Goldberg, E. D. 1954 *J. Geol.* **62**, 249–265.
Goldschmidt, V. M. 1954 *Geochemistry.* Oxford: Clarendon Press.

Krauskopf, K. B. 1956 *Geochim. cosmochim. Acta* **9**, 1–32B.
Ku, T.-L. 1966 *Uranium series disequilibrium in deep-sea sediments.* Ph.D. thesis, Columbia University, New York.
Ku, T.-L. 1976 *A. Rev. Earth planet. Sci.* **4**, 347–379.
Moore, W. S. & Sackett, W. M. 1964 *J. geophys. Res.* **69**, 5401–5405.
Nozaki, Y. & Nakanishi, T. 1985 *Deep Sea Res.* **32**, 1209–1220.
Nozaki, Y., Turekian, K. K. & von Damm, K. 1980 *Earth planet. Sci. Lett.* **49**, 393–400.
Rosholt, J. N., Emiliani, C., Geiss, J., Koczy, F. F. & Wangersky, P. J. 1961 *J. Geol.* **69**, 162–185.
Sackett, W. M. 1960 *Science, Wash.* **132**, 1761–1762.
Shimmield, G. B., Murray, J. W., Thomson, J., Bacon, M. P., Anderson, R. F. & Price, N. B. 1986 *Geochim. cosmochim. Acta* **50**, 2499–2507.
Spencer, D. W., Bacon, M. P. & Brewer, P. G. 1981 *J. mar. Res.* **39**, 119–138.
Sverdrup, H. U., Johnson, M. W. & Fleming, R. H. 1942 *The oceans.* New York: Prentice-Hall.
Turekian, K. K. 1977 *Geochim. cosmochim. Acta* **41**, 1139–1144.
Yang, H.-S., Nozaki, Y., Sakai, H. & Masuda, A. 1986 *Geochim. cosmochim. Acta* **50**, 81–89.

Discussion

W. Roether (*University of Bremen, F.R.G.*). I am somewhat surprised by the high margin–interior exchange timescale (*ca.* 700 years) that Dr Bacon derives. Would Dr Bacon's data, considering data uncertainties, be consistent also with appreciably smaller values (i.e. by a factor of two or three), or might it be an artefact of restricting oneself to essentially a two-box model approach?

M. P. Bacon. The parameter V/v_{ex} is a measure of the time for fluid exchange between interior box and margin box. With $V/v_{ex} = 710$ years, and with the volumes of the interior and margin boxes equal to $0.65V$ and $0.20V$, respectively, the value $V/v_{ex} = 710$ years means that water in the interior box is replaced every 460 years by water from the margin box and that water in the margin box is, in turn, replaced every 140 years by water from the interior box. The result obtained in the calculations is sensitive to the assumed concentration difference between the two boxes. My suspicion is that the Guatemala Basin concentration values may be atypically low, even for ocean margins, and that I therefore may have over-estimated the concentration difference and hence the exchange time. A value lower by a factor of two or three would be within the present range of uncertainty. Additional concentration data would considerably improve the constraints we have on the exchange time.

H. Elderfield (*Department of Earth Sciences, University of Cambridge, U.K.*). I was very interested by Dr Bacon's examination of the relation between Th and Pa deposition fluxes in the ocean interior and on margins and ocean mixing rates in terms of the evidence it might give for palaeo-oceanographic changes. Of course, other factors will affect the contrast in radionuclide fluxes, such as changes in productivity. For example, the effect of more rapid mixing could be offset by increased productivity of ocean margins.

M. P. Bacon. Dr Elderfield is correct in stating that more than one variable could be involved in governing the Th/Pa ratio deposited to the sediment, and a unique interpretation of downcore variations in the ratio would not be possible without additional information. The determination of past changes in the rates of oceanic processes is difficult, and the interpretation of other indicators can also be ambiguous. So I believe it is worthwhile to explore the development of other possible tools, such as the one I am suggesting here. A logical first step is to find out whether a systematic glacial-to-interglacial variation in the ratio can be observed in pelagic and hemipelagic sediment cores.

The use of Cs and Sr isotopes as tracers in the Arctic Mediterranean Seas

By H. D. Livingston

Woods Hole Oceanographic Institution, Woods Hole, Massachusetts 02543, *U.S.A.*

The Arctic Mediterranean Seas constitute an oceanic region in which the thermohaline circulation has a strong advective component and deep ventilation processes are very active relative to other oceanic areas. Details of the nature of these circulation and ventilation processes have been revealed through use of Cs and Sr isotopes from bomb-fallout and nuclear-waste sources as ocean tracers. In both cases, their regional input is dominated by advective supply in the Norwegian Atlantic Current and Norwegian Coastal Current, respectively. The different temporal, spatial, and compositional input patterns of these tracers have been used to study different facets of the regional circulation. These input differences and some representative applications of the use of these tracers are reviewed. The data discussed derive from samples collected both from research vessels and from Arctic ice camps. The topics addressed include: (*a*) the role of Arctic Intermediate Water as source, supplying recent surface water to North Atlantic Deep Water via the Denmark Strait overflow; (*b*) deep convective mixing in the Greenland Sea; (*c*) circulation or recirculation of Atlantic water in the Arctic basins; and (*d*) the role of Arctic shelfwaters in the ventilation of intermediate and deep water in the Eurasian and Canadian basins.

1. The Arctic Mediterranean Seas

The term Arctic Mediterranean Seas has been collectively applied to the seas lying north of the Greenland–Scotland ridge (see figure 1). In terms of global oceanic circulation, they represent the major northern source of ventilation of the world ocean (Mantyla & Reid 1983). The early studies of Helland-Hansen & Nansen (1909) laid the groundwork for an understanding of the nature of the circulation of the seas south of Fram Strait, the Nordic Seas. The understanding of this regional circulation has recently taken on a new dimension in the light of the critical link it represents between the atmosphere and the abyssal ocean. Accordingly, the pace of circulation studies has quickened and now tracer tools, based on anthropogenic sources, have been added to the more traditional suite of physical and chemical methods available. The circulation in the Arctic basins north of Fram Strait is less well studied because of the obstacle to sampling access represented by the ice cover. The most recent review of the circulation of the Arctic Mediterranean Seas both to the north and south of Fram Strait is contained in a paper by Aagaard *et al.* (1985). The traditional site of deep-water formation in the region is in the Greenland Sea. It is only the intermediate waters there and in the Iceland Sea that contribute to the dense overflows over the Greenland–Scotland ridges that ventilate the deep water to the south. The deep waters of the Greenland Sea are too dense to pass over these southern ridges and hence are forced to circulate in the deep basins of the Nordic Seas and the Arctic Ocean, through Fram Strait, the deep connection between Greenland and Spitzbergen.

The major feature of the upper-water circulation in the region is driven by the Norwegian–

FIGURE 1. Arctic sampling locations in Atlantic inflow and outflow regions around Fram Strait and interior Arctic.

Atlantic Current, a northward extension of the Gulf Stream. This current, as the West Spitzbergen Current, separates into several branches in the Fram Strait region. The westernmost waters recirculate southward in the cyclonic Greenland Sea gyre. The eastern branches enter the Arctic as the West Spitzbergen Current (Perkin & Lewis 1984) and drive the main circulation of Atlantic water in the Arctic basins. There the Atlantic water underlies the less-saline polar waters that, with contributions from the Bering Sea, form the water masses shallower than the sharp Arctic halocline (Aagaard *et al.* 1981) that eventually exit through western Fram Strait as the East Greenland Current.

The traditional view (Helland-Hansen & Nansen 1909) held that deep convective mixing in the Greenland Sea drove the deep regional circulation. Greenland Sea Deep Water (GSDW) was then supposed to flow into the Norwegian Sea to form Norwegian Sea Deep Water (NSDW). This water mass in turn was thought to feed the deep Arctic basins by inflow through eastern Fram Strait (figure 1). A return flow of Arctic Ocean Deep Water (AODW) left the deep Arctic basins on the western side of Fram Strait, along the Greenland margin (figure 1). This scheme has proven to be oversimplified. The deep inflow has been shown to consist of more than pure NSDW. Occasional contributions of GSDW have been noted (Swift *et al.* 1983). The deep Arctic has too high a salinity to be fed from the south. In the water-mass progression, GSDW–NSDW–EBDW (Eurasian Basin Deep Water) – CBDW (Canadian Basin Deep Water), the salinities and temperatures increase steadily. The solution to this paradox appears to lie in a supply of a brine-enriched shelf water end-member to the deep Arctic (Aagaard *et al.* 1985). As will be discussed later, anthropogenic tracer evidence supports this proposition in the Eurasian Basin. However, in the Canadian Basin, the deep waters studied in this and previous studies lack a measurable anthropogenic tracer signal (Aagaard & Reed 1987; Östlund 1982; Östlund *et al.* 1987; Top 1984; Wallace & Moore 1985) and renewal rates of *ca.* 700 years (Aagaard & Reed 1987; Östlund *et al.* 1987). This slow rate may explain the lack of a recent shelf tracer signal from the locations that have been studied. They do not include any so far that are close to the basin boundary. Current and future efforts should help to address this question more fully.

In addition to the use of nutrients as classic chemical tracers in the Arctic seas (Jones & Anderson 1986), use has been made of both naturally occurring chemical tracers, e.g. Zn, Cd, Cu and Al (Moore 1981), radium isotopes (Moore 1980), and anthropogenic tracers, e.g. tritium (Östlund 1984; Östlund *et al.* 1982; Moore *et al.* 1983; Smethie *et al.* 1986; Peterson & Rooth 1976), carbon-14 (Östlund *et al.* 1982; Östlund *et al.* 1987; Smethie *et al.* 1986; Peterson & Rooth 1976), krypton-85 and argon-39 (Smethie *et al.* 1986), chlorofluoromethanes (Bullister & Weiss 1983; Smethie in Swift 1984*c*; Wallace & Moore 1985) and artificial radioisotopes from nuclear-weapons testing fallout and European nuclear-waste disposal (Aarkrog *et al.* 1983; Bowen & Sugihari 1964; Holm *et al.* 1983; Kautsky 1980; Livingston *et al.* 1982*a*, *b*; Livingston *et al.* 1984, 1985; Swift *et al.* 1983; Vakulovskiy *et al.* 1985). These various tracers include both transient and steady-state tracers, cover a range of chemical reactivities with respect to particle association and operate over a range of timescales from several years to thousands of years. Because their mode of introduction to the region varies with each tracer or tracer family, they find use in different ways in respect to the processes that they may trace. It is the purpose of this paper to review the ways in which the artificial radionuclides from nuclear-weapons testing fallout and European coastal nuclear-waste disposal have been used as tracers in regional circulation studies.

2. Cs and Sr input patterns

Detailed discussion of the temporal and spatial input pattern of Cs and Sr isotopes from bomb fallout and nuclear-fuel reprocessing discharges have appeared at some length elsewhere (Kupferman *et al.* 1979; Livingston *et al.* 1982*a*; Livingston 1985). In the context of the present paper, it is convenient to summarize the important features of these input patterns and various characteristics of their behaviour or composition that have bearing on their use as tracers in the Arctic regions.

Whether introduced to the oceans from bomb fallout or coastal nuclear-fuel reprocessing discharge, the association of both Cs and Sr isotopes with particles is small enough that their removal via particle association may be disregarded for practical purposes. Their chemistries, as an alkali metal and an alkaline earth, respectively, assure that their oceanic behaviour closely resembles that of the other conservative members of these families, and hence they act as tracers of water transport, or, at least, of its dissolved salt content. The radioactive half-lives of the longest-lived fission-product nuclides of these elements are ^{90}Sr (28 years) and ^{137}Cs (30 years). Because of the close similarity in half-life, their activity ratios to each other are not changed very significantly by radioactive decay over the timescales presently studied. Therefore, because no other process causes them to fractionate with respect to each other, their relative composition (or activity ratio) is a water-mass characteristic linked to one or more of their input sources. Other than input-source composition effects, the only other process that effects a change in their relative composition is riverine input. In terrestrial systems, ^{137}Cs exhibits strong particle reactivity and is associated largely with freshwater sediments and soils; ^{90}Sr, in contrast, is mobilized in freshwater systems and, like tritium, is delivered via rivers to the oceans at relatively high concentrations. The effect of this is to reduce the ratio $^{137}Cs:^{90}Sr$ in coastal waters. In the Arctic this has been noted in low-salinity surface waters of the Polar Water layer by a reduction in this ratio to about 1, as opposed to the value of 1.5 that characterizes open ocean waters containing these isotopes only from bomb fallout (Livingston *et al.* 1984).

(*a*) *Nuclear-weapons testing fallout*

As noted above, an important source identification diagnostic of ^{137}Cs and ^{90}Sr from fallout sources in open ocean water is the constancy of the activity ratio $^{137}Cs:^{90}Sr$ at 1.5 (Bowen *et al.* 1974) in proportion to their fission yields. The input of fallout ^{137}Cs and ^{90}Sr to the Arctic Ocean regions needs to be viewed as having two principal components. The first, direct input of fallout to the surface ocean, came at two intervals: (1) an early, relatively small input in the mid-1950s; and (2) a larger input in the early 1960s. However, because the Arctic regions are north of 60° N and because fallout delivery rates were much higher at mid-latitudes than at high latitudes (Kupferman *et al.* 1979), a major continuing input to the surface waters of the Norwegian and Greenland Seas came in the form of the advective influx from the south via the Norwegian Atlantic current. Between 1960 and 1981 the ^{137}Cs concentrations of this water mass rose to a maximum in 1963–1964 and declined rapidly thereafter, reaching slowly declining levels in the 1970s and 1980s at around 20–25% of peak values (Livingston *et al.* 1985). This input is primarily a tracer for the high-salinity Atlantic water flowing into the Arctic regions, in contrast to the following source.

(*b*) *Nuclear-fuel reprocessing discharges*

The major contributor to this source is the low-level discharge of liquid waste from the British Nuclear Fuels Plant at Sellafield (formerly Windscale) on the northeast coast of the Irish Sea. Much smaller discharges come from the French plant at Cap de la Hague on the Normandy coast, or from the Scottish plant at Dounreay. The time history of the amounts, composition, and advective transport of these discharges into the Arctic Mediterranean Seas has been described in earlier reports (Livingston *et al.* 1982*a*, *b*). The features of greatest relevance in their use as regional tracers are as follows.

(1) Discharges of ^{137}Cs began to increase in 1970, peaked in 1975 and have since declined steadily to pre-1970 levels.

(2) Over much of the history of releases during the high ^{137}Cs discharge period, the input was characterized by a $^{137}Cs:^{90}Sr$ ratio of around 10, which is substantially higher than that characterizing bomb fallout.

(3) The discharges were carried northward in the shelf circulation and appeared in the northern Norwegian Sea about five years after their release to the Irish Sea.

(4) As they travel northwards, the discharges are diluted continuously by mixing with inflowing Atlantic water carrying ^{137}Cs and ^{90}Sr at fallout levels and ratios. As this mixing occurs, the ratio of $^{137}Cs:^{90}Sr$ is reduced substantially. For example, dilution of the releases during the peak of ^{137}Cs discharges was such that, on arrival in the northern Norwegian Sea, the ratio of $^{137}Cs:^{90}Sr$ was reduced to the range 3–4 (Casso & Livingston 1984).

(5) Because of their coastal origin, these discharges serve to trace the coastal circulation and shelf-water contributions to the larger-scale circulation in the Arctic regions.

(6) In contrast to ^{137}Cs and ^{90}Sr from fallout, the nuclear-waste source has an association with a relatively lower salinity signal, derived from its low-salinity coastal end member. This low-salinity signal is scarcely detectable as high-salinity Atlantic water, mixed with coastal contributions, recirculates in the Greenland Sea gyre or circulates in the Eurasian or Canadian basins of the Arctic Ocean.

3. Deep convection in the Greenland Sea

Despite considerable effort, most recently during the winter cruise of C. S. S. *Hudson* in 1982 (Clarke *et al.* 1986), direct observation of deep convective events has not been made. Perhaps the nearest to direct observations were those of Malmberg (1983) who found uniform properties from 500 m to the bottom in the central Greenland Sea in February. Killworth (1979) pointed out that the convection events could be both short-lived and localized and easily escape observation. The main evidence for the existence of the process is the finding that deep water in the central Greenland Seas has properties suggesting very recent connection with the sea surface. Various anthropogenic tracers have supported this finding and have led to a series of estimates of deep Greenland Sea residence times with respect to exchange with surface water, namely 24–34 years (Smethie *et al.* 1986), 26–31 years (Peterson & Rooth 1976), and 35–42 years (Bullister & Weiss 1983).

In addition to deep convection, Aagaard *et al.* (1985) have shown that a deep flow from the Eurasian Arctic Basin, relatively warm and salty, enters the northwest Greenland Sea through Fram Strait (figure 1) and propose that this mixes with Greenland Sea water around the outer edges of the Greenland Sea gyre, eventually entering the Norwegian Sea through deeper sections of the mid-ocean ridge. An updated and expanded study of this process has been made very recently (Swift & Koltermann 1988). Smethie *et al.* (1986) reported that tracer and hydrographic observations from the 1981 Transient Tracers in Oceanography (TTO) cruise in the Norwegian–Greenland Seas also support this proposition and obtained similar residence times for NSDW with respect to exchange with GSDW (19–30 years) to those reported by Bullister & Weiss (1983).

Data for Cs and Sr isotopes are also relevant to the question of the ventilation of the deep Greenland and Norwegian Seas. The measurements described here cover the time period 1972–1981. This is a significant interval because it spans the transition created when input from the nuclear-waste discharges began to add to the pre-existing bomb-fallout distribution of ^{137}Cs and ^{90}Sr. The evolution of this transition has been described elsewhere (Livingston 1985). In summary, between 1972 and 1981 the following changes were noted in tracer properties of central Greenland Sea deep water. Fallout concentrations of ^{137}Cs and ^{90}Sr remained essentially unchanged, i.e. the rate of supply from the surface being balanced by the rate of loss through exchange to the deep Norwegian Sea. Deep tritium concentrations increased by more than 50%, but the reason for this increase relative to ^{137}Cs and ^{90}Sr was believed to reflect parallel changes in surface-water composition over the previous two decades.

The most remarkable change in the deep-water properties of these isotopes came from a comparison of the 1972 and 1979 inventories and concentrations. For ^{137}Cs, there was an average increase of 45% over the 1972 fallout levels, the increase essentially all being attributable to the new nuclear waste source. The presence of the fresh input was verified by the finding of a deep-water ratio of $^{137}Cs:^{90}Sr$ of 1.7, up from the fallout value of 1.5 noted in 1972. What this tells us is the rapidity with which the deep water is responding to changes at the surface. It was noted in another report (Livingston *et al.* 1985) that the nuclear-waste ^{137}Cs signal could not have been in central Greenland Sea surface waters before 1978. The deep-water increase noted in 1979 seems to offer evidence of either continuing active annual ventilation or unusually intense deep convection during the 1978–1979 winter. In any event, it serves to illustrate the action of a ventilation process that keeps the deep Greenland Sea in contact with the surface ocean over very short timescales.

Perhaps the most compelling indirect evidence that deep ventilation occurs in the central Greenland Sea and then spreads laterally, comes from a comparison of surface and deep-water tracer concentrations in the Nordic Seas in 1979–1982 (Livingston 1986). In addition to measurements made in my laboratory, tritium data by H. G. Östlund (University of Miami) and tritium and ^{137}Cs data reported by P. Schlosser (University of Heidelberg) were used in this comparison. For both tritium and ^{137}Cs, a consistent reversal exists between surface and deep concentration patterns (see figure 4). In surface waters, concentrations are *highest* in the periphery of the Greenland Sea gyre and *lowest* in the interior. In deep waters, concentrations are *lowest* in the gyre periphery and *highest* in the interior. This does not argue against the deep inflow of Arctic Ocean water through the western side of Fram Strait referred to earlier. Rather, it serves to emphasize the classical view of the primary role in ventilation of the regional deep waters that takes place in the centre of the Greenland Sea gyre.

4. Connections to North Atlantic Deep Water

The two dense overflows into the North Atlantic that pass over the Greenland–Scotland sills have long been recognized as the vehicles that propagate surface-ocean properties into the deep North Atlantic. In both cases the outflows have been shown not to be outflows of deep water from the north, but rather of intermediate waters formed at the surface in the Greenland and Iceland Seas in winter (Swift 1986; Swift *et al.* 1980). The two overflows, Iceland–Scotland and Denmark Strait, eventually combine to form North Atlantic Deep Water (Swift 1984*a*, *b*, *c*). The detection of a plume of tritium in the deep water of the western boundary current at 30° N in 1977 served to highlight the route and rate of the southward-spreading ventilation process (Jenkins & Rhines 1980). The TTO cruise in 1981 presented an opportunity to study the nature of this process in greater detail. By 1981 the new nuclear-waste signal had been in the overflow waters long enough to permit it to be propagated southward and used to set some further constraints on the dynamics of the connection between North Atlantic Deep Water and its northern source waters.

Livingston *et al.* (1985) described the detection of the nuclear-waste signal in deep overflow water along a section just south of Denmark Strait. The location included a reoccupation of GEOSECS station 11, and hence there was the chance to compare the changes in tracer properties that had taken place between 1972 and 1981. All of the radionuclide-tracer concentrations reported (tritium, ^{137}Cs and ^{90}Sr) had increased between 1972 and 1981. This increase paralleled a shift in salinity and temperature of northern North Atlantic deep waters during the same interval (Swift 1984*a*, *b*, *c*; Brewer *et al.* 1983). All the tracers were inversely correlated with salinity in the overflow water with the tracer maxima corresponding to the salinity

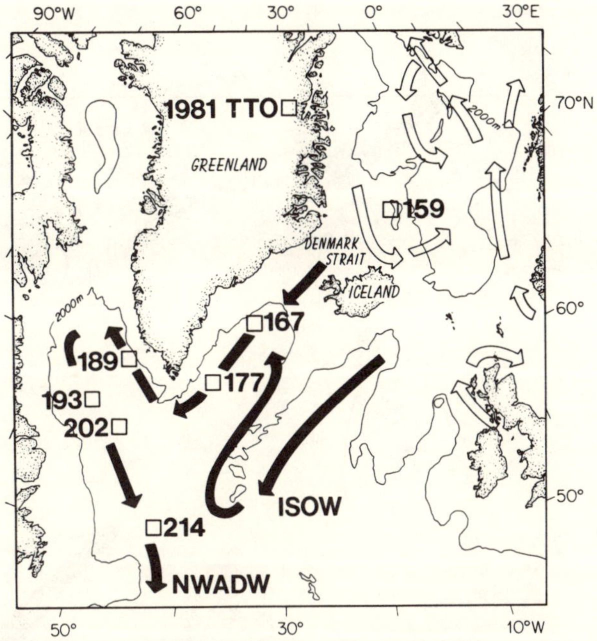

FIGURE 2. Sampling locations south of Denmark Strait during the 1981 Transient Tracers in Oceanography – North Atlantic Study. Open arrows represent the surface North Atlantic current inflow (as the Norwegian Atlantic Current). The solid arrows represent the deep outflow from the Nordic Seas, the Denmark Strait Overflow Water and the Iceland–Scotland Overflow Water. Water depths at these stations were: 159–2168 m, 167–2339 m, 177–3180 m, 189–2970 m, 193–3522 m, 202–3630 m and 214–4382 m.

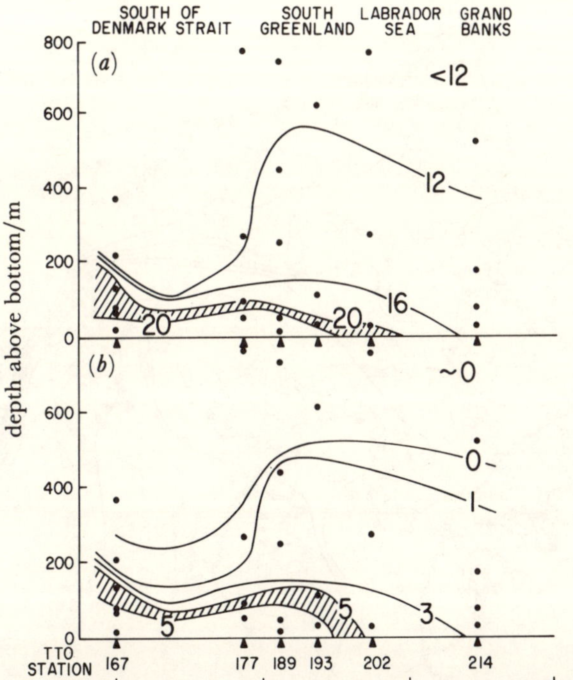

FIGURE 3. Distributions of ^{137}Cs in Northwest Atlantic Bottom Water, Denmark Strait–Grand Banks section. (*a*) Total ^{137}Cs concentration contours. (*b*) Nuclear-waste ^{137}Cs concentration contours (resolved from total ^{137}Cs data by method described by Livingston (1985)).

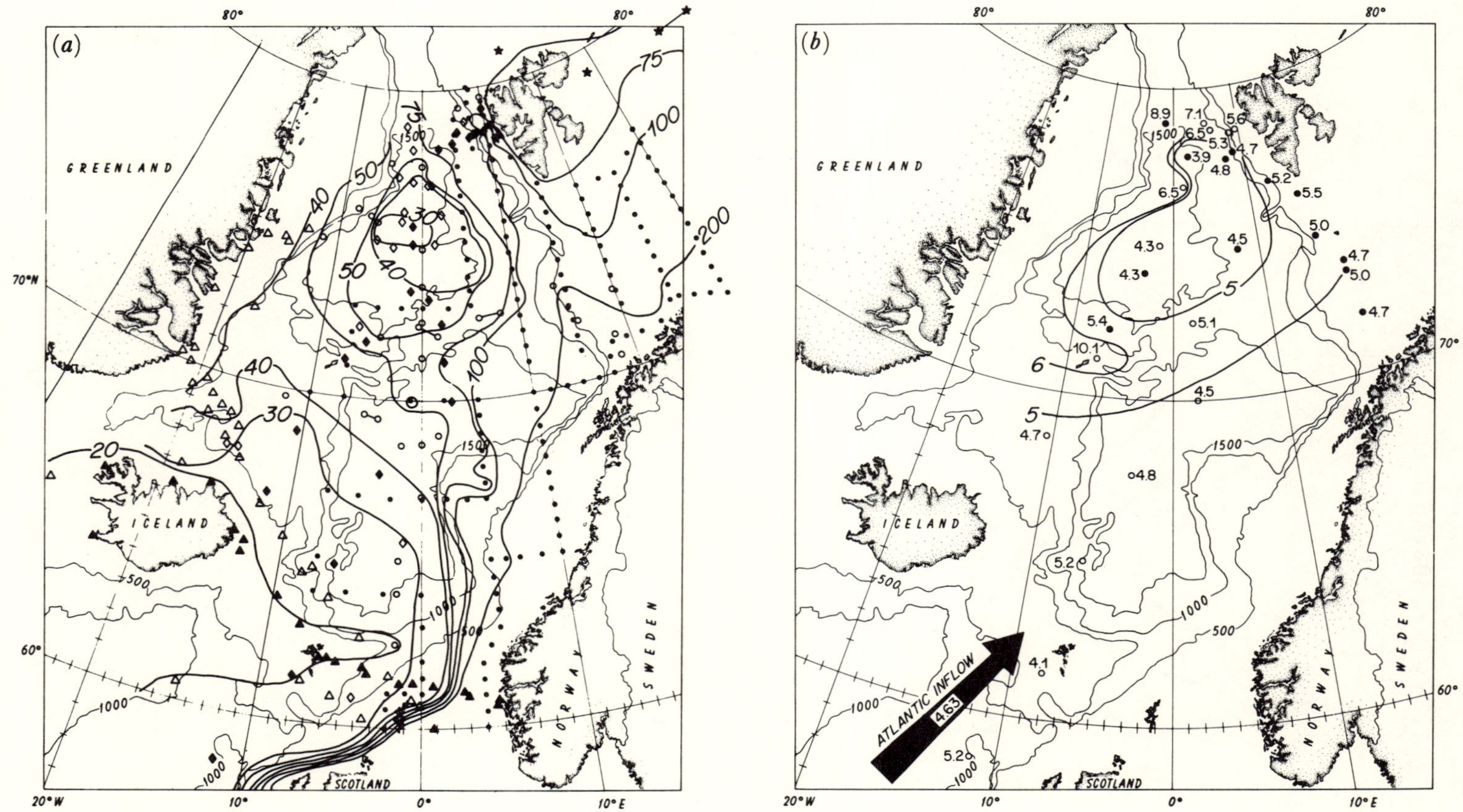

FIGURE 4. (*a*) Surface distributions of ^{137}Cs (disintegrations per minute per 100 kg, as of 1 January 1981) in the Norwegian–Greenland Seas. Solid symbols are stations occupied in 1981 by the ships indicated; open symbols are for 1982. Symbols: ◆, *Knorr*; ●, *Cirolana*; ▲, *Smyril*; ★, *Lance*; ◇, *Hudson*; ○, *Meteor*; △, *Nella Dan.* (1 disintegration per minute (d.p.m.) = 60 Bq.) (*b*) Surface distributions of ^{3}H (tritium units, as of 1 January 1981). Open circles are for stations occupied by R. V. *Knorr* in 1981, and solid circles for R. V. *Hudson* in 1982.

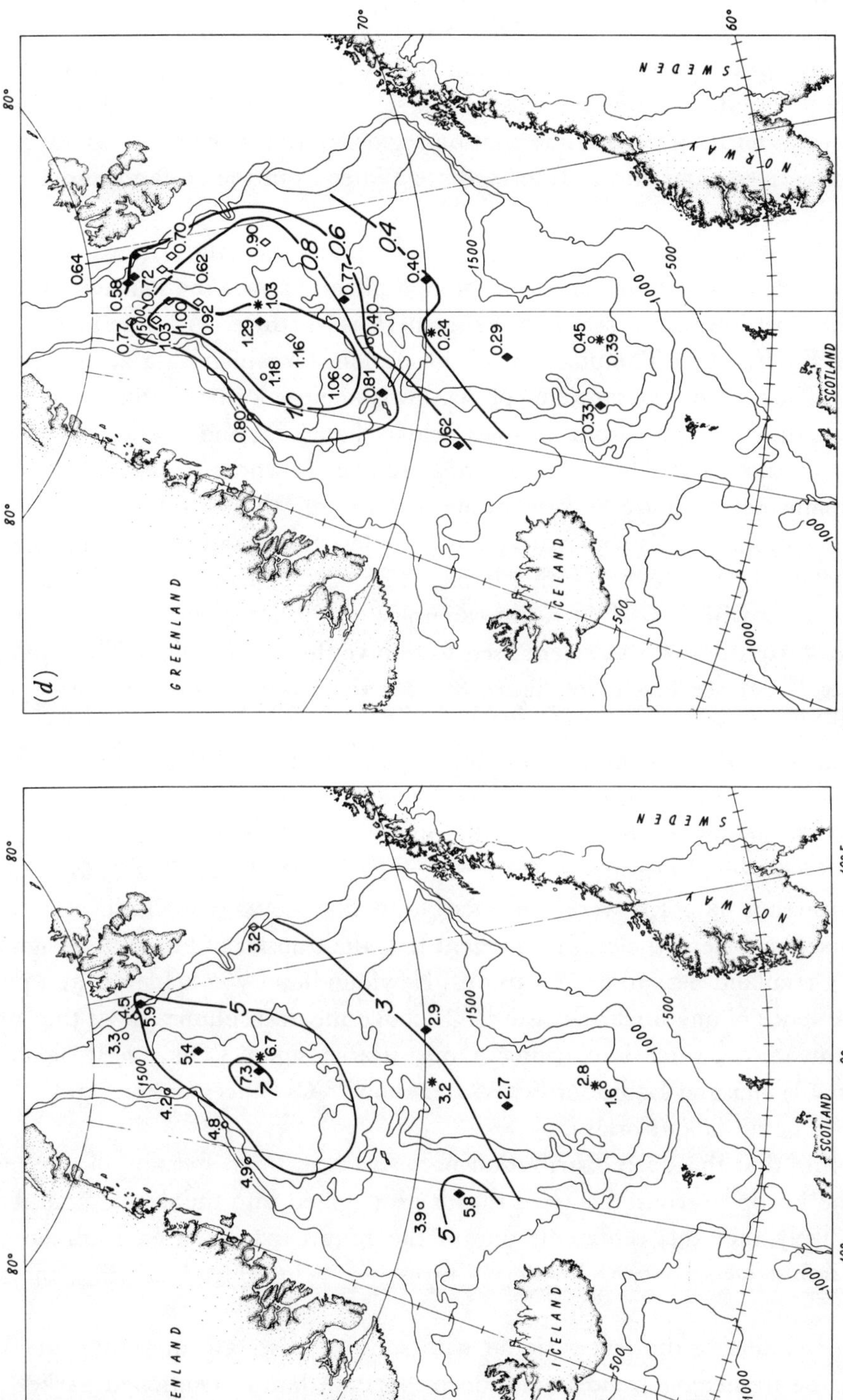

FIGURE 4. (*c*) Average ^{137}Cs concentrations (disintegrations per minute per 100 kg, as of 1 January 1981) in deep water, and (*d*) average ^{3}H concentrations (tritium units as of 1 January 1981) in deep water. Symbols denote ship and year of collection: ✱, *Meteor* 1979; ◆, *Knorr* 1981; ◇, *Hudson* 1982; ○, *Meteor* 1982.

minima. The sharp increase noted for ^{137}Cs was attributed to the presence of ^{137}Cs from the nuclear-waste source on the basis of the elevated ratio ^{137}Cs:^{90}Sr and the presence of ^{134}Cs (both diagnostic indicators). At the core of the tracer plume, the major fraction of the ^{137}Cs present could be attributed to bomb fallout and the nuclear-waste component comprised a little over 25% of the total. This nuclear-waste ^{137}Cs signal presence confirmed the rapidity of transport of surface water from northern formation regions to the overflow. It was argued that the source waters derived from Arctic Intermediate Water from the Iceland and Greenland Seas and that transport took two years or less.

Additional data are presented in this paper that extend this study downstream of the overflow-water trajectory. The data come from samples in and above the southward extension of Denmark Strait Overflow water, referred to elsewhere as Northwest Atlantic Bottom Water, (NWABW) (Lee & Ellett 1967). The positions of the stations for which data are presented are shown in figure 2. They extend the sampling of NWABW from near Denmark Strait to the Grand Banks. The only data shown are from samples colder than 3 °C and fresher than 34.95‰. Complete data from these and other stations are available elsewhere (Casso *et al.* 1988).

In the upper panel (*a*) of figure 3 are contoured the total ^{137}Cs concentrations along the section. The tongue of high ^{137}Cs water, inversely correlated with salinity, extends nearly all the way to TTO station 214 off the Grand Banks. The core of the low-salinity plume was characterized by water with ^{137}Cs:^{90}Sr ratios ranging up to 2, clear evidence of the nuclear-waste component. Strontium-90 data were used to resolve the bomb fallout ^{137}Cs signal from the nuclear-waste ^{137}Cs (see legend of figure 3) and the latter is contoured in the lower panel (*b*).

It is evident that the speed of southward propagation of the new ^{137}Cs signal exceeds the sampling strategy employed. No samples were collected for Cs and Sr isotope measurement south of TTO 214. Although the intensity of the plume was upstream of TTO 214, the deep samples there still contained measurable traces of nuclear-waste ^{137}Cs. It is unfortunate that no further downstream samples were collected to define better the limit of southward penetration at that time. Nevertheless, it is evident that the transfer of this surface signal from the Iceland and Greenland Seas to deep water off Newfoundland was achieved in a period of two years. The absence of any nuclear-waste ^{137}Cs above the deep plume shows that only the Denmark Strait overflow water is carrying a measurable signal. The overlying Iceland–Scotland Overflow Water and Labrador Sea Water show ^{137}Cs concentrations and ^{137}Cs:^{90}Sr ratios indicative of bomb-fallout origin.

It is worth noting that this deep distribution pattern traces the dispersion of any material originating on the European coastal shelves that is propagated into the deep North Atlantic. Therefore, it is likely that this represents one of the routes by which environmental contaminants from industrial activity in Western Europe are being introduced into the abyssal ocean.

The inverse correlation of the tracer signal with salinity raises an interesting question in connection with the freshening trend in the deep North Atlantic mentioned earlier in this section. Swift (1984*b*) speculated that this phenomenon might have been caused by wind-forcing shifts, allowing the incorporation of relatively more low-salinity surface water into the overflows. He suggested the East Greenland Current as a potential candidate. Given the finding that the low-salinity plume is accompanied by a ^{137}Cs-rich tracer signal, rather than by a tritium or ^{90}Sr-rich signal, which would signal an East Greenland Current component, the possibility emerges that we are witnessing a freshening originating in proportionately more low-salinity shelf water from the Norwegian side mixing into the water overflowing the sill at Denmark Strait. The absence of nuclear-waste ^{137}Cs in the Iceland–Scotland Overflow Water may be also relevant to this point but may reflect the longer surface–deep water path lengths, dilution patterns, and transport times for the formation of this water mass.

5. Atlantic water circulation in the Arctic basins

The Atlantic water in the Arctic basins is typically characterized by a temperature maximum. It is generally viewed as being fed by the warm and salty inflow of the West Spitzbergen current through Fram Strait. This current has a complex input, splitting into several branches on the basis of the bathymetry of the Yermak Plateau (Perkin & Lewis 1984), and recirculation with short-circuited return flows southward in the Fram Strait region (Quadfasel *et al.* 1987). Those branches of the current that enter the Arctic basin are subsequently found to be below the southward-bound, fresh and cold polar water that feeds the East Greenland Current. The core of the Atlantic water, as defined by the temperature maximum, is found to be at progressively greater depths and with lower temperature maxima as its transit through the Arctic Basin is followed. This is generally believed to follow a broad anticlockwise trajectory, ending up at the western side of Fram Strait as relatively old water relative to its time of inflow to the Eurasian Basin.

If this was the only mechanism whereby Atlantic water circulated in the basin, one would expect that a tracer maximum would be found to correspond with the temperature maximum for those types of tracer having an input with the Atlantic water. For Cs and Sr isotopes, this should be seen as an older fallout-characterized signal, followed by a later nuclear-waste derived signal. This scenario seems to be met at locations near the inflow, e.g. at the stations PS-325 (1984) and FRAM-4 (1983) in figure 1. At these, as seen in figure 5*a*, *b* (and at other stations in the area (Casso *et al.* 1988)) the inflowing water is characterized by a high ^{137}Cs concentration and a high $^{137}Cs:^{90}Sr$ ratio at the temperature maximum.

Only four stations (figure 1) have been occupied at interior Arctic locations where the lower temperature maxima imply a significant time interval since their Atlantic cores passed north of Fram Strait. The stations are as follows.

(1) LOREX (Lomonosov Ridge Experiment), an ice station near the North Pole occupied in 1979, at which the first Arctic Cs and Sr isotope profile was measured (Livingston *et al.* 1984). Situated at the Lomonosov ridge, it is expected to be a location at which one would find Atlantic water that had travelled around the basin boundary and back across the mid-ocean ridge.

(2) CESAR (Canadian Expedition to Study the Alpha Ridge), another ice station (1983) over the Alpha Ridge where the low-temperature maximum led to an expectation of relatively old Atlantic water, especially with its location being away from the dynamically active basin boundaries.

(3) AIWEX (Arctic Internal Wave Experiment). This 1985 ice station in the Canadian Arctic was sampled for Cs and Sr isotopes in the expectation that it would act as a representative of a location where the ventilation predated the nuclear-waste ^{137}Cs signal, an erroneous expectation as will be noted.

(4) PS-331. This 1984 station from the POLARSTERN Arktis II-3 cruise sampled very old

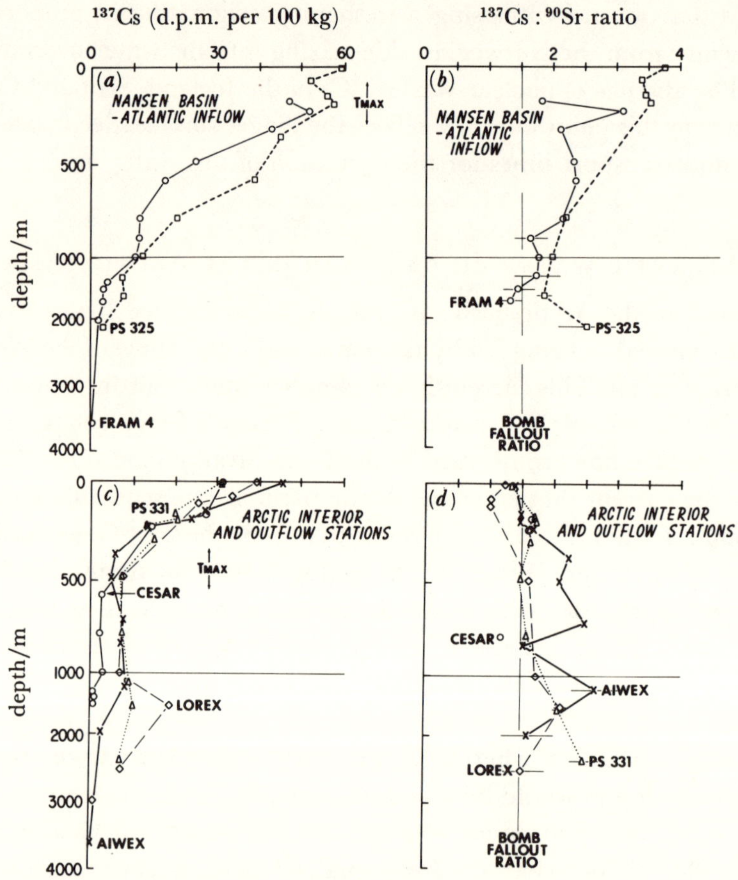

FIGURE 5. (*a*) Concentration profiles of ^{137}Cs at Atlantic inflow stations POLARSTERN-325 and FRAM-4. (*b*) Ratios of ^{137}Cs:^{90}Sr at above station. (*c*) Concentration profiles of ^{137}Cs at interior Arctic and Atlantic outflow stations. (*d*) Ratios of ^{137}Cs:^{90}Sr at above stations. Where error bars are not indicated in the figure, measurement errors are comparable with, or less than, symbol size.

Arctic basin water, assumed to have taken 'the Grand Tour' around the Arctic basins before moving southwards through Fram Strait.

At none of these four stations is there any suggestion of a tracer maximum corresponding to the temperature maximum (figure 5*c*). This trend also characterized other anthropogenic tracers when they were measured at these stations (Östlund 1982, 1985; Östlund *et al.* 1987; Smethie in Swift 1984*c*; Top 1984). In fact, when the patterns of the ^{137}Cs:^{90}Sr ratio with depth are examined (figure 5*d*), there is a trend of increase in ratio with depth in the lower Atlantic water and deeper layers at the AIWEX, LOREX and PS-331 stations. Only at the AIWEX station is the Atlantic core characterized with water with ^{137}Cs:^{90}Sr ratios higher than the typical fallout value of 1.5. The discussion of these various features properly belongs in the final section.

6. SHELF VENTILATION OF THE ARCTIC BASINS

In recent years there have been various suggestions that the formation and export of cold and saline water by brine rejection during freezing on the broad Arctic shelves may be an important process in regard to ventilation of the basin interior. Initially this process was proposed for halocline ventilation (Aagaard *et al.* 1981). Subsequently, this idea was extended to the deeper water masses (Aagaard *et al.* 1985). In fact, the possibility of deep-water penetration by shelf waters was raised by the failure of a model developed for the Arctic halocline to simulate properties below 200 m (Killworth & Smith 1984). Aagaard *et al.* (1985) noted that in the Eurasian Basin the shelf waters are already saline because of their Atlantic origin and can become denser through cooling alone, without the need to invoke a brine rejection mechanism. Both processes may contribute to a range of dense waters available for off-shelf ventilation and the Barents Sea has been cited as a potential site for ventilation water formation. Midttun (1985) has described the observation of bottom water more saline than 34.95‰ and near the freezing point in channels north of Novaya Zemlya that connect with channels to the Arctic Ocean from the Barents and Kara Seas. Samples collected in September 1985 in this same region confirm that water formed there is carrying a strong tracer signal from the southern nuclear-waste source. In figure 6, ^{137}Cs concentrations measured in bottom-water samples are plotted. The corresponding surface values are somewhat higher (Casso *et al.* 1988). Both the high values and the ratios of ^{137}Cs : ^{90}Sr in the 3–5 range reinforce the expectation that this signal will provide a unique and readily detectable tracer for water of shelf origin found in the deep Arctic basins. Hence this offers a powerful tool for studying the process of deep Arctic ventilation from the Eurasian shelf. The pattern of Cs and Sr distribution and their ratios to each other at the various locations indicated on figure 4 speak to this ventilation hypothesis.

The observations made in 1979 at the LOREX ice-station near the North Pole (Livingston *et al.* 1984) provided the first clue that shelf ^{137}Cs and ^{90}Sr signals were being rapidly transmitted into the Arctic Ocean interior. At that time, the sampling was limited and the only unequivocal signal seen was the high ^{137}Cs concentration and ratio of ^{137}Cs : ^{90}Sr at 1500 m, on the Eurasian side of the Lomonosov ridge. The LOREX data are reproduced in figure 5*c*, *d* and serve as a contrast and comparison with the data for the other Arctic interior or outflow stations.

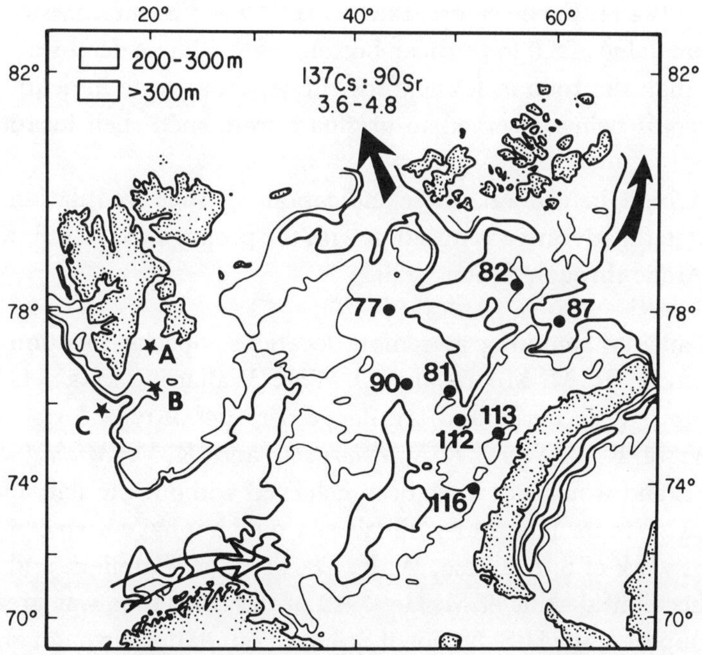

FIGURE 6. Concentrations of ^{137}Cs in September 1985 in bottom water in the eastern Barents Sea (in units of d.p.m. per 100 kg).

As noted in the previous section, there is no evidence at these stations of a nuclear-waste signal advected directly with the core of the Atlantic water, as defined by the temperature maxima. With the exception of the data from the CESAR station, at all three other locations there is a significant trend of increase in $^{137}Cs:^{90}Sr$ ratio with increasing depth below the Atlantic core (figure 5*d*). At the Canadian Basin station (AIWEX), 500 km north of Prudhoe Bay, there is also a trend of increasing ^{137}Cs concentration with depth, at least to about 1200 m. The absence of any nuclear-waste signal below 1500 m suggests that the Lomonosov ridge sill at that depth may have prevented deeper water, bearing the tracer signal, from moving into the Canadian Basin from the Eurasian side. Otherwise, the pattern with depth is consistent with that observed previously at the LOREX station. The data from all the profiles sampled at the AIWEX station suggested that this location was characterized by water with fairly high dynamic properties with decadal scale renewal times at mid-water depths (Swift *et al.* 1987). The CESAR station data show very low Cs and Sr isotope signals and no indications of any source except bomb fallout. The location on the Alpha Ridge must represent a central basin situation where the ventilation rate is relatively slow.

Finally, the data from the POLARSTERN Arktis II-3 station 331 are relevant to the shelf-ventilation question. The water sampled at this station had properties in the Atlantic layer that implied long residence in the Arctic Basin (Swift 1984*c*). At greater depths both the hydrographic properties and the chlorofluoromethan tracer data (Smethie *et al.* 1987) were indicative of Arctic Ocean Deep Water before its proposed passage into the northern Greenland Sea to act as a parent water mass for Norwegian Sea Deep Water (Swift & Koltermann 1988) after mixing with Greenland Sea Deep Water. The ^{137}Cs concentrations at this station (figure 5*c*) reflect bomb-fallout origins in the Atlantic core, as evidenced by their $^{137}Cs:^{90}Sr$ ratios at these depths (figure 5*d*). Below this depth, ^{137}Cs concentrations pass through a maximum at about 1500 m thus at a depth where the salinity of 34.925‰ is evidence of the presence of Arctic Ocean Deep Water. Furthermore, the ratio of $^{137}Cs:^{90}Sr$ increases across this depth and reaches a maximum value of 2.5 in the near-bottom water. These deep-water trends, especially at depths greater than the 1500 m Lomonosov sill depth, provide indications that ventilation of these deep layers is being observed to originate from such shelf locations as the Barents Sea.

In the Arctic Ocean in general, these indications of shelf ventilation seem to be more pronounced than lateral advective ventilation with the propagation of inflowing Atlantic water coming into the Arctic through Fram Strait.

Much effort went into sampling at remote locations. Special mention and thanks go to R. M. Moore (LOREX), R. M. Moore and D. W. R. Wallace (CESAR), G. C. Anderson and J. H. Swift (AIWEX), P. J. Stern and A. Baggeroer (FRAM-IV), Lars Midttun and the officers and crews of R. V. *Knorr*, R. V. *Polarstern*, and R. V. *Michael Saars*. These were major group efforts and would not have been achieved without the skill and dedication of all these people. Likewise, in the laboratory, the work would not have been completed without the expertise and hard work of S. A. Casso, W. R. Clarke, J. M. Palmieri, and L. D. Surprenant, as is true for the secretarial skills of M. Hess. All of this assistance was greatly appreciated.

The financial support of the U.S. National Science Foundation through grant number OCE-8402849 and the U.S. Office of Naval Research through contract number N00014-85-C-0715 is acknowledged with thanks. This is contribution number 6478 of the Woods Hole Oceanographic Institution.

References

Aagaard, K., Coachman, L. K. & Carmack, E. C. 1981 On the halocline of the Arctic Ocean. *Deep Sea Res.* **28**, 529–545.

Aagaard, K., Swift, J. H. & Carmack, E. C. 1985 Thermohaline circulation in the Arctic Mediterranean Sea. *J. geophys. Res.* **90**, 4833–4846.

Aagaard, K. & Reed, R. 1987 Fram Strait-Exchange and Dynamics. *Eos, Wash.* **68**(9), 124–125.

Aarkrog, A., Dahlgaard, H., Hallstadius, L., Hansen, H. & Holm, E. 1983 Radiocesium from Sellafield effluents in Greenland waters. *Nature, Lond.* **304**, 49–51.

Bowen, V. T. & Sugihari, T. T. 1964 Fission product concentration in the Chukchi Sea. *Arctic* **17**, 198–203.

Bowen, V. T., Noshkin, V. E., Volchok, H. L., Livingston, H. D. & Wong, K. M. 1974 Cesium-137 to strontium-90 ratios in the Atlantic Ocean, 1966 through 1972. *Limnol. Oceanogr.* **19**, 670–681.

Brewer, P. G., Broecker, W. S., Jenkins, W. J., Rhines, P. B., Rooth, G. G., Swift, J. H., Takahashi, T. & Williams, R. T. 1983 A climatic freshening of the deep Atlantic north of 50° N over the past 20 years. *Science, Wash.* **222**, 237–239.

Bullister, J. L. & Weiss, R. F. 1983 Anthropogenic chlorofluoromethanes in the Greenland and Norwegian Seas. *Science, Wash.* **221**, 265–268.

Casso, S. A. & Livingston, H. D. 1984 Radiocesium and other nuclides in the Norwegian-Greenland Seas 1981–82. Technical report WHOI-84-40. (101 pages.) Massachusetts: Woods Hole Oceanographic Institution.

Casso, S. A., Clarke, W. C., Livingston, H. D., Palmieri, J. M. & Surprenant, L. D. 1988 Cesium and strontium isotopes in the northwestern North Atlantic and Arctic Ocean, 1981–1985. Woods Hole Oceanographic Institution technical report (in preparation).

Clarke, R. A., Reid, J. L. & Swift, J. H. 1986 The Greenland Sea in Winter. CM1986/C:32, unpublished document. Copenhagen: International Council for the Exploration of the Sea.

Helland-Hansen, B. & Nansen, F. 1909 The Norwegian Sea, its Physical Oceanography Based Upon the Norwegian researches 1900–1904. Report on Norwegian Fishery and Marine Investigations, vol. 2, part 1, no. 2. (Oslo): Mallingske, Christiana.

Holm, E., Persson, B. R. R., Hallstadius, L., Aarkrog, A. & Dahlgaard, H. 1983 Radio-cesium and transuranium elements in the Greenland and Barents Seas. *Oceanol. Acta* **6**, 457–462.

Jenkins, W. J. & Rhines, P. B. 1980 Tritium in the deep North Atlantic Ocean. *Nature, Lond.* **286**, 877–880.

Jones, E. P. & Anderson, L. G. 1986 On the origin of chemical properties of the Arctic Ocean Halocline. *J. geophys. Res.* **91**, 10759–10767.

Kautsky, H. 1980 Distribution of radioactive fallout products in the water of the North Atlantic and the Barents Sea during the year 1972. In *Isotope marine chemistry* (ed. E. Goldberg), ch. 2, pp. 9–23. Tokyo: Uchida Rokakuho.

Killworth, P. D. 1979 On 'chimney' formations in the ocean. *J. phys. Oceanogr.* **9**, 531–554.

Killworth, P. D. & Smith, J. M. 1984 A one-and-a-half dimensional model for the Arctic halocline. *Deep Sea Res.* **31**, 271–293.

Kupferman, S. L., Livingston, H. D. & Bowen, V. T. 1979 A mass balance for ^{137}Cs and ^{90}Sr in the North Atlantic Ocean. *J. mar. Res.* **37**, 157–199.

Lee, A. & Ellett, D. 1967 On the water masses of the northwest Atlantic Ocean. *Deep Sea Res.* **14**, 183–190.

Livingston, H. D., Bowen, V. T. & Kupferman, S. L. 1982*a* Radionuclides from Windscale discharges I: non-equilibrium tracer experiments in high latitude oceanography. *J. mar. Res.* **40**, 253–272.

Livingston, H. D., Bowen, V. T. & Kupferman, S. L. 1982*b* Radionuclides from Windscale discharges II: their dispersion in Scottish and Norwegian coastal circulation. *J. mar. Res.* **40**, 1227–1258.

Livingston, H. D., Kupferman, S. L., Bowen, V. T. & Moore, R. M. 1984 Vertical profile of artificial radionuclide concentrations in the central Arctic Ocean. *Geochim. cosmochim. Acta* **48**, 2195–2203.

Livingston, H. D. 1985 Anthropogenic radiotracer evolution in the central Greenland Sea. *Rit Fiskideildar* **9**, 43–54.

Livingston, H. D., Swift, J. H. & Östlund, H. G. 1985 Artificial radionuclide tracer supply to the Denmark Strait overflow between 1972 and 1981. *J. geophys. Res.* **90**, 6971–6982.

Livingston, H. D. 1986 Radiotracer studies of deep water formation in the Nordic Seas. CM1986/C:3, unpublished document. Copenhagen: International Council for the Exploration of the Sea.

Malmberg, S.-A. 1983 Hydrographic investigations in the Iceland and Greenland Seas in late winter 1971 – 'Deep Water Project'. *Jo Kull* **33**, 133–140.

Mantyla, A. W. & Reid, J. L. 1983 On the abyssal circulation of the world ocean. *Deep Sea Res.* **30**, 805–833.

Midttun, L. 1985 Formation of dense bottom water in the Barents Sea. *Deep Sea Res.* **32**, 1233–1241.

Moore, R. M. 1981 Oceanographic distributions of zinc, cadmium, copper and aluminum in waters of the central Arctic. *Geochim. cosmochim. Acta* **45**, 2475–2482.

Moore, R. M., Lowings, M. G. & Tan, F. C. 1983 Geochemical profiles in the central Arctic Ocean: their relation to freezing and shallow circulation. *J. geophys. Res.* **88**, 2667–2674.

Moore, W. S. 1980 Radium isotopes in sub-Arctic waters. *Earth planet. Sci. Lett.* **49**, 329–340.

Östlund, H. G. 1982 The residence time of the freshwater component in the Arctic Ocean. *J. geophys. Res.* **87**, 2035–2043.

Östlund, H. G., Top, Z. & Lee, V. L. 1982 Isotope dating at Fram III. *Geophys. Res. Lett.* **9**, 1117–1119.

Östlund, H. G. 1984 Arctic Ocean water mass balance from isotope data. *J. geophys. Res.* **89**, 6373–6381.

Östlund, H. G. 1985 R. V. Polarstern Arktis II/3 Cruise 1984, Northwestern Section, Tritium Samples. University of Miami Tritium Laboratory Data Release 85–43. Unpublished document.

Östlund, H. G., Possnert, G. & Swift, J. H. 1987 Ventilation rate of the deep Arctic Ocean from carbon-14 data. *J. geophys. Res.* **92**, 3769–3777.

Perkin, R. G. & Lewis, E. L. 1984 Mixing in the West Spitzbergen Current. *J. phys. Oceanogr.* **14**, 1315–1325.

Peterson, W. H. & Rooth, C. G. H. 1976 Formation and exchange of deep water in the Greenland and Norwegian Seas. *Deep Sea Res.* **23**, 273–283.

Quadfasel, D., Gascard, J. C. & Kolterman, K. P. 1987 Large scale oceanography in Fram Strait during MIZEX 84. *J. geophys. Res.* **92**, 6719–6728.

Smethie, W. M., Östlund, H. G. & Loosli, H. H. 1986 Ventilation of the deep Greenland and Norwegian Seas: evidence from krypton-85, tritium, carbon-14 and argon-39. *Deep Sea Res.* **33**, 675–703.

Smethie, W. M., Chipman, D. W., Swift, J. H. & Koltermann, K. P. 1988 Chlorofluoromethanes in the Arctic Mediterranean Seas: evidence for formation of bottom water in the Eurasian Basin and deep water exchange through Fram Strait. *Deep Sea Res.* (In the press.)

Swift, J. H., Aagaard, K. & Malmberg, S.-A. 1980 The contribution of the Denmark Strait overflow to the deep North Atlantic. *Deep Sea Res.* **27A**, 29–42.

Swift, J. H., Takahashi, T. & Livingston, H. D. 1983 The contribution of Greenland and Barents Seas to the deep water of the Arctic Ocean. *J. geophys. Res.* **88**, 5981–5986.

Swift, J. H. 1984*a* The circulation of the Denmark Strait and Iceland-Scotland overflow waters in the North Atlantic. *Deep Sea Res.* **31**, 1339–1355.

Swift, J. H. 1984*b* A recent θ–S shift in the deep water of the northern North Atlantic. In *Climate processes and climate sensitivity* (ed. J. E. Hansen & T. Takahashi), vol. 5 (Geophys. Monogr. 29), pp. 39–47. Washington, D.C.: American Geophysical Union.

Swift, J. H. 1984*c* Preliminary Data Report, R/V Polarstern Cruise Arktis II/3, 19 July–7 August 1984. (46 pages.) University of California, Scripps Institution of Oceanography, La Jolla, CA. Unpublished document.

Swift, J. H. 1986 In *The Nordic seas* (ed. B. G. Hurdle), ch. 5. (777 pages.) New York: Springer-Verlag.

Swift, J. H. & Koltermann, K. P. 1988 The origin of Norwegian Sea deep water. *J. geophys. Res.* (In the press.)

Swift, J. H., Livingston, H. D. & Östlund, H. G. 1987 Hydrographic sampling during AIWEX. *Eos, Wash.* **68**(7), 100.

Top, Z. 1984 CESAR tritium and C-14 results. University of Miami Tritium Laboratory Data Release 84–30. Unpublished document.

Vakulovskiy, S. M., Nikitin, A. I. & Chumichev. V. B. 1985 Pollution of Arctic Seas by radioactive wastes of West European atomic plants. *Atomn. Energ.* **58**, 445–449.

Wallace, D. R. W. & Moore, R. F. 1985 Vertical profiles of CCl_3F (F-11) and CCl_2F_2 (F-12) in the central Arctic Ocean basin. *J. geophys. Res.* **90**, 1155–1166.

Discussion

P. D. KILLWORTH (*Hooke Institute, University of Oxford, U.K.*). The sinking of dense brine from central Arctic shelves presents a great problem for dynamicists. No 'plume' model has succeeded in moving water even a fraction of the total depth.

Conservative tracers and the ocean circulation

By G. R. Bigg[1] and P. D. Killworth[2]

[1] *School of Environmental Sciences, University of East Anglia, Norwich NR4 7TJ, U.K.*
[2] *Hooke Institute for Atmospheric Research, University of Oxford, Clarendon Laboratory, Parks Road, Oxford OX1 3PU, U.K.; Institute of Oceanographic Sciences, Wormley, Godalming, Surrey GU8 5UB, U.K.*

Inclusion of chemical tracers in inverse schemes for calculating the ocean circulation leads to a better determined system of constraints to the flow and may result in increased knowledge about mixing processes as well as current systems. A type of tracer suggested by Broecker, namely a conservative combination of nitrates or phosphates and oxygen ('NO' or 'PO') uses standard observed quantities. It is shown that although these tracers do add information to more conventional inverse schemes, the present sampling strategy is inadequate for determining the ocean circulation quantitatively. The principal reasons for this are the variability of the ocean and the limited spatial coverage of present hydrographic surveys.

1. Introduction

One of the reasons for interest in chemical tracers in the ocean is the hope that they may provide orthogonal information to the more traditional tracers of density, temperature and salinity when investigating the general circulation. Much qualitative insight and a degree of quantitative data has been gained from analysis of various chemical species' distribution, detailed well in Broecker & Peng (1982). In this paper we will investigate the degree of success with which one of the most comprehensive oceanic chemical surveys, GEOSECS (see references in Broecker & Peng, pp. 670–672), can be integrated into the methodology of inverse methods of determining the ocean circulation. Two different inverse methods will be used, namely the β-spiral method (Schott & Stommel 1978) and the Bernoulli method (Killworth 1986), and in addition to the conventional conservative tracer, density, we will consider potential vorticity and the chemical combinations of oxygen, nitrates and phosphates suggested by Broecker (1974).

2. The data

The GEOSECS data are far from ideal for use with many inverse methods for determining the general circulation of the ocean, as they provide information mostly along long sections, making estimation of the full set of derivatives required by the inverse methods difficult. The box-type inverse technique pioneered by Wunsch (1978), which we will not be considering, partly does away with this problem. It still, however, requires the sections to enclose a region of ocean and the GEOSECS cruises, shown in figure 1, do not generally meet this requirement except on very large scales. Nonetheless, we have been able to consider three regions for this paper, marked on figure 1. They include a point in the central North Atlantic (station 30)

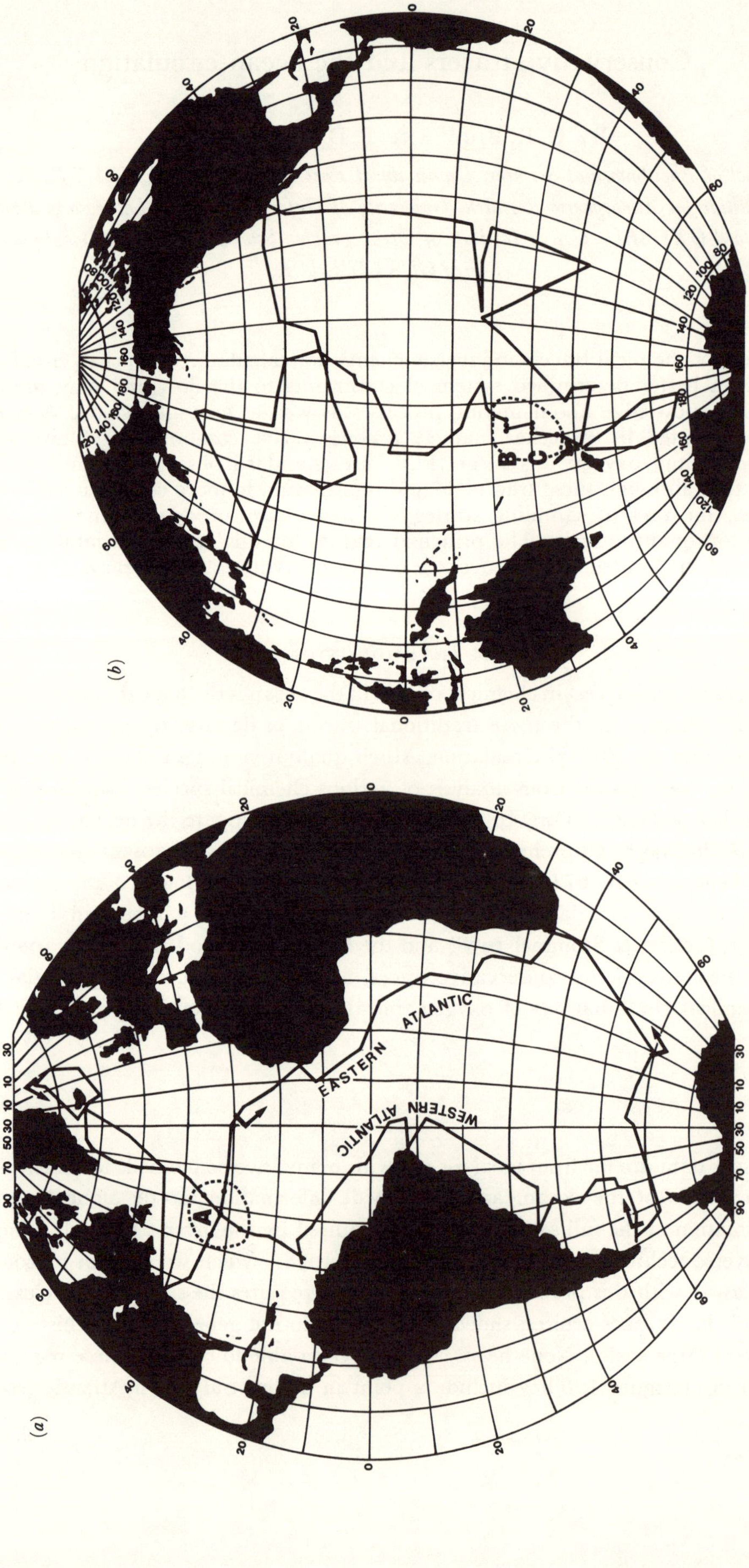

FIGURE 1. GEOSECS (*a*) Atlantic and (*b*) Pacific sections. A, station 30; B, Tongan Trench section; C, Kermadec Trench section. The ends of the sections shown in figure 2 are marked on (*a*).

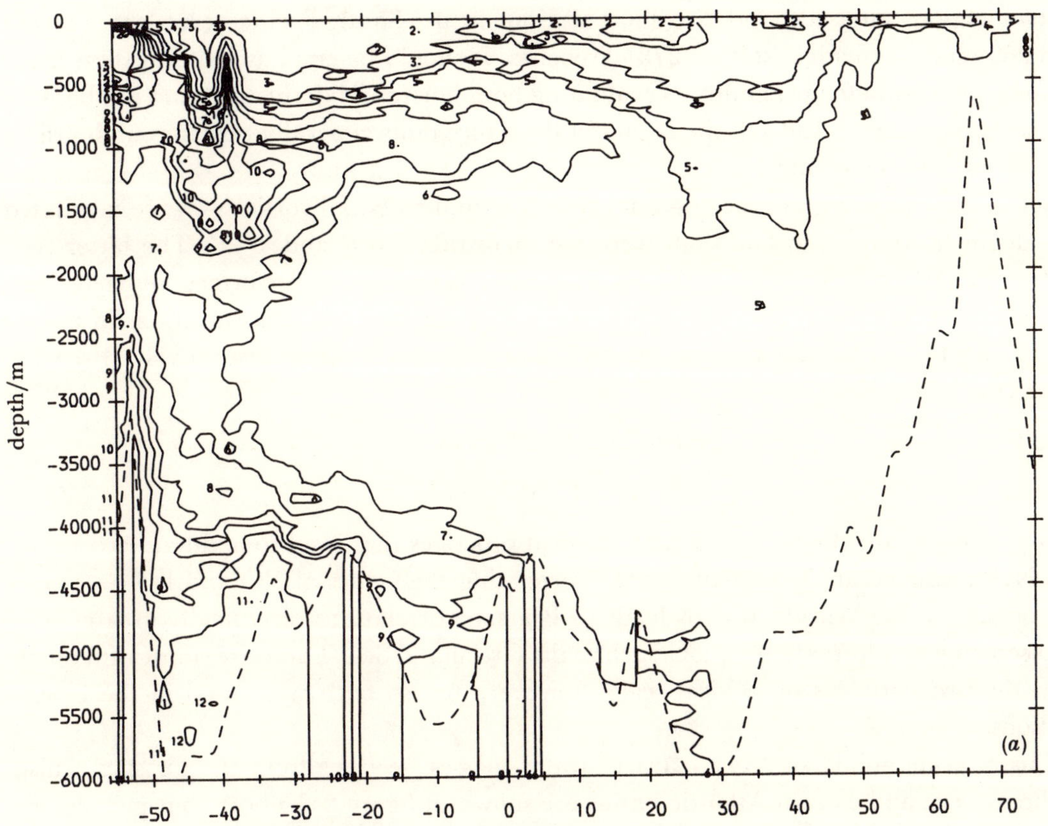

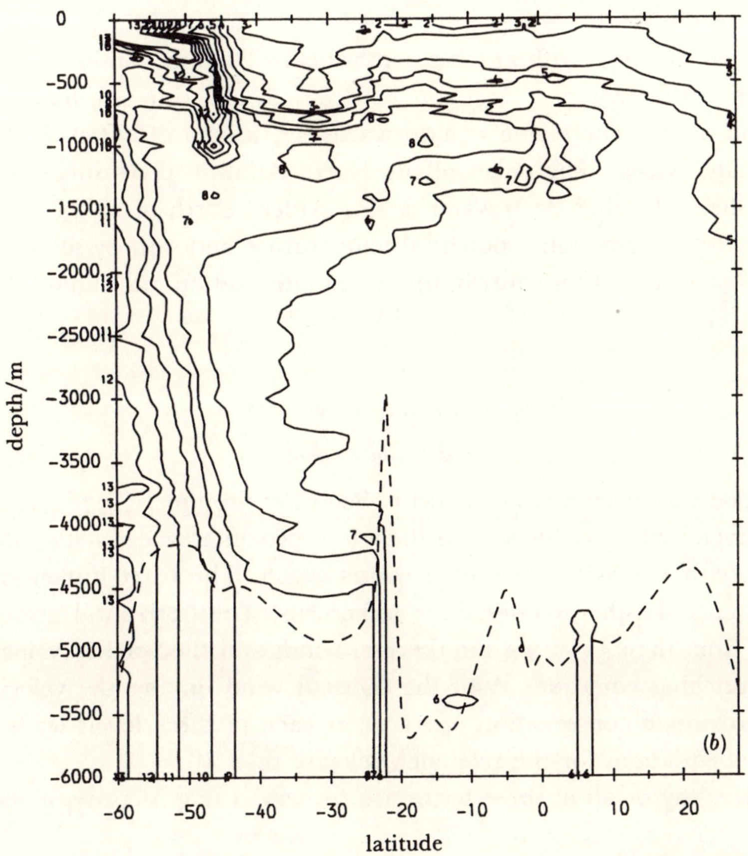

FIGURE 2. NO sections in (*a*) the western Atlantic and (*b*) the eastern Atlantic. The sections are marked on figure 1. The contour intervals are *n* micromoles per kilogram: 1–4, $n \times 100$; 5–6, $400 + (n-4) \times 50$; 7–8, $500 + (n-6) \times 25$; 9–13, $550 + (n-8) \times 10$. The broken line follows the deepest sampling depth.

where two sections intersect, and two short sections in the South Pacific across the Tongan (station 269) and Kermadec (station 273) Trenches. It should be emphasized that these areas were chosen because of the possibility of evaluating both horizontal derivatives and not because of their scientific interest. Indeed, the layout of the data points generally possesses undesirable features for inverse modelling.

In each of these three regions the potential density σ_θ and potential vorticity q were calculated at each depth horizon as, in addition, were the quantities 'NO' and 'PO'. The latter two, derived, variables are conservative tracers essentially representing preformed nutrient concentrations (Broecker 1974) and are defined as combinations of the concentrations in micromoles per kilogram of nitrate and oxygen and phosphate and oxygen respectively (Broecker *et al.* 1985) as

$$\left.\begin{aligned} NO &= 11[NO_3]+[O_2], \\ PO &= 175[PO_4]+[O_2]. \end{aligned}\right\} \qquad (1)$$

The two quantities are closely linked by biological processes that tend to maintain a constant ratio of N:P in the ocean (Goldman 1980). Correlating the values of NO and PO over large regions of each of the World's oceans leads to Pearson correlations of typically greater than 98%, except in water formation regions such as the Greenland Sea. There are subtle differences between the two variables on a local scale, however, and so both were used in the inverse calculations.

To illustrate the ability of NO to 'trace' water masses, sections from the GEOSECS cruises along the western and eastern Atlantic basins are shown in figure 2. In both, but particularly the western basin (figure 2*a*), we can clearly see the intermediate NO waters (high oxygen but low nitrate) formed in the North Atlantic north of Iceland, and subducted below relatively NO-poor surface water, moving south at most depths below 2000 m. Similarly, the very cold, dense Antarctic Bottom Water with a large nutrient content and thus still higher NO values, is seen on both basin floors penetrating equatorwards, as, around 1000 m, is the less dense Antarctic Intermediate Water. The effect of the North Atlantic Gyre mixing the low-NO surface waters into the NO-richer deep water is also evident north of 20° N. Comparison of these diagrams with the corresponding potential temperature and salinity sections of GEOSECS (Bainbridge 1980) shows that NO is contributing extra information and should therefore be a useful additional tracer.

3. The inverse methods

(*a*) *The β-spiral method*

The β-spiral method was originally described in Schott & Stommel (1978). It assumes that in the locality of a given hydrographic station there is approximate steady-state thermal-wind balance and tracer (either density, NO, etc.) conservation. The three-dimensional velocity vector at some reference depth provides three unknowns. The horizontal velocity at other depths is computed from these by using the thermal wind, and the vertical velocity by using the linear vortex stretching equation. With the thermal wind linking the velocities at each depth horizon, approximate conservation of tracer at each of these levels leads to an overdetermined system of equations for the reference velocity (u, v, w).

If it is assumed that any or all of three terms are conserved (e.g. density, q and NO) then we can combine all the resulting equations for the reference-level velocity to give a highly over-determined problem. This enables us to allow horizontal and vertical diffusion coefficients to be introduced as unknowns at each level and still retain an over-determined problem. In essentially all calculations with diffusion coefficients the unknowns were found to be not significantly different from zero by using the error analysis intrinsic to least-squares problems (Press *et al.* 1986).

(*b*) *The Bernoulli method*

This method is developed in Killworth (1986). It assumes a steady-state ocean in approximate geostrophic and hydrostatic balance that conserves mass, density and chemical tracer. Linear vorticity $q = f\rho_z$ is thus conserved (Welander 1971), where f is the Coriolis parameter, ρ the density and z a coordinate vertically upwards. The Bernoulli function $B = P + \rho g z$ is also conserved, where P is the pressure and g is the acceleration due to gravity. With several quantities all being approximately conserved we can consider one (e.g. B) as being a function of others. So we have considered $B = F(\rho, q)$ (as in Killworth 1986) or $B = F(\rho, \mathrm{NO})$. In both these systems it is B that is not uniquely defined as the surface pressure is not known. Once B is known the horizontal velocity may be evaluated from the pressure field.

The problem then is to find B. To do this we take a set of stations and examine them pairwise, seeking for a depth (below the mixed layer) at one station at which ρ and q (or ρ and NO) match the values at some depth at the other station. As B must also match, this gives a linear equation connecting the differences in surface pressure at the two stations. With there being typically a number of matches between any pair of stations a large over-determined system of equations for the surface pressure is produced that can be solved by singular-value decomposition in the least-squares manner (as in the β-spiral method).

(*c*) *Gauging the success of inverse methods*

One of the biggest problems with the use of inverse methods to infer the ocean circulation is to estimate how well they capture the desired flow. Many different oceanic locations have been examined in the literature but it is rare to have velocity measurements with which to compare the inversions. One way around this problem is to seek bounds to the flow rather than its actual value. Wunsch (1984) has had some success with this approach.

In previous work (Killworth & Bigg 1987) we have addressed this problem by examining the performance of the two techniques used here, plus a simple version of Wunsch's (1978) box-type inverse method, on data obtained from an eddy-resolving general circulation model of Cox (1985). We therefore had the 'real' velocities with which to compare the inversions of that study. The reader is referred to Killworth & Bigg (1988) for a full description; it suffices here to say that in general the Bernoulli and box methods performed well, whereas the β-spiral method was to a large degree unsatisfactory, producing large residuals to its composite equations. The β-spiral method has been retained for this investigation because of the different quantities (e.g. NO) used in the present study and because the model and real oceans are perforce not identical in their dynamics. Another result of Killworth & Bigg (1988) worth reporting here is that for accurate thermal winds (i.e. well within 1 cm s^{-1} top-to-bottom of the model values) horizontal derivatives needed to be evaluated over distances of less than 100 km. This has important implications for most oceanographic surveys.

Apart from the error estimates accompanying least-squares problems there are two means by which we will check for 'realistic' inversions. One is how well the data fit the equations, i.e. the residuals. The other is to choose a standard water column for the calculations and study the reproducibility of the inverted velocity from case to case, and between considering the whole of the water column and sub-sections of it.

4. The results

(a) *Thermal-wind calculations*

At each of the three sites considered, four different estimates of the thermal-wind shear from 3000 to 400 m were made. Three employed σ_θ as the key variable in the thermal-wind calculation

$$\left.\begin{aligned} u_z &= g(\sigma_\theta)_y/f\rho_0, \\ v_z &= -g(\sigma_\theta)_x/f\rho_0, \end{aligned}\right\} \tag{2}$$

while one used geopotentials (see Gill 1982). The differences between the σ_θ estimates came from the way in which the derivatives at each depth level were evaluated. One method assumed that the density was of the form $a+bx+cy$ at a given level and gave a least-squares fit to the linear coefficients, so giving the derivatives. The others assumed that the stations used to calculate horizontal derivatives by finite differences lay on a longitude–latitude cross. One method let the stations on, say, the longitude arm be unevenly spaced, whereas the other assumed that they were equally spaced. Both of the two calculations gave similar results and will not be separately considered. The geopotential horizontal derivatives were calculated by using the uneven spacing.

The Pacific sites show much greater thermal-wind shear than the Atlantic site and have larger differences between the three methods. This is shown in figure 3 for the Tongan Trench

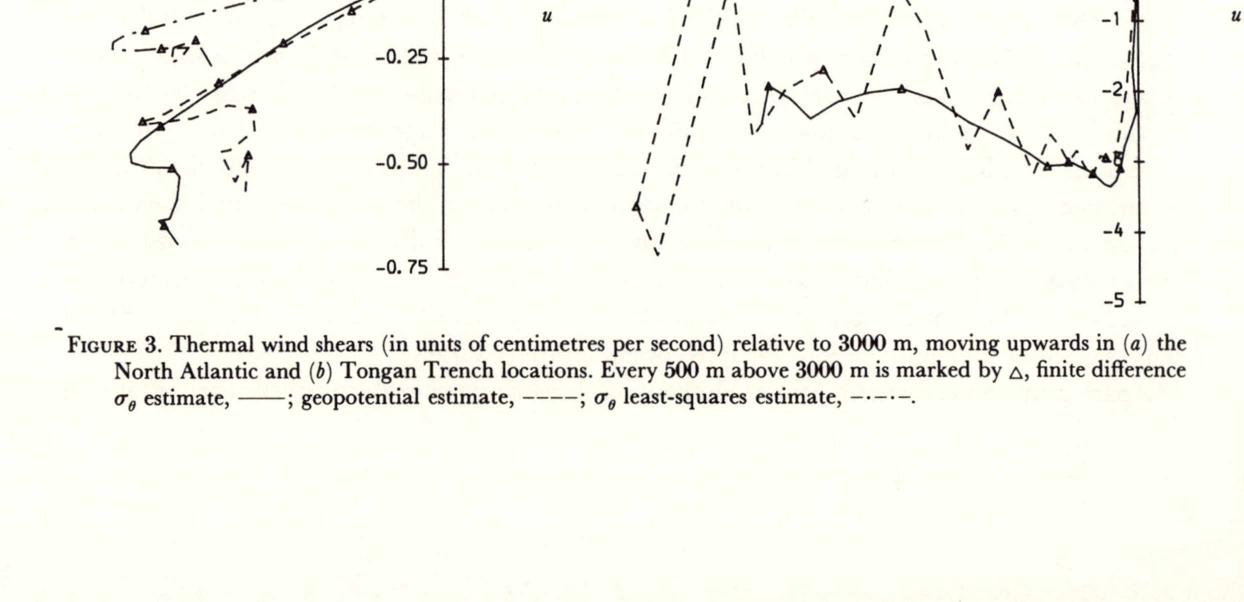

FIGURE 3. Thermal wind shears (in units of centimetres per second) relative to 3000 m, moving upwards in (*a*) the North Atlantic and (*b*) Tongan Trench locations. Every 500 m above 3000 m is marked by $\triangle$, finite difference σ_θ estimate, ——; geopotential estimate, ----; σ_θ least-squares estimate, -·-·-.

(or northern section, B) in the Pacific and the North Atlantic site (A). The σ_θ and geopotential finite difference derivative cases are similar at all three locations, with the geopotential thermal wind always being noisier, because of the level-by-level method of calculation (as opposed to the depth-integrated form of calculation for σ_θ estimates). However, the least-squares σ_θ estimate, although roughly similar to the others at the Atlantic location, is wildly different for both Pacific sites. This is largely because of the special nature of the Pacific sites. The zonal sections are much shorter than the meridional scale, and including both zonal sections in a calculation seriously distorts the x-derivative fit.

So it is found that calculating the thermal wind, before trying to apply any inverse methods, does not lead to a unique answer. Instead, it is strongly dependent on the method, and site, of calculation. Fortunately, the inverse methods are robust to the significant noise level of the geopotential thermal wind and give similar results to the equivalent σ_θ estimate. Use of this thermal wind will therefore not be considered further. It remains now to examine the degree of internal consistency of the inverse methods.

(*b*) *Inverse calculations*

At each location we carried out three inversions from both methods. These were for different depth ranges: a 'full depth' calculation (400–3500 m), an upper range (400–1500 m) and a lower range calculation (1600–3500 m). The GEOSECS data were interpolated to give σ_θ, etc., at every 100 m so that the least over-determined β-spiral inversion had eleven equations and three unknowns. The three velocity components were always resolved but the addition of unknown mixing coefficients raised the rank only slightly and so these components were not resolved. The Bernoulli inversions were characteristically an order of magnitude more over-determined with well over 100 $\rho - q$ (or $\rho -$NO) crossings and five unknown surface pressures. For consistency, only the results for the finite difference σ_θ estimates of thermal wind are shown, except for the North Atlantic Bernoulli runs that used the least-squares fits (see the previous section for comments). In all the following diagrams, the velocity at 1500 m only will be examined. The thermal-wind profiles of figure 3 enable the reader to gauge the ramifications of the discussion for other depths.

In figure 4, the 1500 m Bernoulli velocities for all three depth ranges and the two variable combinations (ρ and q, ρ and NO) are shown for each of the three locations. Comparing the magnitudes of the scatter in the solutions with the thermal wind of figure 3 shows that the variability in the inverse velocity profile is considerable. There tends to be grouping of the two solutions for a given depth range but even this is not uniformly true.

The Bernoulli method assumes conservation of several variables simultaneously. In the 'snap-shot' data set with which we are working it may be that this is asking too much of the ocean. The β-spiral calculations, however, are very diverse, letting individual quantities be conserved, or collections of them. It performed poorly in the sensitivity tests of Killworth & Bigg (1988) but it was thought that a major reason for this was the strong influence of numerical diffusion in the model. With the GEOSECS inversions, diffusion was found not to improve the solution, with the extra unknowns being unresolved. In figures 5 and 6 we show the 1500 m velocities for the solutions giving the best and worst fits respectively of the trials for the Atlantic and Tongan Trench locations. The Kermadec Trench site gives similar results. As in figure 4 the three different depth-range solutions are shown.

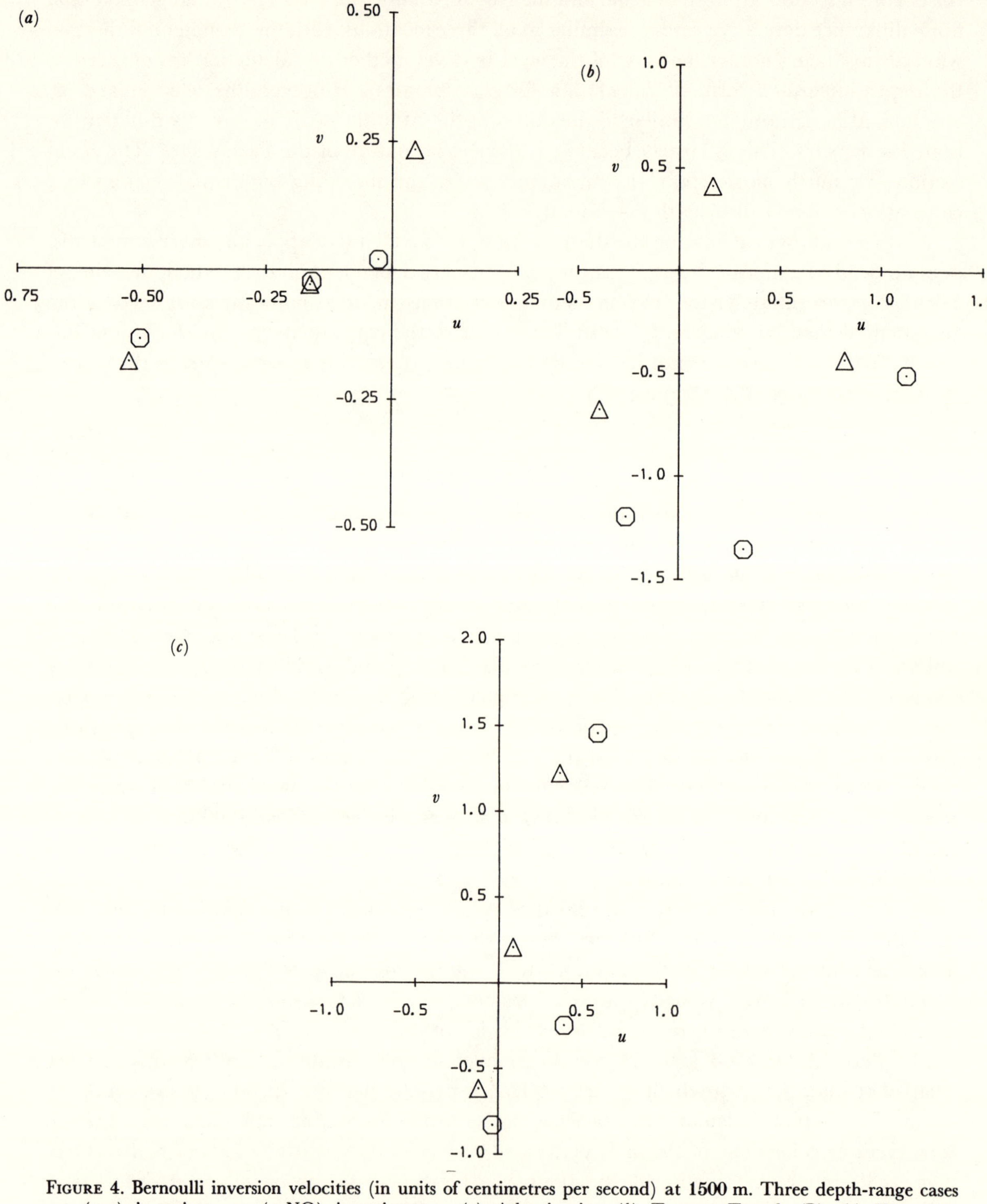

FIGURE 4. Bernoulli inversion velocities (in units of centimetres per second) at 1500 m. Three depth-range cases (ρ, q) inversions, ○; (ρ, NO) inversions, △. (*a*) Atlantic site, (*b*) Tongan Trench (B), (*c*) Kermadec Trench (C).

The inconsistency between the various solutions and depth ranges is manifest. The inversions combining conservation of ρ, q and NO tend to give the largest residuals (and hence the worst fits) whereas the ρ-only results tend to give the smallest residuals. There are significant exceptions to these statements. In particular, the good fits of PO in the Atlantic illustrate the compatibility of the PO construction with more conventional conserved quantities.

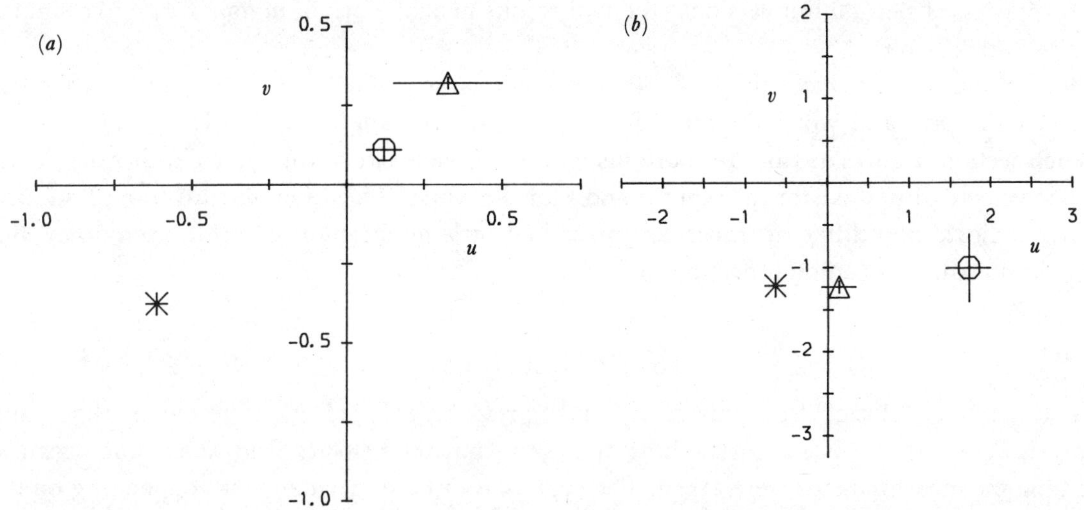

FIGURE 5. The best fit β-spiral velocities (centimetres per second) at 1500 m. Three depth ranges: full depth, △; upper range, ○; lower range, ×. (a) Atlantic site (variable runs providing best fit: △, PO; ○, PO; ×, ρ). (b) Tongan Trench (B) (variable runs providing best fit: △, ρ; ○, ρ; ×, ρ).

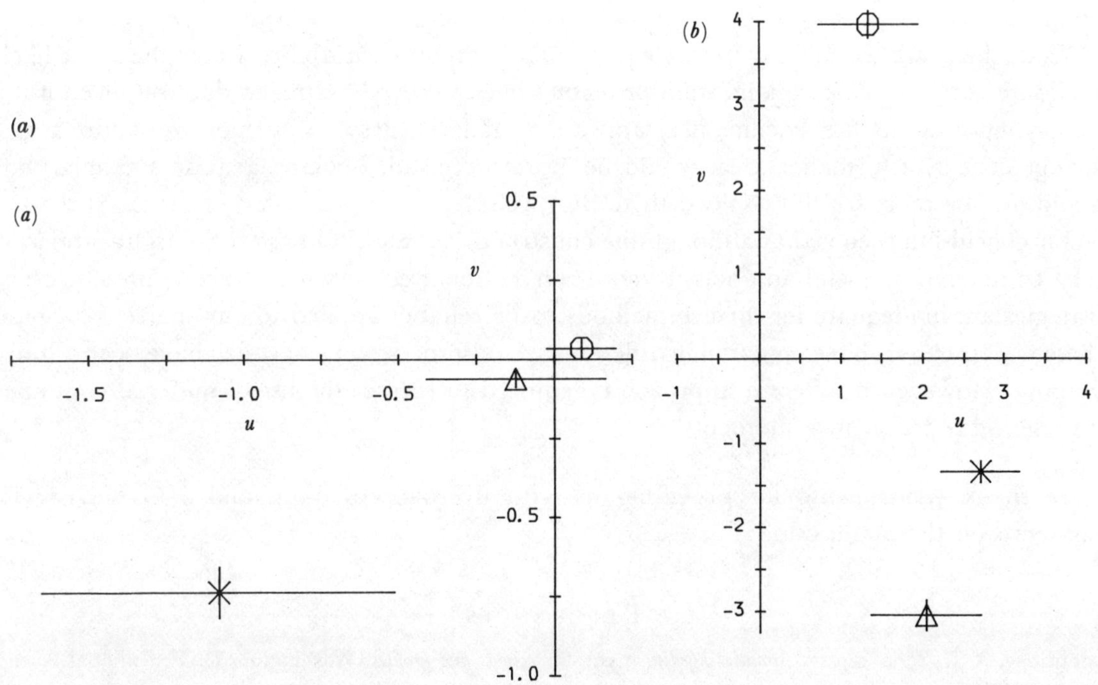

FIGURE 6. As for figure 5, but for worst fits. (a) Atlantic site (variable runs providing worst fit: △, $\rho+q+$NO; ○, q; ×, NO). (b) Tongan Trench (B) (variable runs providing worst fit: △, $\rho+q+$NO; ○, $\rho+q+$NO; ×, $\rho+q+$NO).

The upper-Atlantic depth range is the only set where both best and worst solutions fall within the respective error bars (interestingly the density inversion is not the best fit here). Note the smallness of the velocity predicted. The information from the deeper regions at this location seems to be completely at variance with the consistency of the upper region. Examining the thermal wind (figure 3*a*) shows a significant *u* reversal around 1500 m. Presumably the dynamics of this transition region violate the model equations, perhaps through nonlinearities. The upper-level Bernoulli inversions (the two results near the origin in figure 4*a*) give similar results to the β-spiral.

Both Pacific sites behave similarly, that is, very poorly. There is considerable difference between the best and worst fits and also between depth ranges. Curiously, in the Tongan Trench section (figures 5*b* and 6*b*) both fits tend to agree between the three depth-range cases for one velocity, but it is *u* for the best fits and *v* for the worst! The different runs (not all shown) clearly indicate that different tracers are providing orthogonal information to each other, but they do not share a common origin.

5. Conclusions

The major conclusion from our many and varied inversions with different methods and depth ranges is that it is extremely difficult to say what the velocity field was in the ocean at the time the measurements were taken. The variability in the thermal wind, depending on the method of calculation, was to some degree lost in inversion. However, the inconsistency between different inversions leads one to conclude that there are some important discrepancies between the GEOSECS ocean and the dynamical models used. Only at the North Atlantic site, over a limited depth range where velocities were less than 0.5 cm s^{-1}, can consistency be claimed for the inversions.

What, then, are we missing in our equations? Temporal variability seems the most likely candidate. Trying to use conservation principles over a sparsely sampled data set, even if it is near synoptic as in the Pacific sites, appears to fail. Possibly using more data, in a grid arrangement over a smaller area, would be more successful. Looking at time-averaged data should also be more fruitful (Killworth & Bigg 1988).

Our conclusion then is that although the constructs NO and PO possess the right properties to be considered as useful adjuncts to more conventional conservative tracers, present cruise strategies are inadequate for inverse methods to be reliably applied to our sparse data base. These tracers clearly have potential in different circumstances, some of which have been alluded to above. However, a different approach is required to extract the useful material contained in present-day tracer measurements.

We thank J. Sarmiento for providing us with the GEOSECS data, and J. G. Harvey for comments on the manuscript.

References

Bainbridge, A. J. 1980 *GEOSECS Atlantic expedition*, vol. 2, *sections and profiles*. Washington, D.C.: National Science Foundation.

Broecker, W. S. 1974 'NO', a conservative water-mass tracer. *Earth planet. Sci. Lett.* **23**, 100–107.

Broecker, W. S. & Peng T.-H. 1982 *Tracers in the sea*. (690 pages.) Lamont–Doherty Geological Observatory, New York: Eldigio Press.

Broecker, W. S., Takahashi, T. & Takahashi, T. 1985 Sources and flow patterns of deep-ocean waters as deduced from potential temperature, salinity and initial phosphate concentrations. *J. geophys. Res.* **90**, 6925–6939.

Cox, M. D. 1985 An eddy resolving numerical model of the ventilated thermocline. *J. phys. Oceanogr.* **15**, 1312–1324.

Gill, A. E. 1982 *Atmosphere–Ocean dynamics.* (662 pages.) London: Academic Press.

Goldman, J. C. 1980 Physiological processes, nutrient availability and the concept of relative growth in marine phytoplankton ecology. In *Primary production in the sea* (ed. P. G. Falkowske), pp. 179–193. New York: Plenum.

Killworth, P. D. 1986 A Bernoulli inverse method for determining the ocean circulation. *J. phys. Oceanogr.* **16**, 2031–2051.

Killworth, P. D. & Bigg, G. R. 1988 An intercomparison of inverse methods using an eddy resolving general circulation model. *J. phys. Oceanogr.* (In the press.)

Press, W. H., Flannery, B. P., Teukovsky, S. A. & Vetterling, W. T. 1986 *Numerical recipes.* (818 pages.) Cambridge University Press.

Schott, F. & Stommel, H. 1978 Beta spirals and absolute velocities in different oceans. *Deep Sea Res.* **25**, 961–1010.

Welander, P. 1971 The thermocline problem. *Phil. Trans. R. Soc. Lond.* A**270**, 415–421.

Wunsch, C. 1978 The general circulation of the North Atlantic west of 50° W determined from inverse methods. *Rev. Geophys.* **16**, 583–620.

Wunsch, C. 1984 An eclectic Atlantic ocean circulation model, I. The meridional flux of heat. *J. phys. Oceanogr.* **14**, 1712–1733.

Purposefully released tracers

By A. J. Watson[1] and J. R. Ledwell[2]

[1] *Marine Biological Association, The Laboratory, Citadel Hill, Plymouth PL*1 *2PB, U.K.*

[2] *Lamont-Doherty Geological Observatory of Columbia University, Palisades, New York* 10964, *U.S.A.*

It has been recognized since the late 1970s that it should be possible to use the release of conservative tracers into the deep ocean to measure vertical mixing rates. Diapycnal mixing (i.e. vertical mixing across density strata) is rather crudely described by the present generation of ocean models, usually being taken to be constant irrespective of depth. Model circulations are, however, sensitive to quite minor changes in this assumption. Technology and experimental methods in the open ocean with perfluorinated tracers have now advanced to the state where direct-tracer experiments are feasible and a pilot study in a small basin has given encouraging results.

The design of open-ocean experiments is discussed. Consideration of the problem of sampling a streaky distribution leads to the conclusion that the period between release and sampling should be several times the eddy-rotation time.

Although any number of tracers could be used for such experiments, sulphur hexafluoride seems to have unique advantages over other candidates: it has been demonstrated to be conservative at least on a timescale of several months, and can be detected with better sensitivity than other possible tracers.

Perfluorinated tracers show promise in other areas besides deep-sea oceanography: they are well suited to studies in the coastal ocean and could be used to measure dispersion and air–sea exchange properties at the surface.

1. Introduction

The concept of large-scale experiments with a deliberately released tracer to measure the rate of mixing in the ocean thermocline is now nearly ten years old. It was first suggested by J. Shepherd and W. S. Broecker who proposed an experiment christened 'DOTREX' (Deep ocean tracer release experiment). The problem to be solved was this: diapycnal mixing (vertical mixing across density strata) is a poorly measured process that must be parametrized in all ocean-circulation models, from the simplest geochemist's box-flux diagram to the most complex global circulation model. Not even the high-resolution models, which should be available by the end of the century, will be able to treat explicitly the physical processes responsible for vertical mixing, for these have scales of tens of metres at the most.

Neither can estimates of true diapycnal mixing be made by using geochemical tracers such as tritium, fluorocarbons, etc. The source of these transient tracers is generally the wind-mixed layer, which contacts the entire range of densities found in the ocean as one goes from the Equator towards the Poles. The tracers are therefore injected onto all densities simultaneously. Because isopycnal mixing (along the nearly horizontal density strata) is perhaps six or seven orders of magnitude faster than the diapycnal variety, a transient tracer may penetrate deep into the interior of the ocean by spreading down the gently sloping isopycnal horizons without undergoing appreciable diapycnal mixing. Measurement of apparent vertical-mixing coefficient by such tracers therefore gives only an upper bound on the true diapycnal mixing rate.

The importance of getting not only the right number but also the correct variation with depth, has been illustrated by Gargett (1984). She points out that if the coefficient of vertical mixing, K_v, increases with depth as stratification decreases, then upwelling velocities may also be expected to increase with depth beneath the thermocline. Convergence of the vertical velocity then dictates equatorward flow of water at mid-depth, in contrast to the Stommel–Ahrons view of the abyssal circulation in which these waters move poleward. It seems therefore that the nature of the K_v variation with depth radically influences the simplest attempts to describe the global circulation.

Shepherd and Broecker's original plan called for the deliberate release of a tracer onto a chosen isopycnal in the main thermocline. The rate at which the injection spread in the vertical would be an unambiguous measure of diapycnal mixing rate. DOTREX spawned lively discussions and quite a mass of correspondence, but the experiment appeared to be impractical for two reasons: theorists said that it would not work because the release would give rise to a streaky distribution that would be a nightmare to sample. Worse, no tracer could be found that was sensitively enough detected yet sufficiently conservative to make a large-scale experiment possible. The only candidate, helium-3, was already present in the ocean and proving valuable as a 'non-deliberate' tracer. No one wanted to risk confusing the ^3He signal by a deliberate release.

The DOTREX oceanographers were unaware that large-scale tracer experiments in the atmosphere were in fact quite common. These mostly used perfluorinated compounds detected by electron-capture gas chromatography. When, at J. E. Lovelock's suggestion, consideration was given to these compounds it soon became clear that they were strong candidates for a DOTREX-style experiment. In 1983 funding was obtained in Britain and America to begin development work. Results to date have been encouraging and a feasible and useful experimental plan to tackle the diapycnal mixing problem can now be advanced, although the emphasis has moved away from the idea of a single, huge release. We will describe such a plan later (§4*b*) after reviewing some properties of these tracers, their potential fields of use in marine studies, and the results from a pilot experiment in the Santa Monica Basin.

2. Properties and uses for perfluorinated tracers

The ideal purposeful tracer for marine experiments would be fully conservative in the sea and continuously quantifiable at extreme dilution, by simple apparatus. It would have no appreciable environmental effects and no toxicity, a consideration that rules out radioactive substances such as tritium. Perfluorinated tracers come close to this ideal in many respects. It is useful to compare them with rhodamine dyes, which are presently the 'standard' method of water tracing for most applications. Below is a summary of the relative advantages and disadvantages of rhodamine and two perfluorinated tracers, sulphur hexafluoride (SF_6) and perfluorodecalin (PFD).

Useful detection limit: for rhodamine dyes this is set by the background fluorescence of natural substances in seawater, and is of the order 10^{-12} g g^{-1} in the deep ocean (Ewart & Bendiner 1974). SF_6 has a background in the sea because of dissolution from the atmosphere where it is present at 1–2 parts per trillion (by volume).† This generates concentrations of about 5×10^{-17} g g^{-1} in surface waters, declining to zero in the deep ocean. The instrumental detection limit for SF_6 is about one tenth of the surface-water value (Watson & Liddicoat 1985).

PFD has no measurable background in the sea, so its detection limit is set by the performance of the instrument. The minimum detectable concentration is of the order 10^{-16} g g^{-1}.

One tonne is probably a practical upper limit on the amount of any of these tracers that could be released in a given experiment. The detection limits, then, imply that rhodamine could be used for experiments on scales up to 100 km linear dimension, PFD for scales up to 1000 km and SF_6 for oceanic scales.

Conservative behaviour: rhodamine has a tendency to be absorbed on suspended particulates, and is also destroyed by photodegradation in sunlit water (Carter 1974). Its lifetime in the sea will therefore depend on the suspended particle load, but is probably more than a month and less than a year in typical ocean conditions. Perfluorinated compounds are biologically and chemically almost inert in the environment. However, PFD appears to have some affinity for surfaces and we have some evidence that transport on settling particles marginally affected the distribution five months after release during our pilot experiment.

Sulphur hexafluoride would be expected on theoretical grounds to be the least surface active of these three tracers. It seems unlikely that particle scavenging significantly affected the concentrations in the pilot experiment but it may be premature to assume that its lifetime in subsurface water is indefinite.

PFD and SF_6 will not be conservative in the surface layer of the ocean. Because the perfluorinated tracers are sparingly soluble, non-ionic compounds they will tend to escape to the atmosphere from surface waters, their residence time depending on the depth of the surface mixed layer and the air–water gas-exchange rate. They are of little use in rivers, for example, because from such shallow, turbulent water they will vent to the atmosphere within hours. But in the coastal ocean they should last for weeks or months and can be used to good effect provided the experimenter recognizes that the tracer is not conservative (Watson *et al.* 1987).

Toxicity and environmental impact: SF_6 and PFD are both very safe compounds to handle, with negligible toxic effects. Rhodamine is not highly toxic but may be carcinogenic. It is also rather messy because it visibly colours water at concentrations above 10^{-8} g g^{-1}.

Response time and ease of measurement: whereas the fluorometer used for rhodamine determination is a robust instrument that gives a continuous measurement and can be used *in situ*, the perfluorinated tracers are analysed by gas chromatographic methods for which samples of seawater must be brought on board the survey ship. Although, at 2–3 min, the analysis is fast for a chromatographic method, it is two orders of magnitude slower than fluorometry and incapable of resolving fine detail. For near-surface measurements where a pumped supply of seawater is available, continuous sampling with the analyses giving time-integrated concentrations can be arranged (Watson *et al.* 1987). It is interesting to note that Thomson & Lovelock (1976) have described a device for measuring sulphur hexafluoride on a continuous basis in the atmosphere at levels of around 5×10^{-14} g ml^{-1}. With some modification this instrumentation should work equally well for determinations in seawater and could be used to monitor the distribution in the early stages of a marine experiment when

† In this paper, 1 trillion represents 10^{12}.

concentrations are high, but pending the development of an *in situ* instrument (one that could be lowered into the water column) it would not alleviate the requirement that water samples be pumped up to a surface ship.

To summarize, perfluorinated tracers score high for the exquisite sensitivity of the analysis and for conservative behaviour beneath the surface layer, but the analysis is relatively slow and they will vent to the atmosphere from surface waters. Thus, for experiments in very shallow waters (rivers, estuaries, etc.) rhodamine dyes remain the most useful tracer. Short-duration experiments (i.e. a few days or less) are also better tackled with rhodamine because rapid continuous measurement is usually desirable. Longer releases require the greater sensitivity of perfluorinated tracers. Even in tidally active coastal waters experiments having durations of a month or more are possible, although atmospheric venting must be corrected for.

In the deep ocean, where water cannot be continuously pumped to the ship, the slow rate of sample recovery means that the horizontal advection of the tracer can be studied only imperfectly and there is a strong possibility that the release will be lost unless it is tracked by independent means such as isopycnal drifting floats. The release of clusters of such floats will by itself give a good idea of horizontal eddy dispersion so there is probably no need to release a tracer if horizontal transport is to be the focus of the study. For vertical mixing, however, a tracer release seems to be the only practical approach.

3. The Santa Monica Basin experiment

A release of SF_6 and PFD to investigate the practicality of deep-ocean experiments was performed in the Santa Monica Basin off southern California in August 1985. Figure 1 shows the bathymetry of the basin, which is a depression in the sea floor about 50 km long by 30 km wide, connected via a narrow saddle to the San Pedro Basin to the southeast. The two basins together form a single feature with the deepest sill at 737 m. The injection was made below sill depth at about 800 m, where the buoyancy period was 53 ± 4 min. The advantage of making the release in a small, nearly closed basin was that no floats were needed to track the tracers. The disadvantage is that care needs to be exercised in extrapolating the results to the open ocean.

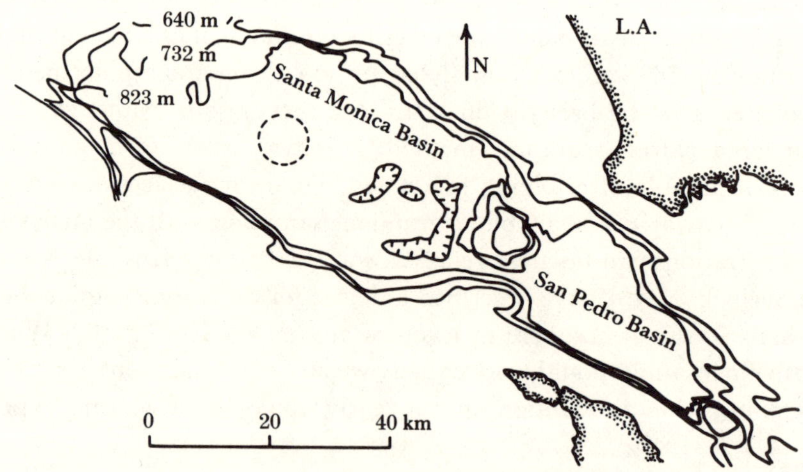

FIGURE 1. Bathymetry of the Santa Monica and San Pedro basins. The broken circle shows the region in which the tracers were released.

Preliminary results of the experiment have been reported by Ledwell *et al.* (1986). A complete account awaits the reduction of all the hydrographic data. Here we will concentrate on a few important points of technique that made for a successful experiment, and on the results up to 50 days from release.

(*a*) *Injection*

It was considered important to inject the tracers into a constant-density surface rather than simply at constant depth, and to disperse the initial injection widely to ensure that the distribution was affected by a large number of mixing events from the start. Isopycnal injection was accomplished by constantly adjusting the depth of the injection apparatus in response to the output of a conductivity–temperature–depth unit (CTD) to keep the release within 0.01 °C of the 5.15 °C isotherm. (In practice, constant temperature could be substituted for constant density because a close temperature–salinity relation exists in the basin.) The r.m.s. amplitude of internal waves at the injection depth was about 12 m (similar to the value that would be expected in the open ocean at the same stratification). Had the injection been performed at a constant depth therefore, the initial condition would have been substantially broader than necessary.

For success in these experiments it is essential that the injection procedure is accurate and faultless: if buoyancy effects cause the tracer to fall or rise after release the experiment is ruined and cannot be repeated at that site until the faulty release has been diluted to insignificant levels. To verify the initial condition therefore, measurements of the tracers were made for fourteen days following the release. These showed no evidence for a less-than-adequate injection; averaged over this period, the equivalent width of the profile of concentration against depth was 9 m for SF_6 and 7 m for PFD. The dispersion was, in part, caused by natural mixing during the time that had elapsed since injection, and in part by the effect of the turbulent wake of the injection apparatus.

(*b*) *Sampling*

The early DOTREX discussions focused attention on the difficulty of sampling a distribution that exhibits fine-scale streakiness. Instead of a well-behaved, continuous and reasonably isotropic concentration field, dispersion from a point will soon come to resemble a plate of spaghetti, with greatly elongated streaks of tracer, wound up by eddies and interspersed with clear water. Sampling such a distribution with Niskin bottles would be a frustrating experience, because all except a small minority of the samples would contain no tracer. Instead, in the Santa Monica Basin, syringes that filled slowly with water were used to collect samples that had been integrated over several kilometres of horizontal path. An array of the samplers was vertically spaced on a hydrographic wire as shown in figure 2, together with a CTD unit that was usually at the centre of the array. Once the array was in position with the CTD at the isotherm on which the injection had been made, the samplers were made to begin filling and the whole array was towed through the water with the CTD being 'flown' along the target isotherm in the same way as for the injection. This procedure does not result in the individual samplers following isotherms; rather they are at fixed heights above the injection isotherm, leading to a slight overestimate of the true vertical spread. In the Santa Monica Basin, the integrating samples were no longer needed after five months: as the streakiness of the distribution faded, spot samples from Niskin bottles became adequate. For open-ocean experiments integration would presumably be required for longer periods.

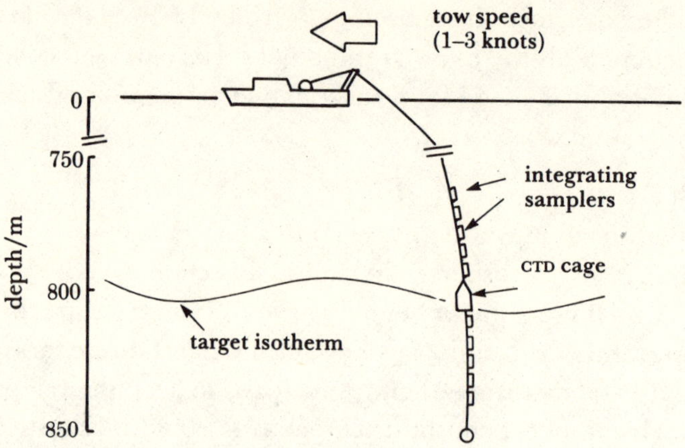

FIGURE 2. Sampling array used to average over streaks of tracer. The integrating samplers fill with water slowly as the array is towed through the water. The CTD is maintained on the target isotherm by adjusting the length of wire out. (1 knot = 1.85 km h^{-1}.)

(c) *First results*

Figure 3 shows mean interior profiles from the first two sampling cruises at 2–13 d and 47–56 d after injection. Only profiles taken at least 4 km from the walls have been included. The two tracers, SF_6 and PFD, were injected on separate casts but were sampled on the same casts: they thus give quasi-independent estimates of the mixing. PFD was injected in just four streaks whereas SF_6 was introduced in 15 lower-concentration streaks. SF_6 was therefore easier to find than PFD on the earlier cruises because it was more dispersed from the start, so there are more data. Furthermore, the SF_6 measurements had a somewhat better signal: noise ratio than PFD, so the SF_6 data set is generally of superior quality.

In figure 4 the second moments of the mean profiles are plotted against time from the

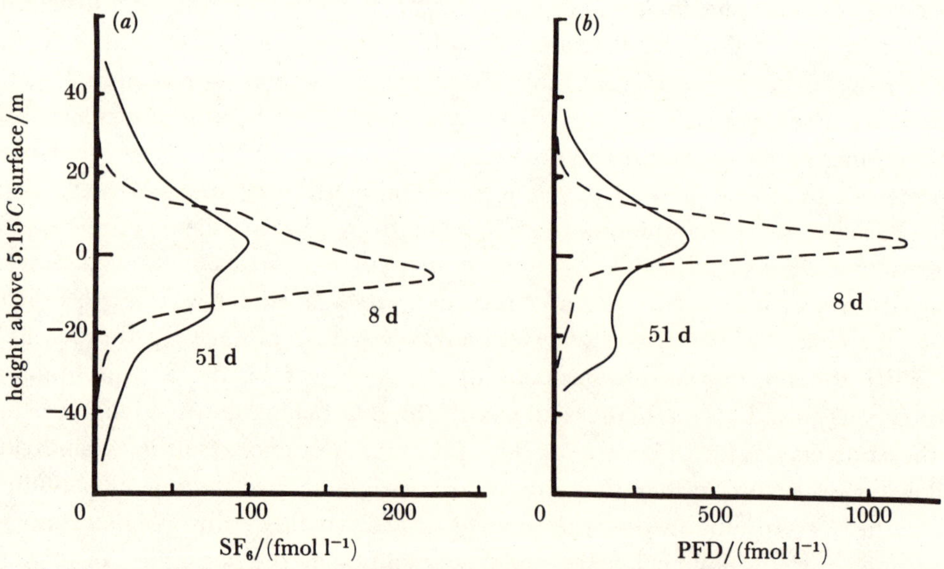

FIGURE 3. Mean vertical profiles of (*a*) SF_6 and (*b*) PFD. The concentrations in the later profiles have been multiplied by factors of 2.4 for SF_6 and 10.0 for PFD to normalize the areas under the curves to those of the earlier profiles.

injection. Assuming fickian diffusion with coefficient K_v the profiles ought to be gaussian with second moments,

$$\sigma^2 = 2K_v t.$$

From the slope of the lines in figure 4 we get $K_v = 0.37\ \text{cm}^2\ \text{s}^{-1}$ (SF_6) or $0.27\ \text{cm}^2\ \text{s}^{-1}$ (PFD). The difference between the two estimates is presumably because of the different time histories of the two tracers, and to the relatively small number of PFD profiles. This explanation is supported by more recent data that indicate that by the time six months had passed, the two values for K_v were closely in agreement.

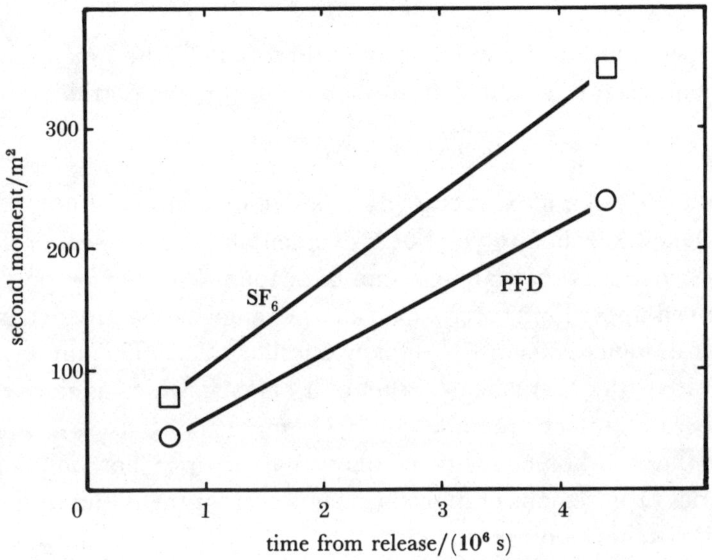

FIGURE 4. The growth of the second moments of the tracer profiles with time from release.

A recent analysis (Gargett & Holloway 1984; Gargett 1984) has suggested that diapycnal diffusivities should scale inversely with buoyancy frequency, N. Gargett proposes the relation:

$$K_v = aN^{-q},$$

where a and q are constants in any given system but will vary between different systems, and q lies between 0.5 and 1. For the open ocean, Gargett's best estimates for a and q are respectively $10^{-3}\ \text{cm}^2\ \text{s}^{-2}$ and 1. The Gargett 'rule' would therefore predict $K_v = 0.51\ \text{cm}^2\ \text{s}^{-1}$ in the open ocean at the same stratification as the experimental release. The order-of-magnitude agreement with the measured value seems to indicate that the presence of the basin walls does not greatly enhance the mixing in the basin.

4. Measuring vertical mixing in the open ocean

(a) *Existing techniques*

A deliberate release such as the Santa Monica experiment gives a measure of diapycnal diffusivity at one particular location and stratification. This goes only a short distance towards answering the needs of the ocean-circulation modeller, who requires a validated parametrization of K_v, perhaps along the lines of the Gargett rule, that will be applicable over most of the oceans. Because we are unable from theoretical considerations to say with any certainty how K_v may vary in space and time, we are faced with the prospect of having to do many direct determinations by tracer release. To minimize the effort involved, it will be important to combine tracer experiments with other techniques giving information on K_v. It should be possible to reduce the number of experiments by using the tracer measurement to calibrate, or to eliminate ambiguities in, other methods.

One well-established technique that has been used to estimate vertical mixing is the measurement of temperature or velocity microstructure. The analysis by Osborn & Cox (1972) suggests that K_v should be related to the molecular diffusion coefficient k by

$$K_v/k \leqslant aC,$$

where a is a dimensionless constant between one and three, and C, the Cox number, is the ratio of the mean square temperature gradient to the square of the mean gradient,

$$C = \overline{(\mathrm{d}T/\mathrm{d}z)^2}/(\mathrm{d}\overline{T}/\mathrm{d}z)^2.$$

There is now quite a body of microstructure data (see, for example, Moum & Osborn 1986; Gregg 1977, 1980; Oakey & Elliot 1977). For instrumental reasons, most of the studies were in the top few hundred metres of the water column but Moum & Osborn report measurements down to 2200 m. Derived upper limits on K_v span a large range with many determinations near 0.01 $cm^2\ s^{-1}$, but a few as high as 30 $cm^2\ s^{-1}$ (Oakey & Elliot 1977). The numbers are generally considerably smaller than the 'canonical' value of 1 $cm^2\ s^{-1}$, which is derived from simple models of 'non-deliberate' tracers (Munk 1966; Gammon *et al.* 1982). The mismatch is probably caused by the non-homogeneity of diapycnal mixing: not only may mixing be enhanced at boundaries or in regions of high sheer, but even in the ocean at large it probably occurs in intermittent patches that are difficult to sample by taking 'snapshots' of the microstructure. Perhaps because of the difficulty of deriving a meaningful average from such a wide scattering of numbers, the microstructure-based estimates of K_v do not seem to be much utilized in numerical simulations. A determination of K_v from a tracer release would be integrated over an ensemble of such events and should therefore be much easier to interpret.

It would seem a good idea for future releases to be accompanied by high-quality microstructure measurements. If the microstructure and tracer measurements agree then our confidence in past microstructure determinations will be greatly increased and the number of tracer experiments required to 'tie down' oceanic K_v significantly reduced. If microstructure gives persistently lower values than the tracer then we must look for explanations, probably to do with the intermittency of mixing events.

A 'heat budget' method has occasionally been used to measure K_v in special locations in the oceans. The best example is that of Hogg *et al.* (1982), who utilized measurements of the flow of Antarctic Bottom Water (AABW) into the nearly closed Brazilian Basin. The measured net inflow below the $\theta = 1$ °C isotherm was set equal to the net upward transport over the basin. In steady state, the downward diffusion of heat required to drive this upwelling can be calculated. Hogg *et al.* (1982) obtained diapycnal diffusivities of 3–4 $cm^2\ s^{-1}$ averaged over the basin. The downward diffusion of heat might occur evenly everywhere, or it might be concentrated near the boundaries and topographic features. An independent measurement of K_v away from these areas would therefore answer the question of whether mixing is or is not a relatively homogeneous process, at the same time as giving a low-stratification point on the curve of K_v against N to test the Gargett 'rule'.

(b) The release of tracer

To determine the variation of K_v on stratification it will be necessary to perform experiments at other sites besides the deep Brazilian Basin. Altogether at least three stratifications must be investigated if we are to have enough data to test the Gargett rule with one degree of freedom. A suitable site for shallower releases would be the interior of a subtropical gyre, away from strong boundary currents, which might tend to advect the tracer an inconveniently large distance. In the North Atlantic, suitable depths that would encompass a wide range of stratifications would be 800 and 2000 m.

About 100 kg of sulphur hexafluoride would be required for each experiment. This would be released over a period of two weeks or so by pumping it at high pressure through fine nozzles into the water. The fine droplets thus produced will dissolve rapidly before settling. The injection apparatus would be towed by a ship and maintained on a target isopycnal in the same manner as during the Santa Monica experiment. The injection streaks would be laid within an area about 100 km × 100 km, so that the horizontal scale would be larger than the mesoscale eddy dimensions from the start. Twenty or more neutrally buoyant floats would be released at intervals along the injection track. These would be used to study horizontal transport processes and to keep track of the tracer patch.

Six months or more after release a sampling ship would steam to the area marked out by the floats. Past float releases (Rossby *et al.* 1986) suggest that at 800 m the domain of occupation of the floats should grow to be of order 500–1000 km during this period. A timescale of six months is dictated by the need to allow sufficient time for the vertical distribution of the tracer to evolve, while not waiting until the patch is spread over too large an area to be adequately covered by the sampling ship on a single research cruise. Table 1 gives the expected vertical thickness of the distribution after six months for various values of diapycnal diffusivity, and two values of the initial dispersion. The experiment will yield an accurate value of K_v, providing it is greater than 0.03 $cm^2\ s^{-1}$, if the initial distribution can be confined to within 5 m of the target isopycnal.

TABLE 1. VERTICAL DISPERSION IN METRES OF A TRACER SIX MONTHS AFTER RELEASE, FOR SEVERAL VALUES OF VERTICAL-MIXING COEFFICIENT K_v AND INITIAL DISPERSIONS OF 5 m AND 10 m

	width at 6 months/m	
$K_v/(cm^2\ s^{-1})$	initial dispersion 5 m	initial dispersion 10 m
0.01	7.5	11.5
0.03	10.9	13.9
0.1	18.4	20.4
0.3	31.2	32.3
1.0	56.3	57.0
3.0	97.3	97.7

(c) Sampling to overcome streakiness

Experience in the Santa Monica Basin suggests that by using slowly filling samplers in a vertical array, the sampling ship could make about four casts per day. Each cast can be towed through a distance of order 6–12 km. For the experiment to be successful, we require that this sampling pattern be sufficient to encounter a large number of tracer streaks during a normal length research cruise of no more than a month's duration.

The streakiness model of Garrett (1983) is helpful in deciding whether the sampling will be adequate. Garrett proposed that streaks should grow exponentially in length at a rate of the order of the r.m.s. strain rate in the eddy field, while their width remained roughly constant at a value set by the balance between longitudinal straining and small-scale isopycnal diffusion. The total length of the streaks at time t is

$$L_{st} = L_0 \exp(S\gamma t),$$

where L_0 is the total length of the streaks at injection, γ is the strain rate, *ca.* $10^{-6}\ s^{-1}$, and S is an unknown scaling constant of order unity.

If these streaks are randomly distributed throughout an area A_p, the domain of occupation of the floats, the average number of streaks encountered by a sampling track of length L_{sa}, also randomly distributed, is

$$N = 2L_{st}\, L_{sa}/(\pi A_p).$$

Figure 5 shows how N evolves with time for various values of the strain rate with the scale constant fixed at 0.4 (i.e. on the conservative side of unity). The area A_p was assumed to grow linearly with time from $10^4\ km^2$ at time zero to $2.5 \times 10^5\ km^2$ at six months. The underlying assumption here is that the domain of occupation will increase in a fickian manner with a horizontal eddy diffusion rate of order $10^7\ cm^2\ s^{-1}$. L_{st} was fixed at 6 km and L_0, the initial total streak length, at 500 km. Values of N approaching unity, meaning that most sampling tracks will cross at least one streak, should ensure the success of the experiment. As the figure shows,

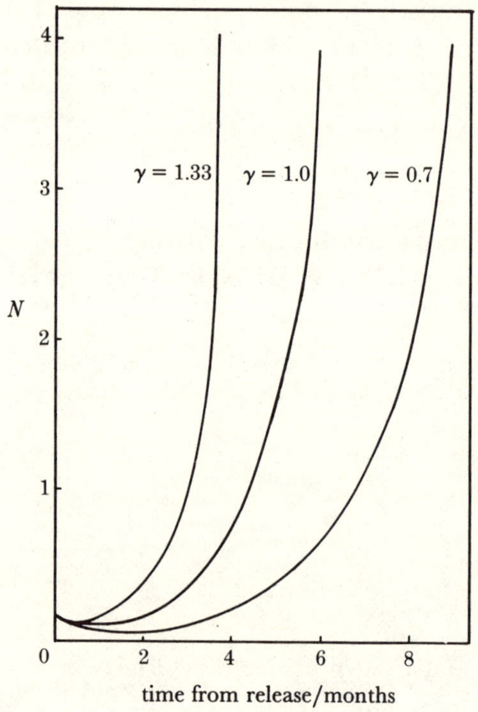

FIGURE 5. The evolution of N, the expected number of tracer streaks crossed by a sampling track, for three values of γ, the r.m.s. strain rate. Values of γ are in units of 10^{-6} reciprocal seconds. Calculated from the model described in the text.

the time taken for the probability of encountering a streak to become high is critically dependent on the exponent $S\gamma t$ in the expression for the streak length. A strain greater than $0.7 \times 10^{-6}\ s^{-1}$ is necessary to ensure success after six months for the particular numbers chosen here, meaning that $S\gamma t \gtrsim 5$. From dimensional considerations the strain rate should be of the same order as a typical angular velocity for the eddies, so a successful experiment requires that several 'eddy-rotation periods' must elapse between release and sampling.

5. Conclusions

With perfluorinated tracers we have the opportunity to perform some elegant experiments that will provide important information about oceanic mixing, particularly in the vertical. They are also likely to be useful in a number of practical applications involving the dispersal of pollutants, etc. The opportunity to use SF_6 as a tracer for large experiments will not last for ever; it will fade as, over the next few decades, the sulphur hexafluoride background in the surface oceans rises by perhaps two orders of magnitude as a result of industrial usage. Perfluorodecalin emulsion may well be used in bulk within a few years as a whole-blood substitute (Mitsuno *et al.* 1982) in which case it too may eventually become unusable as a large-scale tracer. We should take the chance to use these substances for scientific purposes while we may, before they become ubiquitous in the sea.

References

Carter, H. H. 1974 The measurement of rhodamine tracers in natural systems by fluorescence. *Rapp. P-v. Réun. Cons. perm. int. Explor. Mer* **167**, 193–200.

Ewart, T. E. & Bendiner, W. P. 1974 Techniques for estuarine and open ocean dye dispersal measurement. *Rapp. P.-v. Réun Cons. perm. int. Explor. Mer* **167**, 201–212.

Gammon, R. H., Cline, J. & Wisegarver, D. 1982 Chlorofluoromethanes in the northeast Pacific Ocean: measured vertical distribution and application as transient tracers of upper ocean mixing. *J. geophys. Res.* **87**, 9441–9454.

Gargett, A. E. 1984 Vertical eddy diffusivity in the ocean interior. *J. mar. Res.* **42**, 359–393.

Gargett, A. E. & Holloway, G. 1984 Dissipation and diffusion by internal wave breaking. *J. mar. Res.* **42**, 15–27.

Garrett, C. 1983 On the initial streakiness of a dispersing tracer in two and three-dimensional turbulence. *Dynam. Atmos. Oceans.* **7**, 265–277.

Gregg, M. C. 1977 Variations in the intensity of small scale mixing in the main thermocline. *J. phys. Oceanogr.* **7**, 436–454.

Gregg, M. C. 1980 Microstructure patches in the thermocline. *J. phys. Oceanogr.* **10**, 915–943.

Hogg, N., Biscaye, P., Gardner, W. & Schmitz, W. J. Jr 1982 On the transport and modification of Antarctic Bottom Water in the Vema channel. *J. mar. Res.* **40** (suppl.), 231–263.

Ledwell, J. R., Watson, A. J. & Broecker, W. S. 1986 A deliberate tracer experiment in Santa Monica Basin. *Nature, Lond.* **323**, 322–324.

Mitsuno, T., Ohanagi, H. & Naito, R. 1982 Clinical studies of a perflurochemical whole-blood substitute (fluosol-DA). *Ann. Surg.* **195**, 60–69.

Moum, J. N. & Osborn, T. R. 1986 Mixing in the main thermocline. *J. phys. Oceanogr.* **16**, 1250–1259.

Munk, W. H. 1966 Abyssal recipes. *Deep Sea Res.* **13**, 707–730.

Oakey, N. S. & Elliot, J. A. 1977 Vertical temperature gradient structure across the Gulf Stream. *J. geophys. Res.* **82**, 1362–1369.

Osborn, T. R. & Cox, C. S. 1972 Oceanic fine structure. *Geophys. Fluid Dyn.* **3**, 321–345.

Rossby, T., Price, J. & Webb, D. 1986 The spatial and temporal evolution of a cluster of SOFAR floats in the POLYMODE local dynamics experiment (LDE). *J. phys. Oceanogr.* **16**, 428–442.

Thomson, E. & Lovelock, J. E. 1976 A continuous and immediate method for the detection of SF_6 and other tracer gases by ECD in atmospheric diffusion experiments. *Atmos. Environ.* **19**, 917–920.

Watson, A. J. & Liddicoat, M. I. 1985 Recent history of atmospheric trace gas concentrations deduced from measurements in the deep sea: application to sulphur hexafluoride and carbon tetrachloride. *Atmos. Environ.* **19**, 1477–1484.

Watson, A. J., Liddicoat, M. I. & Ledwell, J. R. 1987 Perfluorodecalin and sulphur hexafluoride as purposeful marine tracers: some deployment and analysis techniques. *Deep Sea Res.* **34**, 19–31.

Discussion

D. J. Webb (*Institute of Oceanographic Studies, Surrey, U.K.*). Given that both satellite-tracked floats and computer models could be used to follow the general movement of a patch of released tracer, would it be possible to carry out an open-ocean experiment with little more than the amount of tracer required for the Santa Monica experiment?

A. J. Watson. Very-short-term experiments of a week or two might not take much more tracer than was released in the Santa Monica Basin, but they would probably not be very informative about rates of vertical mixing because they would not integrate sufficiently in space or time. An experiment that lasts long enough to be useful will have to be substantially larger in the open ocean than in a confined basin.

C. Wunsch (*Massachusetts Institute of Technology, U.S.A.*). Can Dr Watson be certain that the fluid that reached the boundaries was in fact seen in the interior measurements? One might imagine that there is a large-scale circulation, confined near the boundaries, such that fluid that was highly mixed never re-entered the interior, but either remained near the boundaries (Rossby radius confinement), or actually had left the Santa Monica Basin before Dr Watson returned.

It is difficult to understand otherwise, what is wrong with ideas about the importance of boundary mixing. Is the internal-wave field in the Santa Monica Basin 'normal'?

A. J. Watson. Although I have not presented the results here, on the second and third sampling cruises we spent about equal time sampling in the boundary region (operationally defined as up to 4 km from the walls) as we did sampling in the interior. We did not see evidence for dramatically enhanced mixing in these observations, so I think the answer is yes, we are reasonably confident that there was no great difference between boundary and interior mixing rates.

But I think we should be cautious about extrapolating to the open ocean. The basin is quite small with dimensions of the same order as the Rossby radius: so in a sense it is all boundary. Perhaps the internal wave field is not 'normal' but is much less effective at promoting mixing than is the open-ocean spectrum, and this compensates for the fact that boundary effects are tending to promote mixing. When the data are all analysed we should get a detailed picture of the internal motions; at the moment all I can say is that the amplitudes are of the same order as one might expect in the open ocean at an equivalent stratification.

Eclectic modelling of the North Atlantic. II. Transient tracers and the ventilation of the Eastern Basin thermocline†

By C. Wunsch

Center for Meteorology and Physical Oceanography, Department of Earth, Atmospheric, and Planetary Sciences, Massachusetts Institute of Technology, Cambridge, Massachusetts 02139, *U.S.A.*

Renewal rates of the waters of the thermocline in the eastern North Atlantic are estimated by combining linear quasi-geostrophic dynamics with steady and transient tracers into a unified eclectic, reservoir model. The two-dimensional model first employed is finally rejected when it is found that it generates oxygen-utilization rates (OUR) that are, by conventional biological wisdom, too high. The three-dimensional model that replaces the two-dimensional one shows that the OUR is indeterminate, with possible ranges from zero to unacceptably high values. The region is flushed primarily from the north and east.

The problem of using transient tracers is mathematically equivalent to that of distributed-system boundary-control theory, the open-ocean boundary conditions playing the role of the unknown control variables. The missing time histories of this new set of unknowns means that tritium and helium-3 distributions are only comparatively weak constraints on the flow field, but do set upper bounds on the vertical exchange with surface waters. Surface Ekman pumping is adequate to explain the interior distributions without additional buoyancy ventilation, although this latter process is possible. Some speculation is made about conditions under which transient tracers might play a more definitive role.

1. Introduction

We have termed 'eclectic modelling' the pooling of many different types of oceanographic observation into internally consistent models of the general circulation. Oceanographic observations seem inevitably disparate in type, duration and sampling strategy, rendering difficult the conventional comparison with model results. In pursuit of procedures for making the type of synthesis that is required, we examined in part 1 (Wunsch 1984*a*) the meridional heat flux of the North Atlantic Ocean by combining a variety of different observations and hypotheses into an eclectic model. The central data were the International Geophysical Year (IGY) zonal hydrographic lines, and estimates of the wind-stress driven Ekman flux. The central hypotheses were geostrophy and a quasi-steady state. Auxiliary observations and hypotheses involved a great many estimates of deep velocities in the ocean, the flux of the Florida Current, Mediterranean outflow, etc. Despite the several hundred constraints that were written down, the system remained underdetermined and the strategy for exploration was to seek upper and lower bounds upon the meridional flux of heat as a function of season and as an annual mean. The major feature of part 1, apart from the specific numerical estimates of flow and heat flux obtained, was the demonstration that it is possible to combine quantitatively the inevitably fragmentary and diverse measurements that characterize oceanographic observations.

The original intention had been that part 2 would combine all the data of part 1 with the

† Part 1 appeared in *J. phys. Oceanogr.* **14**, 1712–1733 (1984).

transient and steady-tracer data that is extant in the North Atlantic to further refine the estimates of heat flux and other flow properties. A number of events have led however, to deferral of that goal. Of most importance, an entirely new data base for the North Atlantic is gradually becoming available, a data base consisting of several higher density, higher quality hydrographic lines (see, for example, Roemmich & Wunsch 1985). These new lines also contain, as the IGY did not, a suite of accurate oxygen and nutrient data, as well as information on some of the transient tracers (tritium, helium-3, chlorofluoromethanes) (see, for example, Brewer *et al.* 1985). Because the labour involved in constructing an eclectic model is very great, it seemed best to delay further computations until this newer and more useful data set becomes fully available.

Further, experimentation with transient-tracer data suggested strongly that there was a great deal to be learned about how to employ data describing time-evolving tracer fields, where the issues of undersampling in time were clearly much more serious than for fields that on some large scale could be regarded as in a steady state. In particular, the combination of radiocarbon data with the eclectic model, as described in Wunsch (1984*b*), was not obviously extracting the full information content of the radiocarbon data.

A first step towards solving the general problem of making inferences from sparse transient-tracer information was taken in Wunsch (1987*a*). There, it was shown that the process involves solving three related problems: the forward problem, the inverse problem, and what was called the regularization problem. In brief, consider an advection–diffusion equation

$$(\partial C/\partial t)+\boldsymbol{u}\cdot\nabla C-\alpha\nabla^2 C = Q, \tag{1}$$

where α is a mixing coefficient and Q a source or sink of tracer C. The forward problem consists of determining C given $\boldsymbol{u}$, α, Q and appropriate (in the Cauchy–Hadamard sense) initial/boundary data. The inverse problem consists of determining $\boldsymbol{u}$, α and Q given C, including the initial/boundary data. In this second problem, in practice, one must estimate the unknowns when the available data for C are fragmentary and noisy. In Wunsch (1987*a*), it was shown that one was led in solving the inverse problem, to solving (1) in unstable directions, backwards in time and/or upstream; it is a forward problem for an unstable equation and hence is an issue of what in the mathematical literature is called 'regularization' (see, for example, Tikhonov & Arsenin 1977).

In this present paper, we seek to employ the general methods set out in Wunsch (1984*a*, 1987*a*) to study the ventilation rates of the eastern Atlantic. This is a region, the β-triangle area described by Armi & Stommel (1983), of considerable present attention as a testing ground for ideas about the processes and rates governing the renewal of near-surface water masses through exchange with the atmosphere. As a consequence, there exists an unusually complete set of estimates for both geostrophic velocity and of a suite of tracers, both steady and transient. The challenge is to combine these so as to exploit the information content of all of them to estimate ventilation rates, and their relative utility as a guide for future expeditions.

2. The data and the model

2.1. *The data*

Jenkins *et al.* (1985) have reported standard hydrographic variables and the tritium–helium-3 pair (^{3}H and ^{3}He) in the region depicted in figure 1. Potential temperature, salinity, oxygen,

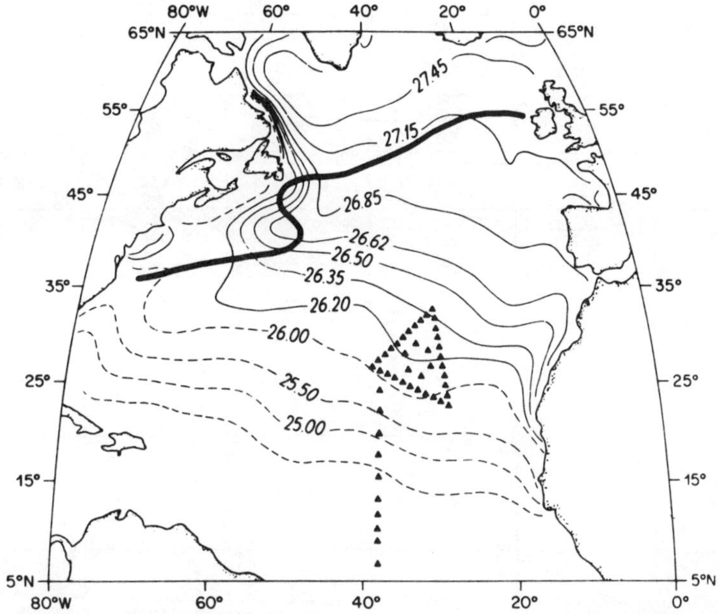

FIGURE 1. Geography of the station data used (taken from Jenkins 1987). The data in the 'β-triangle' stations were used to form a composite meridional section along 38° W as depicted in subsequent figures. The data used combined stations from late 1979 and early 1980.

^{3}H and ^{3}He, and the ^{3}H–^{3}He age are shown in figures 2–6 projected onto the meridional line at 38° W. (I am grateful to Dr Jenkins for making these data available, and for discussions of how best to use them.) The ^{3}H–^{3}He age τ is defined as

$$\tau = \lambda^{-1} \ln\left(1+\frac{[^3\mathrm{He}]}{[^3\mathrm{H}]}\right), \tag{2}$$

where $\lambda = 0.056\ \mathrm{a}^{-1}$ is the decay constant for tritium into ^{3}He. The square brackets mean 'concentration of'. τ is an *apparent* age since the water was in contact with the atmosphere (where $[^3\mathrm{He}] = 0$). The interpretation of τ is simple only if no mixing occurs (Jenkins 1987).

This meridional section will be the focus of the present study, i.e. we will address the oceanic circulation initially as though it were two-dimensional and as though the data were truly representative of the oceans. The two-dimensional model of the region will be shown to fail, in an interesting way, driving us to a fully three-dimensional one. Nonetheless, the two-dimensional model is sufficiently complex that it is worth examining first. In the spirit of eclectic modelling, we wish to bring to bear as much independent information as possible about the behaviour of the ocean in this region.

A number of direct estimates of velocity in this area exist. They include the β-spiral calculations of Armi & Stommel (1983), the climatological β-spiral estimates of Olbers *et al.* (1985), the inversions along zonal lines at 24° N and 36° N of Wunsch & Grant (1982) (from the IGY data) and of Roemmich & Wunsch (1985) for the modern (1981) sections. An examination of all of these estimates, keeping in mind the questions of representativeness of all of them, leads us to take as an *initial* estimate that the velocity at latitude 30° N (a latitude near the centre of the triangle) is as depicted in figure 7*a*, i.e. a linear function of depth, directed

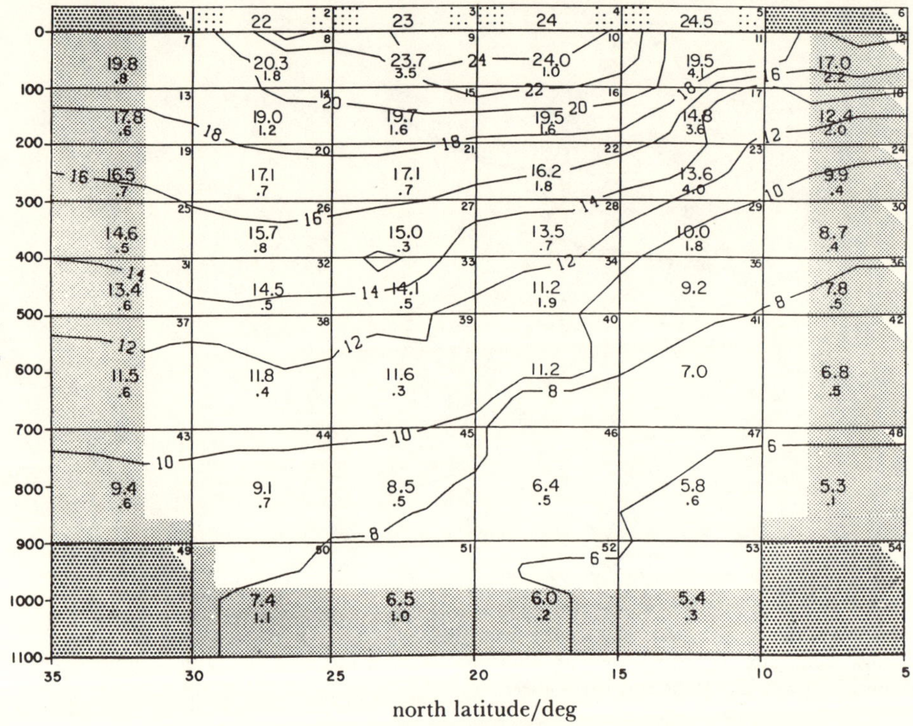

FIGURE 2. Potential-temperature contours along the meridional section. Little smoothing was done by the computer contouring routine (to show the real noisiness of the field). Large numbers in each box are the average temperature assigned to that box for purposes of the model. The smaller number underneath is the estimated standard deviation of the box mean, where the data were adequate to compute one. Values with no standard deviations tend to be interpolations from neighbouring regions. For reference purposes, each box has been given a simple number, in the upper right-hand corner. The boxes fall into five distinct categories. Corners (1, 6, 49, 54) play no role. Stippled boxes (13, 18, ...) are regions used to specify boundary conditions; there are no budgeting constraints applied. The top layer shown (2, 3, ...), is an artifice used to represent the very upper ocean (e.g. an Ekman layer). The thickness of this layer is indeterminate and carries, by way of example, the surface concentration of tritium. Clear boxes are the interior domain of the model for which various balance constraints are applied. The very surface tier, 8–11, is intermediate in that some constraints are imposed there (e.g. salt balance), but others (temperature) are not.

towards the south above 1100 m reaching a value of 1.5 cm s^{-1} at the surface. The formal uncertainty, as a subjective estimate based upon comparison of the above sources, is probably about 10% at any given depth.

Leetmaa & Bunker (1978) estimated the annual average Ekman pumping rate, w_E, in this area. Crudely scaling their figure 7, we estimate w_E in 5° latitude bands at this longitude as shown in figure 7*b*. It is difficult to estimate the uncertainty of these values. A preliminary guess is ±10%, to be changed later if necessary.

The boundary conditions for ^3H involve the description of the atmospheric transfer as a function of latitude and time. Dreisigacker & Roether (1978) and Weiss *et al.* (1979) have compiled estimates of North Atlantic surface concentration from 1952 to 1972. These are shown as a function of latitude and time in figure 8. Because the tracer surveys were made in the autumn of 1979 and late winter of 1980, the curves of figure 8 were linearly extrapolated, based upon the last two years of observation. These extrapolations brought the apparent surface concentrations of ^3H close to zero in 1976 at all latitudes of present concern. No model results are sensitive to this extrapolation.

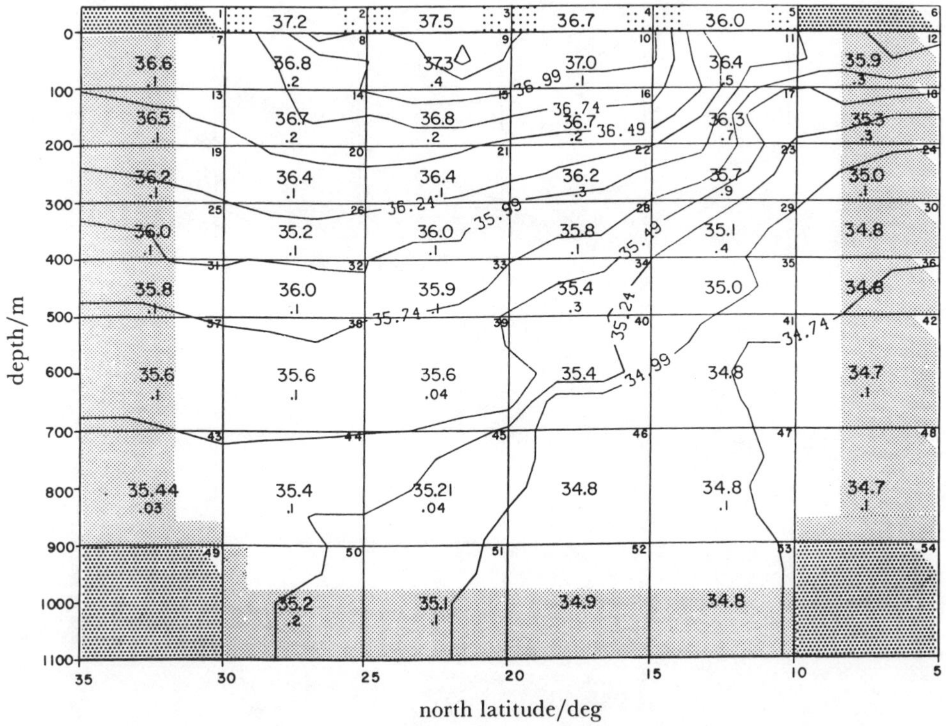

FIGURE 3. As in figure 2, but for salinity.

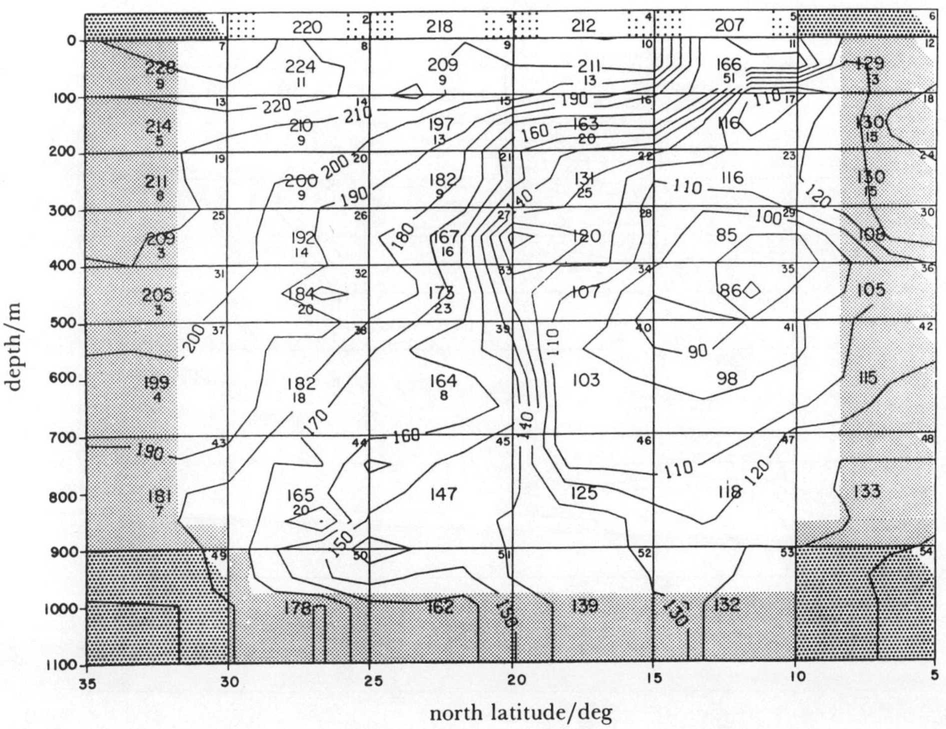

FIGURE 4. As in figure 2, but for oxygen concentration (in units of micromoles per kilogram).

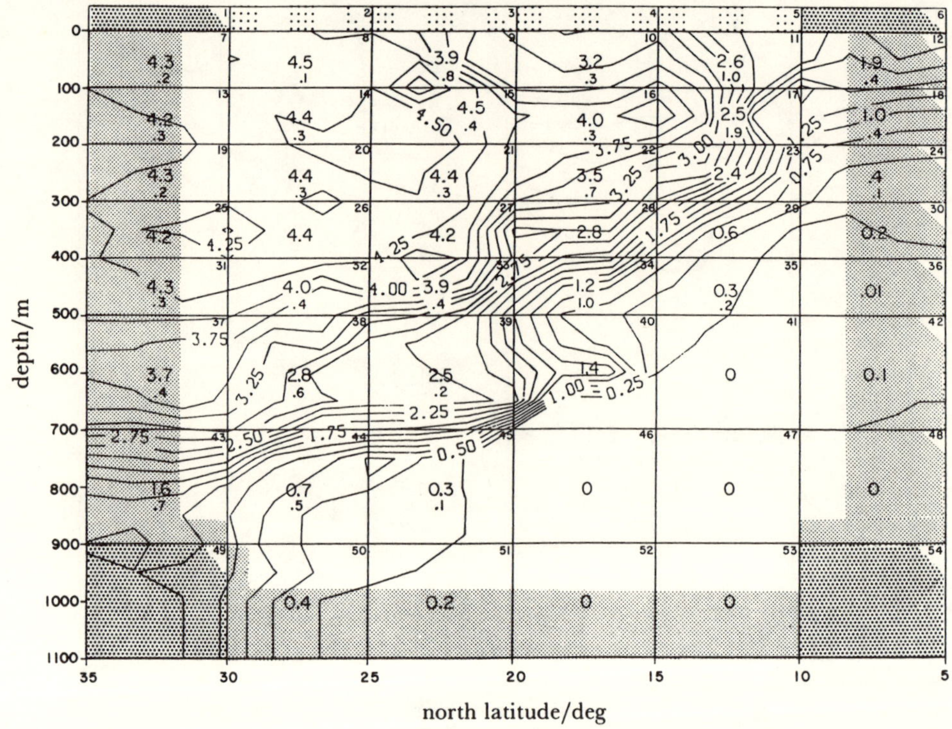

FIGURE 5. As in figure 2, but for the ^{3}H distribution in 1979–1980. Note the very sharp front.

(*a*)

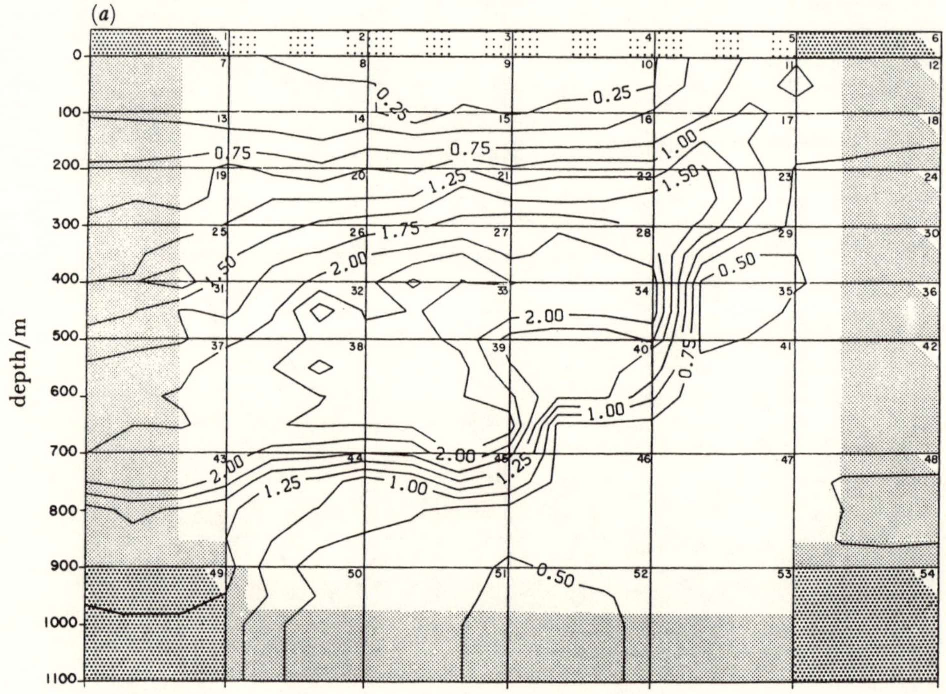

FIGURE 6(*a*). For legend see facing page.

(*b*)

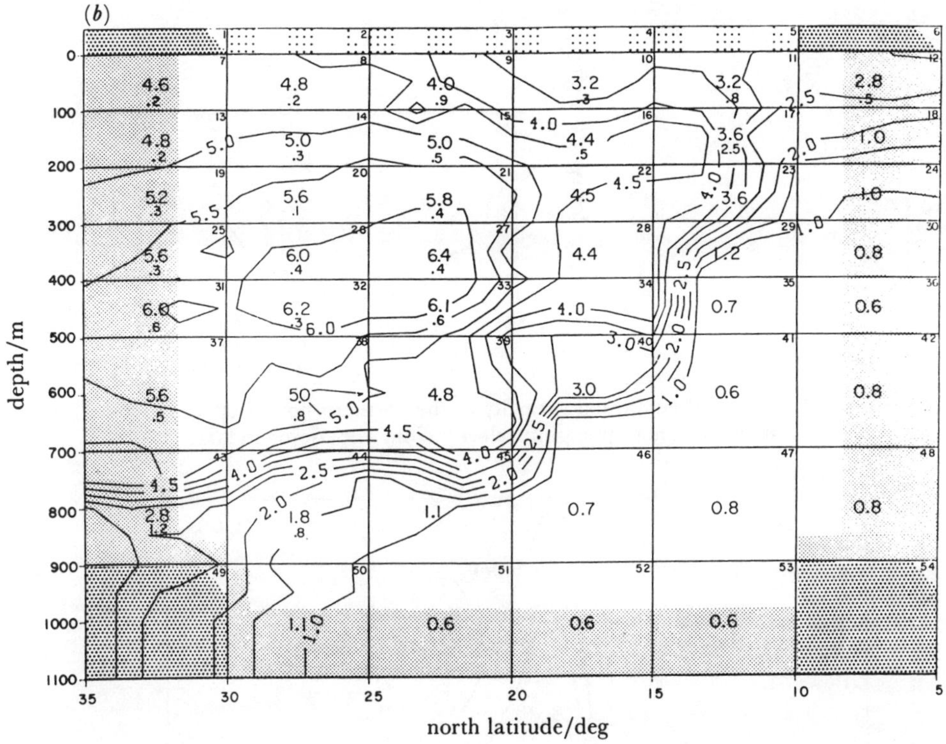

(*c*)

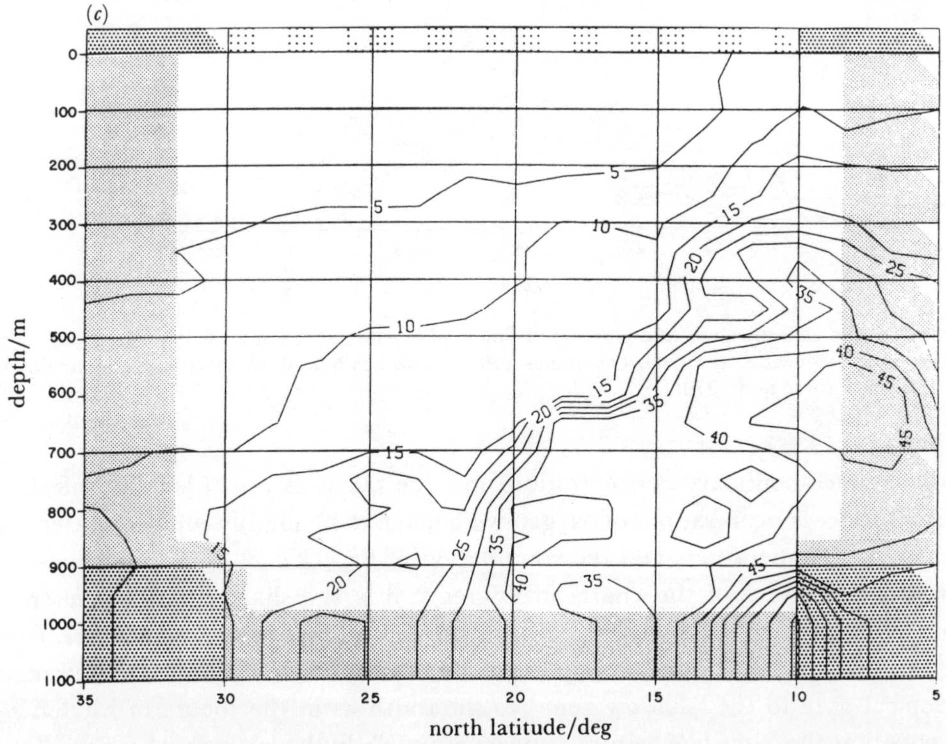

FIGURE 6. (*a*) The estimated ^{3}He distribution in 1979–1980. Box values were not assigned because ^{3}He was used only in the form of stable tritium. (*b*) The stable tritium distribution (see text). (*c*) The ^{3}H–^{3}He age (see text) of the water in the model in 1979–1980.

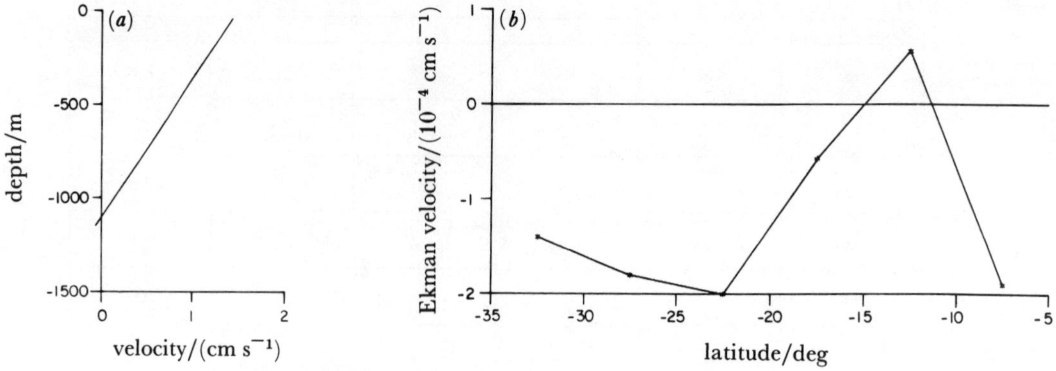

FIGURE 7. (*a*) Estimate of meridional geostrophic velocity at the northern boundary as a function of depth. (*b*) Estimate of Ekman pumping velocity along the top boundary.

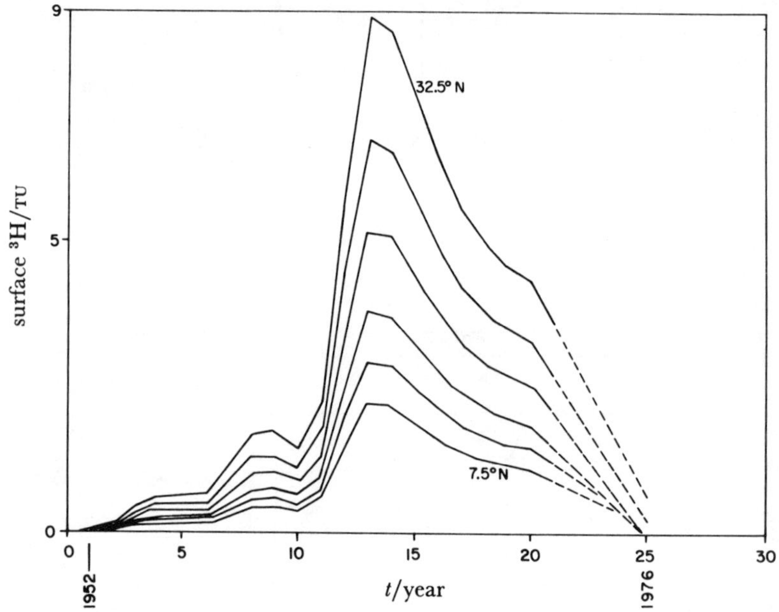

FIGURE 8. Surface tritium concentration as a function of time and latitude as taken from Dreisigacker & Roether (1978). The highest values are at the most northerly latitudes. Broken lines show extrapolation made between the latest published value and 1979.

For ^{3}He, the surface-boundary concentrations may be taken as zero (Jenkins 1987). The annual average surface concentrations of oxygen were taken to be slightly supersaturated, from the observed surface temperatures and the compilation of Broecker & Peng (1982).

The most striking feature of the charts in figures 2–6 is the sharp-front-like outcrop of properties occurring near 15° N discussed by Broecker & Östlund (1979) and others. Cox & Bryan (1984) suggest that the boundary represents the transition from the 'ventilated region' of the sub-tropical gyre to the 'shadow zone' to the south as in the theory of Luyten *et al.* (1983). But note that the front-like feature appears at precisely the latitude where the Ekman velocity estimate of Leetmaa & Bunker (1978) shows a band of Ekman suction rather than the pumping generally characteristic of a sub-tropical gyre. Whether this feature of the wind field is real or not, and if real, whether it is a cause or effect of the property front is indeterminate. We will make a limited effort to understand the extent to which the local wind field controls the tracer properties of the upper ocean in this region.

2.2. *The two-dimensional steady model*

Although the discussion of Wunsch (1988*a*) for the solution of (1) was in terms of a finite difference formulation, we have chosen here to simplify the model even further. Consider the sections in figures 2–6. As displayed, the section was divided into a series of boxes consisting of 5° of latitude and either 100 or 200 m depth intervals. To avoid twice differentiating noisy data we have opted to employ a box-model version of (1). In the chemical-engineering literature such a system is called a 'compartments model' (see Wen & Fan 1975) and is what Keeling & Bolin (1967) described as a 'reservoir model'. Following these latter authors, we abandon the attempt directly to separate advective flows from mixing effects, and simply describe mass transfers between boxes i and j by a single parameter $J_{i,j}$. By choosing to represent the transfer of properties, C, by $C_i J_{i,j}$, we are assuming that this underparametrization of the system will be adequate to describe all the data. We must remain aware that if incompatibilities arise, different properties are surely being mixed at different rates, and it may become necessary to introduce a more complete parametrization. The general philosophy remains as using models as simple as conceivable, until such time as they become in conflict with the data, at which point, more structure is added one step at a time. (Originally the problem was formulated in terms of conventional advection–diffusion as in Wunsch (1988*a*). But then the data reduction is much more complex, involving numerical laplacians, the dimensionality of the mixing field is very high and, in general, there is much more to keep track of. Given the crudity of the data a highly sophisticated form seems unreasonable; cf. the comments by Salmon (1986).)

Depth was used as the vertical coordinate rather than the perhaps more obvious isopycnal one, because it made possible the straightforward employment of the linear vorticity balance described below. Vorticity turns out to be among the more important constraints.

The observed property data, west of longitude 30°, were averaged over the depth range of each box depicted in the figures and this single number was assigned as the particular property value in the box. The variance of each of these averages was also recorded as an estimate of the formal uncertainty where enough data existed in a box to make the calculation useful. Both the box mean and the variances (where they were computable) are depicted in the figures. In addition, a set of surface-layer boxes (of indeterminate thickness), chosen to represent the surface property reservoir was overlain on top of the sections already shown. The obvious uncertainty over the behaviour of the Ekman layer – mixed layer – subduction layer has led us to simply regard the surface properties as defined by a reservoir of unknown physics and vertical exchange through pumping into the box below. The figures display these reservoir values.

2.3. *Conventional tracers*

The traditional tracers of oceanography are temperature, salinity and oxygen. Before grappling with transient tracers, we ask whether the observed fields of these classical tracers can be described consistently. Consider the θ, S and O (potential temperature, salinity and oxygen) fields in figures 2–4. With the time derivative term set equal to zero in (1), the demand is for a flow field $\boldsymbol{J}$ in the box model that satisfies the following constraints for any interior box, labelled i:

$$\text{conservation of mass in the form,} \quad \sum_{j \in N_i} J_{i,j} - \sum_{j \in N_i} J_{j,i} = 0,$$

$$\text{conservation of salt,} \quad \sum_{j \in N_i} S_i J_{i,j} - \sum_{j \in N_i} S_j J_{j,i} = 0,$$

$$\text{conservation of heat,} \quad \sum_{j \in N_i} \theta_i J_{i,j} - \sum \theta_j J_{j,i} = 0,$$

$$\text{conservation of oxygen,} \quad \sum_{j \in N_i} O_i J_{i,j} - \sum O_j J_{j,i} = 0,$$

where $j \in N_i$ denotes the set of indices of the boxes neighbouring box i. All $J_{i,j} \geqslant 0$.

Mass is conserved in each interior box, and for reasons to be made clear presently, an additional statement of mass conservation into the entire interior region was added. Salt conservation was implemented in the same way. Heat conservation was applied only to those boxes below the top two levels.

Finally, oxygen conservation in a steady state was *not* assumed, but a sink term, Q, the oxygen utilization rate (OUR), was introduced for each box as a new set of unknowns. The oxygen boundary conditions were set as slightly supersaturated, as already described.

The final set of relations employed in this initial steady calculation was obtained from the physical estimates of mass flux: the Leetmaa & Bunker (1978) estimates for Ekman pumping of figure 7*b* were imposed at the base of the top layer of boxes. Finally, the geostrophic velocities of figure 7*a* were imposed at the box boundaries at 30° N.

The model was then extended to include the linear vorticity balance

$$\beta v = f \, \partial w / \partial z \tag{3}$$

whose validity in the region is discussed by Armi & Stommel (1983) and Olbers *et al.* (1985). This relation can be applied only approximately. Integrating (3) in the vertical,

$$\beta J'_{i-1,i} = f[J'_{i+m,i} - J'_{i,i-m}], \tag{4}$$

where $J'_{i-1,i}$ represents the net flow into box i through its northern boundary, $J'_{i+m,i}$ the flow from the box below, and $J'_{i,i-m}$ the flow into the box just above. In the top tier of boxes, $J'_{i,i-m}$ was set equal to the Ekman influx from the surface layer; in the northern tier of boxes, $J'_{i-1,i}$ was set equal to the geostrophic flow already described. Each of the J' is of the form

$$J'_{i,j} = J_{i,j} - J_{j,i}, \tag{5}$$

i.e. representing the net flow across the particular boundary into the box.

Wunsch & Roemmich (1985) have questioned the validity of the Sverdrup relation in oceanic modelling, suggesting that evidence for its quantitative applicability was thin. They concluded that Sverdrup balance was unlikely to hold on an Atlantic-wide basis, and that *quantitative* demonstration of a point-wise validity was probably not possible with existing data, but that it could not be disproved either. The present constraint, with w set equal to the Ekman flux, is a demand for approximate Sverdrup balance, and what we are doing is testing it locally. Wunsch & Roemmich (1985) did suggest that at least in the eastern North Atlantic, Sverdrup balance might be plausible in the upper ocean, the chief difficulties coming as one went into the abyss with the bottom-induced w probably becoming dominant. The present data set lies in the thermocline and above, and the relation should work here if it applies anywhere. In the present test, no reason was found to reject (4), although that is a very different matter from proving its validity.

None of the constraints just described can be expected to hold exactly. There are errors expected in every one of them arising from observational inaccuracies (e.g. the geostrophic flow), to model incompleteness (e.g. the assumption of steady two-dimensional flow), or both. Thus for every constraint we make an estimate of the error involved to the extent possible. The bounds used are listed in table 1.

TABLE 1

	row constraints	
mass		
each box	0.05×10^9 kg s^{-1}	
overall	0.025×10^9 kg s^{-1}	
salt		
each box	0.15×10^9 kg s^{-1}	
overall	0.1×10^9 kg s^{-1}	
temperature		
each box	0.5 °C $\times 10^4$ kg s^{-1}	not in boxes 7–12 not in three dimensions
oxygen		
production in boxes 8–11		
consumption in boxes below 8–11		
geostrophic velocity		
northern-boundary boxes		10% of nominal (figure 7*a*)
eastern boxes		10% of nominal (figure 12) three dimensions only
Ekman fluxes		
top tier to second tier		10% of nominal (figure 7*b*)
	column constraints	
all fluxes positive		
all fluxes less than about 10×10^9 kg s^{-1}		

The resulting system consisted then of 128 budgeting, linear-vorticity, and northern-boundary geostrophic constraints in 186 unknowns, $J_{i,j}$, subject to the positivity constraints $J_{ij} > 0$ and an overall constraint that the horizontal velocity equivalent of $J_{i,j} \leqslant 6$ cm s^{-1} in the 100 m thick upper layers (3 cm s^{-1} in the deeper, 200 m thick layers). The resulting system, being underdetermined, was solved by linear programming in the same manner as in Wunsch (1984*a*, *b*). Because of the nature of linear programming solutions, there is a tendency to push all interior-box balances to limits of one sign; to prevent this behaviour, strong constraints requiring overall mass and salt conservation were added. The accounting for constraints is necessarily only approximate, because all constraints have inequalities applied, and others, such as the demands on the Ekman fluxes and on the sign of the OUR, have not been counted.

To make a connection with more conventional models, consider the following. We regard the total flux across an interface as the analogue of the dynamical velocity, u

$$\rho A u = J_{ij} + (-J_{ji}), \tag{6}$$

where A is the area of an interface and ρ is the density (recall that all J_{ij} are positive). The net flux, F, of a tracer, C, across that interface is

$$F_{ij} = C_i J_{ij} - C_j J_{ji}. \tag{7}$$

For those who like to think in terms of eddy coefficients for tracer flux, K, the net flux in a conventional tracer conservation equation is

$$F_{ij} = (\rho C u - \rho K\,(\partial C/\partial n))\,A, \tag{8}$$

where $\partial C/\partial n$ is the normal derivative. Equating (7) to (8) and with the definition (6), we obtain

$$K\frac{\partial C}{\partial n} = \frac{(C_j - C_i)}{2\rho A}(J_{ij} + J_{ji}) \tag{9}$$

having put $C = \frac{1}{2}(C_i + C_j)$. 'Mixing' will be large where $C_j \neq C_i$, and advection will dominate where they are comparable. Minimizing $\Sigma_{i,j}(J_{ij} + J_{ji})$ can be thought of as a search for a minimum mixing solution.

There is a mixing within each box, of course, that in our formulation is sub-grid scale. This mixing includes an implicit component owing to the upstream differencing we are using (Roache 1976). The emphasis in this paper is, however, on integral properties (e.g. the oxygen consumption rates), and we will not focus on distinctions between advective and diffusive transports, much of which can be regarded as largely semantic anyway.

When initially run, the system was 'infeasible', i.e. had no solution at all capable of satisfying all the constraints, despite the considerable under-determinancy. The infeasibility was resolved by permitting the net meridional geostrophic flow from box 7 to 8 to be reduced to $0.5\ \mathrm{cm\ s^{-1}}$. In view of the origins of these estimates, this variation was deemed physically acceptable. A solution to the system is shown in figure 9 for which $\Sigma_{ij} J_{ij}$ was minimized.

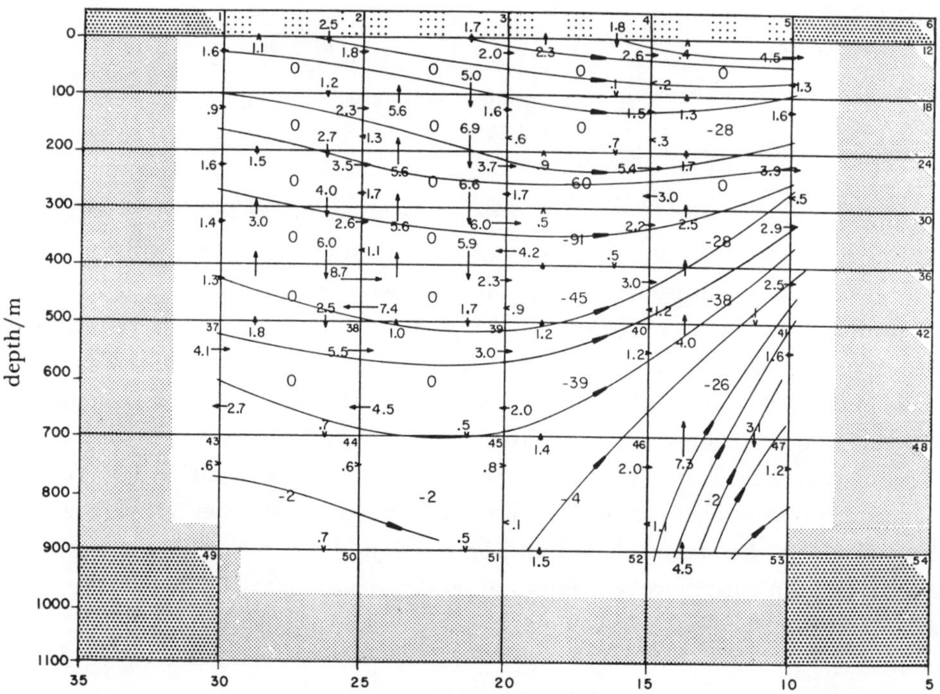

FIGURE 9. Flow in two-dimensional model with objective function minimizing the overall sum of the J and the meridional average OUR as a function of depth. Arrows are labelled with transports in Sverdrups and a schematic interpretation of the net flow (J') across each interface in terms of a stream function is displayed. Contour interval is approximately 1 Sverdrup. Oxygen utilization rate in units of micromoles per kilogram per year is displayed in the centre of each box (negative numbers corresponding to consumption). (1 Sverdrup = $10^6\ \mathrm{m^3\ s^{-1}}$.)

2.4. *First results of the steady model*

To recapitulate, this purely steady model demands consistency with the meridional geostrophic flux at the northern boundary, estimates of the Ekman convergence – divergence at the sea surface, overall mass and salt conservation, mass, heat and salt conservation within the interior boxes, and the linear vorticity balance within each box. For oxygen, it was demanded that there should be a net injection of oxygen into the top tier of boxes, and below that tier, oxygen should only be consumed. Because of various errors of observation, zero oxygen consumption (production) was deemed observationally equivalent to net production (consumption) of up to 0.5 μmol kg^{-1} a^{-1}.

With these constraints as just described, there is no difficulty in finding a solution; indeed, infinitely many solutions exist. Two of them are shown in figures 9 and 10. However, it was quickly realized that all the flows that were found to be consistent with these constraints had very large oxygen consumption below the surface tier, with values reaching higher than 70 μmol kg^{-1} a^{-1} in some boxes, and overall averages below 200 m of 15 μmol kg^{-1} a^{-1}. The values of OUR have been discussed most recently by Jenkins (1987) and Sarmiento *et al.* (1988).

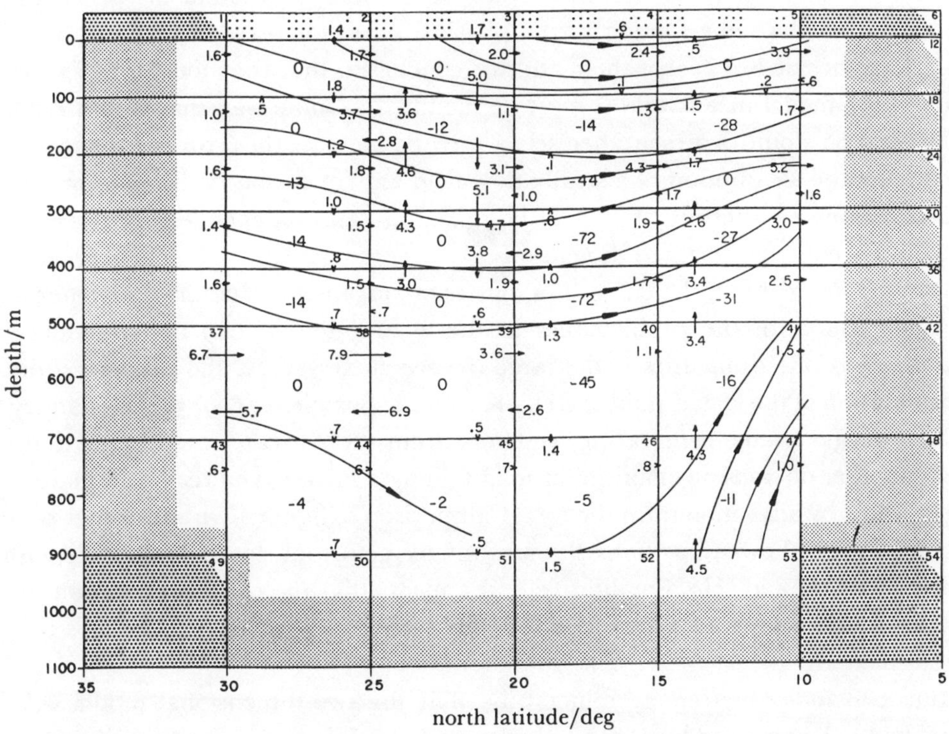

FIGURE 10. As in figure 9, but for a solution in which an attempt was made to minimize the OUR below 200 m.

It would appear that: (a) the actual rates of oxygen consumption (OUR) are very poorly known and, that (b) the biological oceanography community has difficulty in accepting rates even as high as 10 μmol kg^{-1} a^{-1}. Given (a), one might question whether (b) should be given very much weight. However (W. J. Jenkins, personal communication 1987), there do appear to be serious questions as to whether an adequate carbon flux is available to support a very high OUR, and it thus became important to see whether the OUR could be driven down to rates consistent with biological conventional wisdom. This question was explored by making the linear programming objective function (cost function) equal to the average OUR below 200 m and seeking the minimum possible value.

A solution that resulted from the minimum value is depicted in figure 10. The average OUR below 200 m is 14 μmol kg^{-1} a^{-1}. The value of the minimum is unique (up to the constraints imposed), however the flow giving rise to that minimum is not unique. But in some individual boxes the OUR remains over 70 μmol kg^{-1} a^{-1}.

The origins of this very high OUR are unmistakable if figure 10 is compared with the oxygen distribution in figure 4. What is happening is that the geostrophic plus vorticity constraints drive a very strong flow across the oxygen front lying at about 15° N. Thus high-oxygen water is being transported into a region of low oxygen concentration and the only way to maintain the low oxygen concentration is to rapidly consume the incoming flux. Without removing or modifying these dynamical constraints, there is no way to reduce the required consumption.

If we accept the conventional biological wisdom (which says that even Jenkins's (1987) best estimates of 10 μmol kg^{-1} a^{-1} are too high), then the present model must be deemed to have failed. Perhaps the most interesting aspect of the failure is that it only occurs when we couple together the dynamical constraints with a quasi-biological constraint. Either set of constraints by itself would be self consistent. The most troubling question is whether the OUR constraints should be placed on the same footing as the physical ones; for present purposes, with some hesitation, I have decided to accept them and the conclusion that the model must be rejected.

Rejection of the model means that it must be modified in some way that would reduce the high OUR. Several possibilities present themselves: greatly increase the error bars on the physical constraints; relax the assumption of a steady flow field and/or property distribution; increase the parametrization (e.g. permit oxygen to be mixed at rates independent of those for temperature, salinity, etc.); go to three dimensions.

As with any model making, the choice is a matter of judgement. Here my judgement is that the most likely source of the problem lies in the assumption of two dimensionality. The immediate origin of the difficulties is the large oxygen fluxes across the property front. But cursory examination of the charts published by Kawase & Sarmiento (1985) shows a very large tongue of low-oxygen water penetrating westward from the African upwelling region to the east. Here is another obvious mechanism for maintaining a low-oxygen region: replacement of high-oxygen water by advection from the east. Comparison of figure 4 with the other property-distribution diagrams shows that the position of the oxygen front does not coincide with that of either density or ^{3}H and ^{3}He. Qualitatively, it appears that the oxygen minimum lies near the axis of the dynamical front, consistent with the view that low-oxygen water is being fed along it from the east.

It was thus concluded, extremely reluctantly, that the two-dimensional model was indefensible, and had to be replaced by a three-dimensional one. The decision was made reluctantly because the computational load in three dimensions is much greater than in two (especially when time variation is added in §4) and the book-keeping burden grows enormously. Furthermore, the addition of property fluxes in the third dimension increases the parametrization of the model so that the ratio of constraints to unknowns decreases. Nonetheless, the conclusion was that a three-dimensional flow field was inescapable.

3. The three-dimensional steady model

The three-dimensional model was made as simple as seemed plausible while still resolving the OUR contradiction of the two-dimensional flow. The model is shown as depicted schematically in figure 11. The original section is now envisioned as representing a set of three-

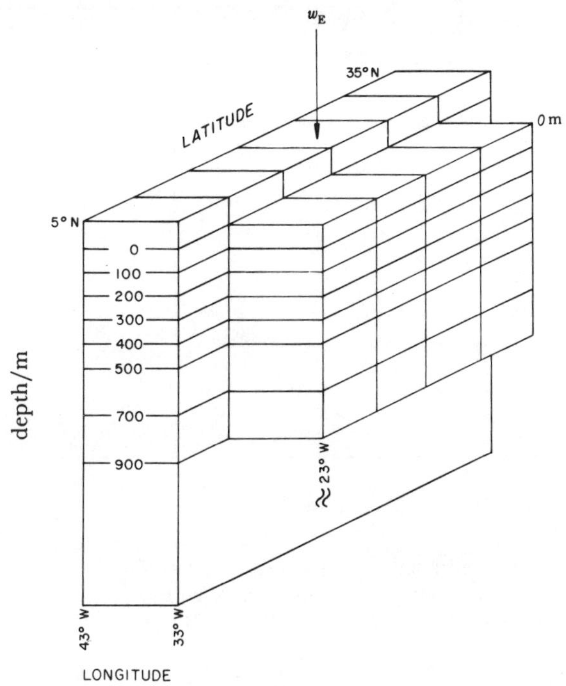

Figure 11. Schematic diagram of the three-dimensional model in which the eastern tier of boxes feeds fluid into the interior boxes to its west at rates determined by the meridional geostrophic shear. A third group of boxes, to the west, is not displayed and serves only as a receiving reservoir for the central tier. No backflow from west to east is permitted.

dimensional boxes sandwiched between the two other such sets of boxes to the east and west. The zonal extent of each of the central boxes was taken as 5° to the east and west of the original section as depicted. The hydrographic data of Jenkins *et al.* (1985) from the region shown in figure 1 were used to estimate the zonal thermal wind relative to the deepest available station pairs as shown in figure 12. A value of the thermal wind was assigned to the zonal flow in each box and was assumed to be the same on both eastern and western boundaries. For each column of boxes, an unknown reference level velocity, b_p, was added to the system. Only flows from east to west were permitted; thus the western tier of boxes is a purely passive recipient of westward flows from the central tier. Divergence in b_p is permitted across the central tier.

The properties assigned to the eastern tier were taken from the charts of Kawase & Sarmiento (1985). These values must be regarded as crude approximations to the true average properties. The constraints of the two-dimensional model were thus modified as follows. Mass, salt and oxygen conservation constraints in each box, and overall, include a zonal flux of properties from the east composed of the known thermal wind plus unknown reference-level velocities. The heat-balance equations were made free (i.e. non-binding) as they are nearly redundant if both mass and salt are conserved. The vorticity constraints are unchanged as (3) does not

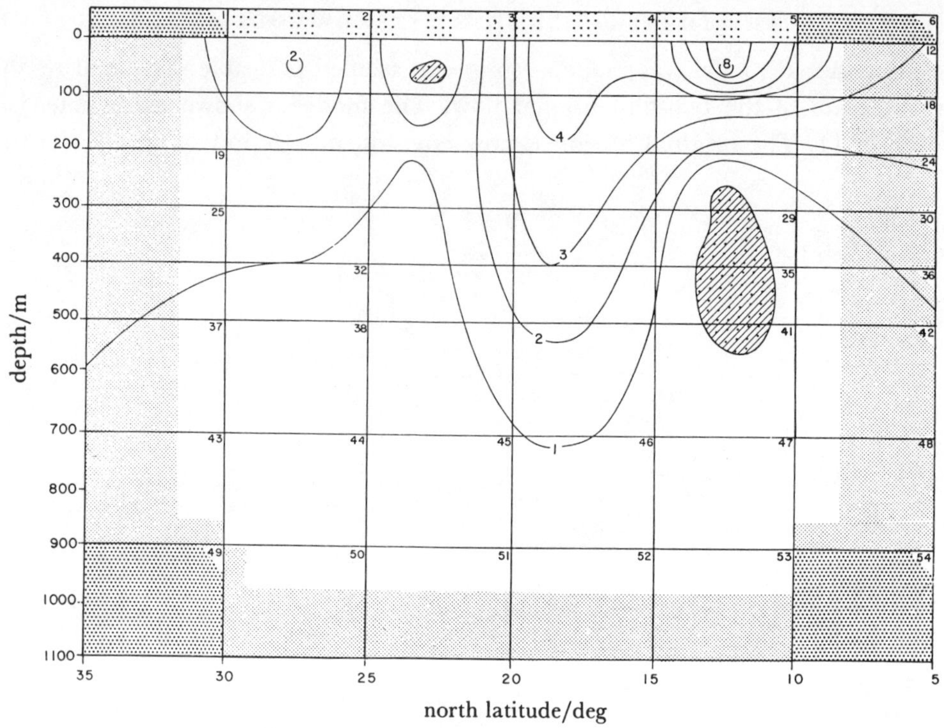

FIGURE 12. Geostrophic flow (in units of centimetres per second) used to drive the east-to-west fluxes. An unknown reference-level velocity is part of the system unknowns in each vertical column of boxes. The reference-level velocity was forced to be sufficiently large so as to drive the small region of eastward flow (stippled) to the west.

involve the zonal flux directly (but the meridional and vertical velocities will be changed from their previous values by the presence of a zonal divergence).

This simplified three-dimensional model nominally contains 181 constraints (each with error bars) in 195 unknowns (all the previous ones plus 8 zonal reference level velocities). When run, it was completely consistent and the problem with the OUR disappeared completely: there are consistent solutions in which there is now no net OUR at depth (within the error bars). No changes had to be made in the northern geostrophic flux; the values in figure 7*a* were used within the stipulated 10% error. This result might be thought of as a lesson in the difficulty of deriving oceanographically useful numbers from models of dimension lower than three.

With this consistent steady-flow model in hand, we now turn temporarily to the transient-tracer problem, returning later to examining the ventilation rates in the eastern North Atlantic.

4. Transient tracers

Consider the original tracer equation (1). If the velocity and mixing coefficients are specified along with the source/sink terms, this equation permits computation of the tracer distribution $C(x, t)$ in any domain if, and only if, appropriate boundary conditions are specified. We could compute the distribution of tracer C in any one of the boxes, i.e. $C_i(t)$ for any time t if all the J_{ij} are known, and if $C_{i0}(t)$ are known in all the boundary boxes, $i_0 \in B$, from which a flow occurs. Furthermore, a set of initial conditions $C_i(0)$, all i is required. That is, the reservoir model is

$$\partial C_i/\partial t + \sum_{j\in N_i} C_i J_{i,j} - \sum_{j\in N_i} C_j J_{j,i} - Q_i = 0, i\in I, \tag{10}$$

$$C_{i0}(t) = \bar{C}_{i0}(t), i_0 \in B, C_i(0) = \bar{C}_i(0), i\in I,$$

where $i_0 \in B$ is the index of all boundary boxes having a flow into the interior, $i\in I$ is the set of all interior boxes, $j\in N_i$ is the set of all neighbour boxes of box i. The forward problem is well-posed, and conventionally solvable, if and only if the governing evolution equation (10), a full set of boundary value time histories, and initial conditions, are available.

Consider a single transient tracer $C = {}^3H$. At the surface of the models in figures 5 and 11, the boundary history is given approximately by the curves in figure 8. Denote these special surfaces boxes as $i\in B_S$, and the remaining boundary boxes as $i\in B_I$. These other boundaries, $i\in B_I$, north, south, east and bottom, are in the open ocean. Apart from a few isolated observations, taken up below, our only available observations are the values on $i\in B_I$ in 1979–1980, plus a set of interior observations, $i\in I$, in that same year. We may reasonably suppose that the initial conditions $C_i(0) = 0$ everywhere, where zero concentration was certainly appropriate before 1950. Whether it was still an appropriate initial condition as late as 1963 (say) is less obvious.

Here is the crux of the transient-tracer problem as defined by a data set like the present one. Even under the drastic and sweeping assumption that all of the $\boldsymbol{J}$ are steady and known exactly, the absence of the open-ocean boundary data (plus the uncertainty in the sea-surface boundary conditions) means that the forward problem is not fully posed and the extent to which inference could be made about the $\boldsymbol{J}$ (which after all is our real goal) from ${}^3H$ is not clear at all.

We now formalize the problem under the assumption that the $\boldsymbol{J}$ are known. Discretize in time as well as space, and write the system as

$$C_i(n+1) = \left[1-\lambda\Delta t - \sum_{j\in N_i} J_{i,j}\frac{\Delta t}{V_i}\right] C_i(n) + \sum_{j\in N_i} J_{j,i}\frac{C_i(n)\,\Delta t}{V_i}$$

or

$$\boldsymbol{C}(n+1) = \boldsymbol{AC}(n) + \boldsymbol{Bq}(n), \quad t = n\Delta t,\ t_f = n_f\,\Delta t, \tag{11}$$

where Δt is the time step, λ is the decay constant (1/17.7 a) for ${}^3H$, V_i is the volume of box i,

$$A_{ii} = \left[1-\lambda\Delta t - \sum_{j\in N_i} J_{i,j}\frac{\Delta t}{V_i}\right], i\in I,$$

$$A_{ij} = \left[J_{ji}\frac{\Delta t}{V_i}\right], j\in N_i,$$

$$\{\boldsymbol{Bq}\}_i = \bar{C}_i(n), i\in B_I \text{ or } B_S.$$

The matrix $\boldsymbol{A}$, when multiplied by the tracer distribution at time n, gives the distribution one time step later if the boundary values vanish (I will ignore the distinction between n and t). The quantity $\boldsymbol{Bq}$ represents the boundary conditions of this problem; this separation of boundary and interior components is made for convenience.

Then, subject to the evolution equations (11), we must find the time histories $\boldsymbol{C}(n)$ for the interior boxes, subject to partial initial/boundary data. In the classification system of Wunsch (1988*b*), this problem is that of 'state estimation'; it is *not* an inverse problem.

By writing the problem in the form (7) we have emphasized a point taken up at length elsewhere (Wunsch 1988*b*), that the transient-tracer problem coincides mathematically with the problem of control of distributed systems, in which the missing data correspond to the need to find a 'control' function. Many of the results of control theory become available for interpreting the present problem. In particular, the tracer observations at time t_f, corresponding to 1979, represent what is called a 'terminal constraint': the time evolution of the system from the initial conditions, under the 'control', or boundary conditions, must result in the tracer concentrations observed at time t_f subject to various demands on the system state $\boldsymbol{C}(n)$ and on the control functions. The formal control problem proceeds by establishing a cost or objective function in general form (see, for example, Luenberger 1979)

$$H = \psi(\boldsymbol{C}_{t_f}) + \sum_{t=0}^{t_f} L[\boldsymbol{C}, \boldsymbol{q}], \tag{12}$$

where L is any function of $\boldsymbol{q}$ and $\boldsymbol{C}$, often quadratic, and ψ is the terminal constraint demand. General discussion of this system is taken up in Wunsch (1987*b*); here the discussion is confined to specifics.

The formal unknowns of this problem are $\boldsymbol{C}(t)$ and $\boldsymbol{q}(t)\ 0 \leqslant t \leqslant t_f$; the solution is rendered unique by driving the cost function to an extreme (as was done above for the steady problem). In the discussion in Wunsch (1988*a*) the problem was formulated by writing the cost function as (typically)

$$\min: H = \sum_{i \in B} \sum_{n=0}^{n_f} q_i^2(n)$$

or

$$\min: H = \sum_{i \in B} \sum_{n} |q_i(n+1) - q_i(n)|.$$

In Wunsch (1988*a*) the entire system of (11) was solved explicitly in the whole time-space domain for all the unknowns at once by using linear programming. This procedure is practical for the one-dimensional (diffusive pipe flow) situation described there. In three dimensions the system size is unwieldly. More troublesome is the huge disparity between the number of calculations required to solve the model and the very small number of actual observations to which we can compare the result. It seems unreasonable that thousands of separate numerical operations must be carried out to test the consistency of a handful of tracer observations from 1979–1980.

We are led to try and reduce the number of degrees of freedom in the model to be more consistent with the number of testable observations.

4.1. *Stability and system size*

The first problem is determining the size of the time step Δt. This value is governed by the stability of the time evolution of (11). The criterion can be derived on simple physical grounds as the demand that in any time step of size Δt, that the tracer concentration in none of the boxes become negative. Examination shows that this requirement in turn demands that the elements of $\boldsymbol{A}$ should be non-negative:

$$1 - \lambda\Delta t - \sum_{j \in N_i} J_{i,j} \frac{\Delta t}{V_i} > 0, \quad \text{or} \quad \Delta t < 1 \bigg/ \left(\lambda + \sum_{j \in N_i} J_{i,j} \frac{1}{V_i}\right). \tag{13}$$

This criterion determines the computational load. For values of J_{ij} found in part 2 above, we find that Δt normally must be less than about one month. If (13) is satisfied, the system (11) is one of non-negative system matrix $\boldsymbol{A}$, permitting the application of powerful theorems (Luenberger 1979) that describe the system behaviour; these will not be pursued here. The paradox of transient tracers is that we have no confidence at all in details of the input on monthly time scales, but it appears that we must nonetheless specify $\boldsymbol{q}$ at monthly intervals and time step the system repeatedly to arrive at the observations in 1979.

To reduce the system size the following procedure was adopted. Let the tracer concentration in boundary box i_0 be unity at the time $t = 0$ for one physical year and zero thereafter. We then compute the evolution of the tracer through the system for all times $0 \leqslant t \leqslant t_f$. Call the result $\boldsymbol{g}_{i_0}(n)$ and make the computation for all boundary boxes. The computation must be carried out for Δt chosen in accord with (11); but the results of the calculation were then averaged year by year to yield the set $\bar{\boldsymbol{g}}_{i_0}(t = n\Delta t') \equiv \boldsymbol{G}_{i_0}(n)$ where $\Delta t' = 1$ year. Two examples are shown in figure 13.

Given $\boldsymbol{G}_{i_0}$ (which depends directly upon the J_{ij}), the general solution of time t_f can be written as

$$\boldsymbol{C}(t_f) = \sum_{i_0 \in B} \sum_{n'=0}^{n_f} \alpha_{i_0}(n')\, \boldsymbol{G}_{i_0}(n-n'); \tag{14}$$

$\boldsymbol{g}_{i_0}$ or $\boldsymbol{G}_{i_0}$ will be recognized as a special form of Green's function; it is related to the solution of the adjoint of system (11) and is the key to system sensitivity testing. The α_{i_0} are the concentration time histories in each boundary box and are the new unknowns.

We now have reduced the formal unknowns of the problem from the complete internal history, $\boldsymbol{C}(n)$ plus boundary history $\boldsymbol{q}(n)$, to just the time histories of the boundary boxes alone. Computation of the $\boldsymbol{G}_{i_0}$ for the present problem suggested that the system tends to flush out all significant memory of previous states in less than about 15 years. If we take 15 years as the duration of the computation and permit the boundary concentrations to vary independently each year and independently in each boundary box, there are 15 years* (18+28 boundary boxes) = 690 unknowns plus the boundary values in the sea-surface boxes (which are partly uncertain). This number accounts for the fluid being fed from the eastern boxes in the full three dimensional calculation. Because the tritium data are represented by 28 interior values in 1979, it still seems unreasonable to have to work with so many unknowns and only 28 constraints. The problem was therefore further reduced in degrees of freedom by writing the concentration time history in each boundary box as

$$C_{i_0}(t) \equiv \alpha_{i_0}(t) = \gamma_{i_0,1} f_1(t) + \gamma_{i_0,2} f_2(t) + \gamma_{i_0,3} f_3(t), \quad t = n\Delta t', \tag{15}$$

with the coefficients γ_{i_0} as constants. For tritium this form still demands 3*(18+28) = 138 unknowns. The three f_i can be chosen as any convenient set of functions. Here they were the coefficients of the Taylor series expansion:

$$f_1(t) = t, \quad f_2(t) = \tfrac{1}{2}t^2, \quad f_3(t) = \tfrac{1}{6}t^3. \tag{16}$$

This reduced form is known as 'parametric control' (see, for example, Stengel 1986) and it clearly restricts the range of optimization of the cost function.

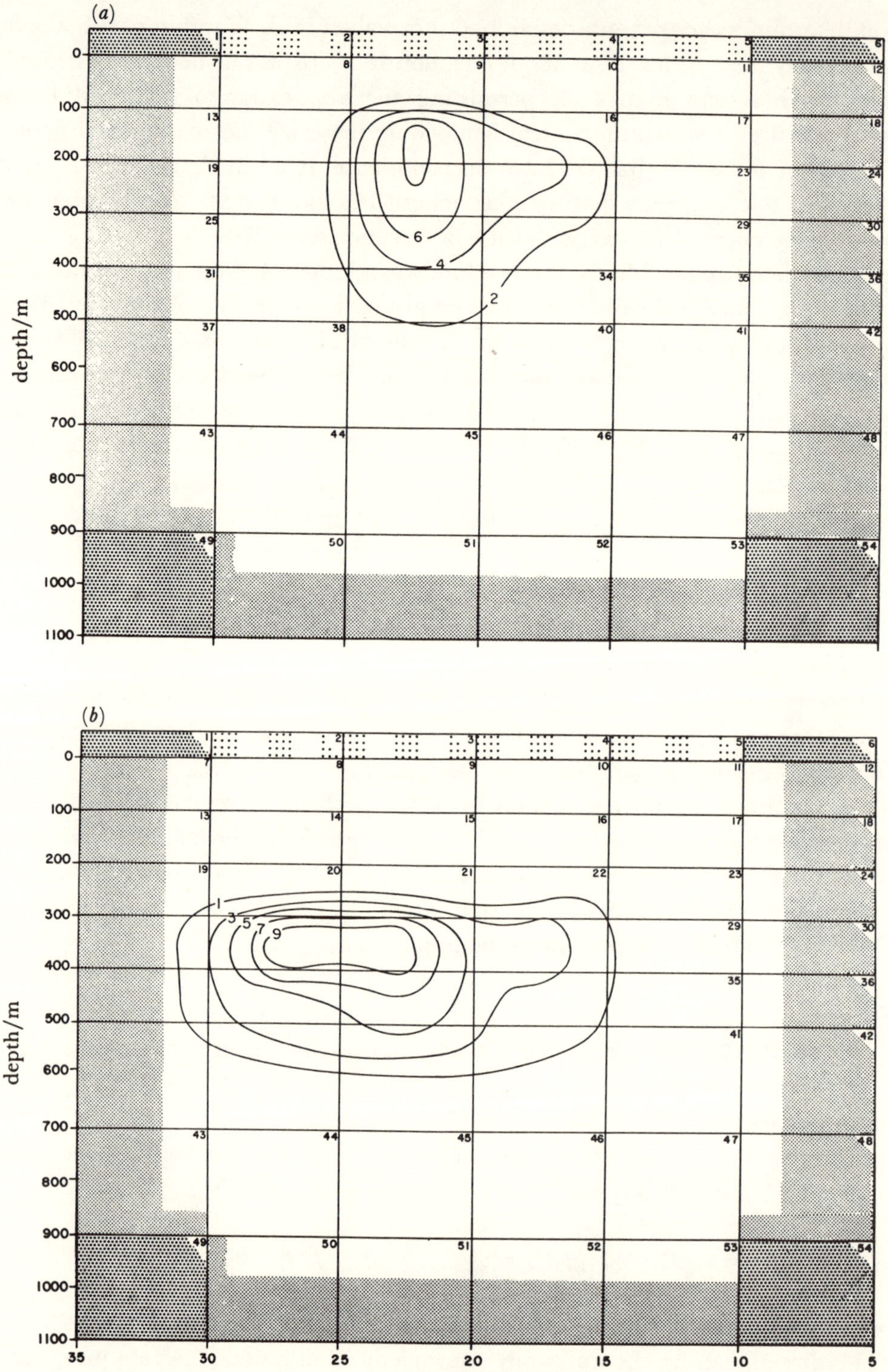

FIGURE 13. Green's function showing the tracer concentration that results when unit value is introduced into boundary boxes 3 (figure 13*a*) and 25 (figure 13*b*) after 5 years, for a given flow field (see figure 16).

In terms of (11), the vector

$$\boldsymbol{q}(t) = \begin{bmatrix} f_1 \\ f_2 \\ f_3 \end{bmatrix},$$

and the matrix $\boldsymbol{B}$ has non-zero rows for $i_0 \in B$, consisting of $[\gamma_{i_0,1}, \gamma_{i_0,2}, \gamma_{i_0,3}]$ plus appropriate zeroes.

What should be apparent at this stage would be the very great underdeterminacy, if the transient-tracer inverse problem were being posed. Discussion of the inverse calculation is deferred to the end, but it is worth making two points at this stage. The transient-tracer equations (11) place demands on the flow field $\boldsymbol{J}$ only if the control variables can be specified, and if the time-derivative terms can be estimated on timescales consistent with equation accuracy and stability requirements. Neither of these requirements is trivial.

4.2. *Transient solution*

In the present case, we chose as objective function the sum of the time history coefficients

$$H = \sum_{i_0 \in B} (\gamma_{i_0,1} + \gamma_{i_0,2} + \gamma_{i_0,3}), \tag{17}$$

and solved the system (14)–(17) by linear programming, which remains convenient because of the non-negativity requirement for tracer distributions.

^3He was included in the system by forming 'stable tritium' ST = $^3H + {}^3He$, a conservative quantity (with ^3He expressed in ^3H units). Appropriate $\boldsymbol{G}_{i_0}$ were computed for ST (its decay constant λ vanishes in (11)), where the surface boundary conditions for ^3He are vanishing concentration so that the ST surface boundary conditions are the same as for tritium. A requirement ST $\geqslant {}^3H$ is also demanded in each boundary box, and the equivalent of (14) is imposed with the stable tritium concentration appearing on the left. The entire concatenated tritium plus stable tritium system was solved simultaneously.

The unknowns of this system are now the coefficients of the three expansion functions in each of the boundary boxes, one set each for ^3H and ST. The nominal knowns are the observed [^3H], [ST] in 1979, including the boundary values themselves. We anticipate, should an infeasibility appear, that we would need to choose between permitting more structure in the boundary time histories, and modifying the transport field $\boldsymbol{J}$. We obviously have much scope for accommodating the available data.

For the steady-flow field depicted in figure 9 (a minimum mass flux solution), a slight infeasibility *did* occur: it took the form of an inability to raise the ^3H and ST concentrations above zero in box 29. Because the concentration in that box is indeed very low, the infeasibility was most simply accommodated by increasing the error bar very slightly to permit zero there. The system was then entirely feasible and for this fixed-flow field and terminal constraint, it became possible to explore the different boundary time histories that are consistent. Two of them are shown in figure 14, for which the cost function (17) was a minimum, and the other a maximum. The isotopic age τ may be expressed as

$$\tau = \lambda^{-1} \ln ([\mathrm{ST}]/[^3\mathrm{H}]) \tag{18}$$

and is displayed in figure 14 for these two time histories.

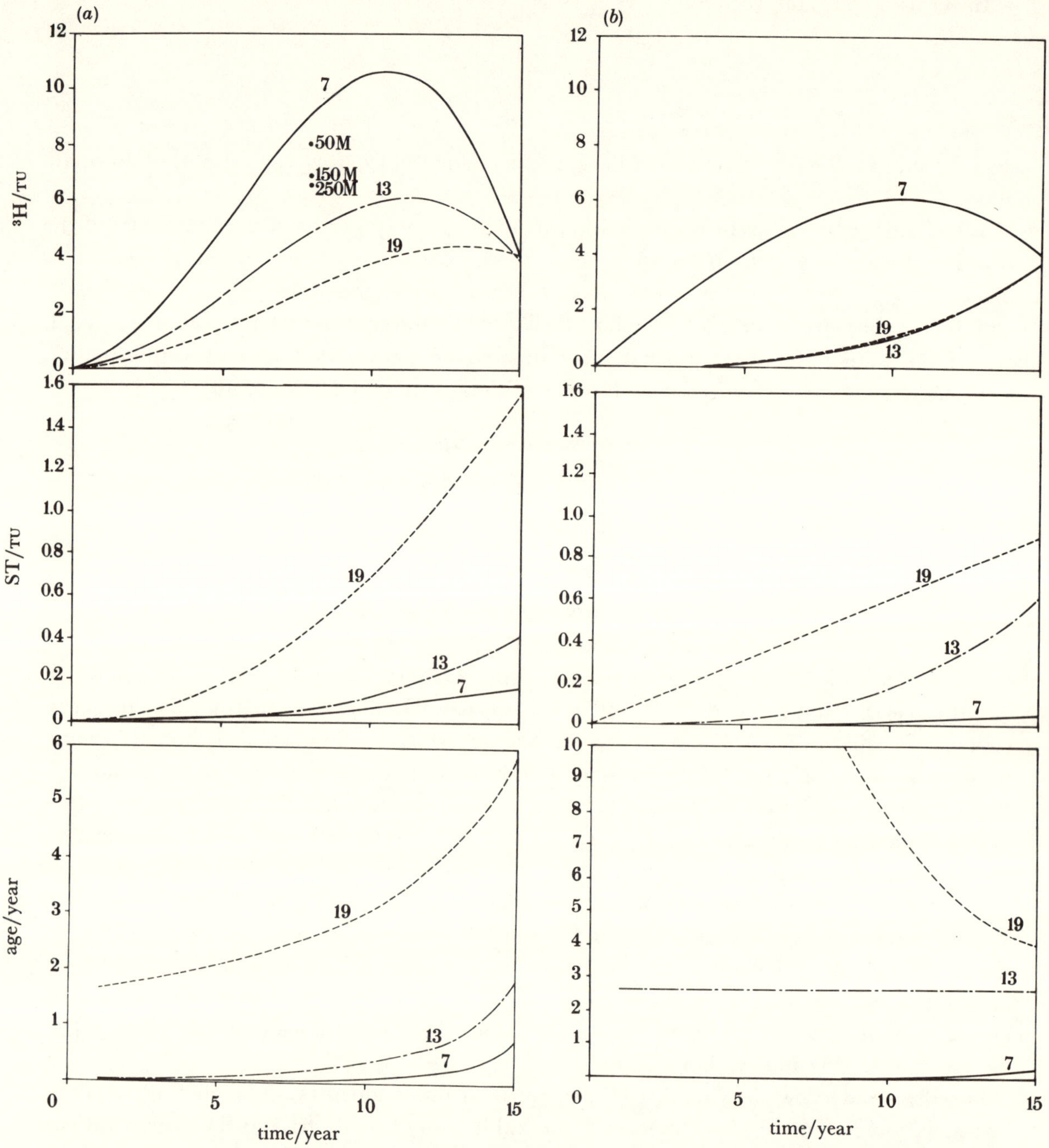

FIGURE 14. (*a*) Time histories in three of the northern-boundary boxes that lead to the final, observed ^{3}H and ST distributions. Box number is shown in the figure. Values determined by Östlund (1984) for 1972 are shown for ^{3}H in upper ocean (top three boxes). (*b*) Another time history in the northern-boundary boxes also leading to the final tracer distributions. The flow field was the same as in (*a*), yet despite the greatly restricted forms of these time histories, they remain underdetermined and put little restriction on the interior flow fields.

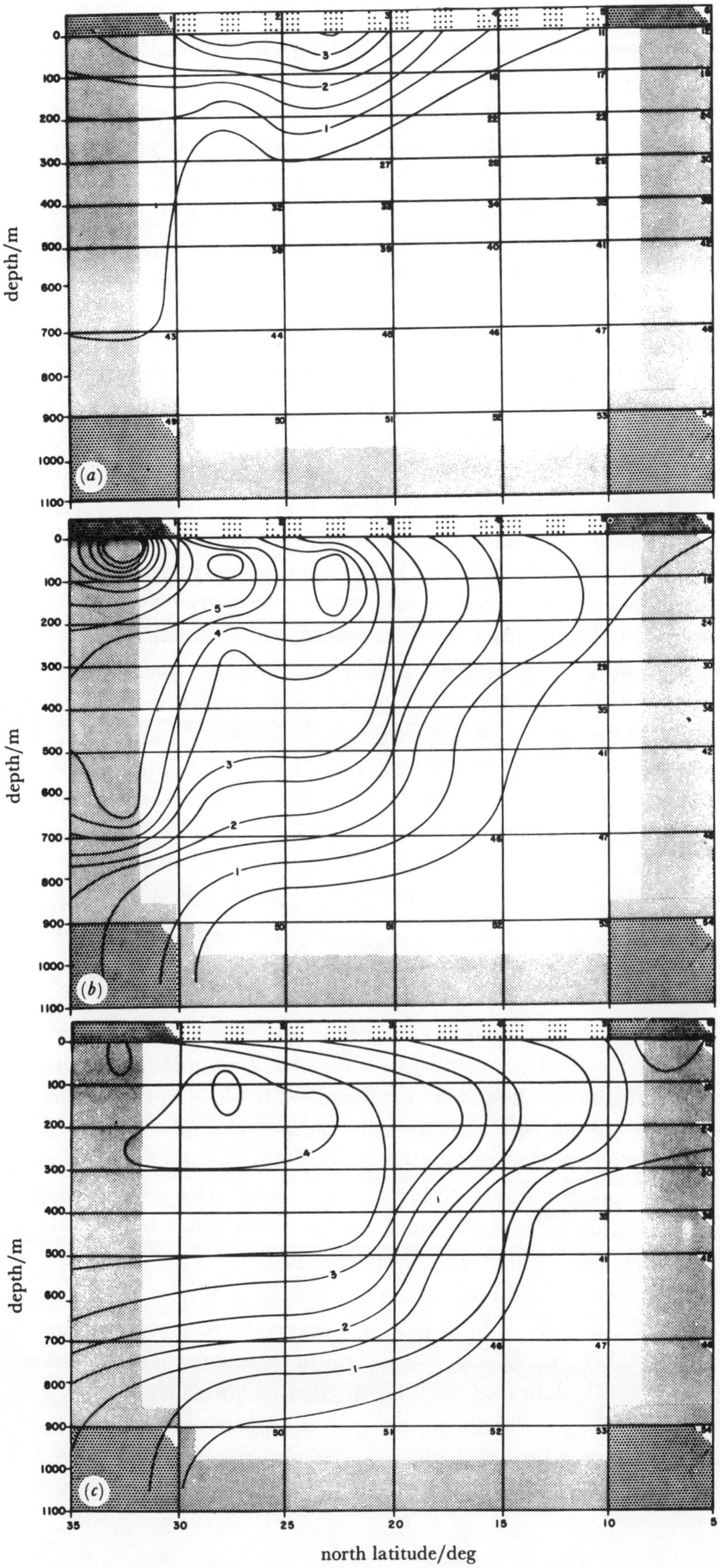

FIGURE 15. Interior distribution of ^{3}H corresponding to figure 14*a* and the flow field of figure 16, for three years (*a*) year 3, (*b*) year 12, (*c*) year 15, including the final year (1979). The initial interior distribution at $t = 0$ was taken to be zero for purposes of this calculation. If non-zero, it would have flushed through before the end. Contours in tritium units.

The actual interior ^{3}H distribution is displayed in figure 15 after 3, 12 and 15 years, the latter corresponding, within error bars, with the terminal constraints.

The role of the pair of transient tracers ^{3}H and ^{3}He is thus seen to be, in this particular problem, a rather weak consistency check on the flow determination previously made from the steady system. Not only does the available information only weakly constrain the $\boldsymbol{J}$, but two rather different time histories (there are infinitely many possible) in the boundary boxes are consistent with the terminal-tracer constraints.

About the only information we gained from the calculation apart from the demonstration of consistency, was that the ^{3}H distribution had to have peaked before 1979 at the northern boundary, because when a completely monotonic rise to the observed values was forced, the system was infeasible. The histories calculated suggest that the maximum should have occurred between about 1969 and 1974, depending upon depth. But this conclusion should not be taken as a very strong one: the width and location of the peak is largely determined by restrictions on the most negative value of the cubic term in (15). More terms in (15) and fewer restrictions on the coefficients would almost surely permit later, sharper maxima.

A certain amount of transient-tracer data in the β-triangle region exists for the time period before 1979. One can ask whether this extra information would make a qualitative change in the degree of constraint placed on the solutions. Figure 14*a* displays the values in the boundary boxes in 1972 (year 8) as inferred from the data reported by Östlund (1984). The 50 m value in 1972 is lower than that calculated from the reference solution, and the values below 50 m are a good deal higher. The question of whether forcing the boundary boxes to these values in 1972 would place strong demands on the system is easily answered (strictly speaking, the 1972 values could have been ascribed to the interior column underneath box 8 (latitude 27.5° N), which is closer to the nominal latitude (30° N) of the Östlund data. But forcing the boundary conditions, with only three parameters governing each box, is much more demanding than forcing the interior boxes, which respond to many parameters distributed over the entire system). The set of constraints that $\boldsymbol{C}(8)$ should equal the Östlund values (only three of which are shown in figure 14*a*) was added to the system (14) and then re-solved. These new observations were easily accommodated, without any difficulty, and thus the interior flow field remains essentially unconstrained.

The Östlund (1984) data also include values that could be assigned either to the southernmost column, or to the southern-boundary conditions. Assignment to the southern boundary adds no information because for the particular flow field being considered, there is no northward transfer into the interior. If the values are assigned to the column at 12.5° N, the necessary extra ^{3}H is easily supplied from the east (the values of the Östlund observations are somewhat higher than shown in figure 15) and again no demand is made on the flow field.

The calculated time histories of the isotopic age are intriguing. If the histories displayed (figure 14) are examined, they suggest a complex variation in the exposures of the waters entering the boxes over time. Should one be able to place restrictions on the age, it holds out hope of writing stronger demands on the $\boldsymbol{J}$ field. However, the water feeding into the box over a decadal period presumably emanates from a variety of 'sources' originating over much of the ocean. Second, the age calculation is very sensitive to slight variations in [^{3}He] at small concentrations and the significance of the apparent source variations seen in figure 14 is unclear, as the water was surely exposed to the surface at different times in the tritium transient and at different locations. Furthermore, as Jenkins (1987) shows, the interpretation of the age is far from straightforward as soon as mixing processes are admitted into the equations.

In what follows then, we return to the steady-box calculations for estimates of ventilation rates, regarding the transient calculation as a check to be made at the very end. It will be seen that the consistency check is not wholly trivial. Some of the more extreme cases generated below will be rejected as inconsistent with the transient tracers. But the point remains that the necessity of regarding the boundary conditions as a part of the problem unknowns weakens the constraining power of the transients.

5. Rates of ventilation

We have show that, in general, we can expect the ^{3}H and ^{3}He distributions to be consistent with any reasonable flow distribution deduced from the steady-tracer distributions (although actual consistency should be checked in each case). We therefore return to the steady box model of §3 and explore the so-called ventilation of this part of the ocean.

'Ventilation' is a term that seems to be more or less synonymous with 'renewal' as used for example by Dietrich *et al.* (1980). More recently, the word has come into prominence through the title of the paper by Luyten *et al.* (1983) in which the context is the rate at which water pumped down from the surface by Ekman-layer convergence reaches the thermocline. This latter process is a special one that is probably best labelled 'Ekman ventilation', to distinguish it from, for example, water being forced downward through buoyancy (convective) effects, perhaps to be called 'convective ventilation'. Almost certainly there are other processes of importance (e.g. small-scale mixing) which also 'ventilate' the interior ocean.

In the present case, the initial motivation had been to focus the question specifically on the rate at which oxygen was being consumed and replaced. But as we have already seen, once the model is carried into three dimensions, OUR can vary between approximately zero, and biologically unreasonable values. In effect, we learn that improving estimates of OUR will depend upon physical constraints that have yet to be measured, or preferably, from a physical oceanographer's point of view, from fundamental biochemical measurements. If the latter were a reasonable prospect, then oxygen distributions could be used to help determine the circulation, rather than vice versa.

For the present, we will define ventilation as the rates at which water moves across the boundaries from outside the volume, with a focus on the northern, and near-surface boundaries.

5.1. *A reference solution*

As a reference state, we minimize the overall fluxes within the region, i.e. we minimize

$$H = \sum_{\text{all } i,j} J_{i,j}$$

in the meridional plane. This solution can be thought of as being the one of maximum simplicity, although this identification is only an analogy and is not rigorous, or as a search for minimum mixing as defined above. A solution that minimized this H (it is not unique) is shown in figure 16*a* in the y–z plane. There are some instructive elements of the solution. Overall, the flow is being forced directly by the geostrophic fluxes from the north; no 'backflow' from the

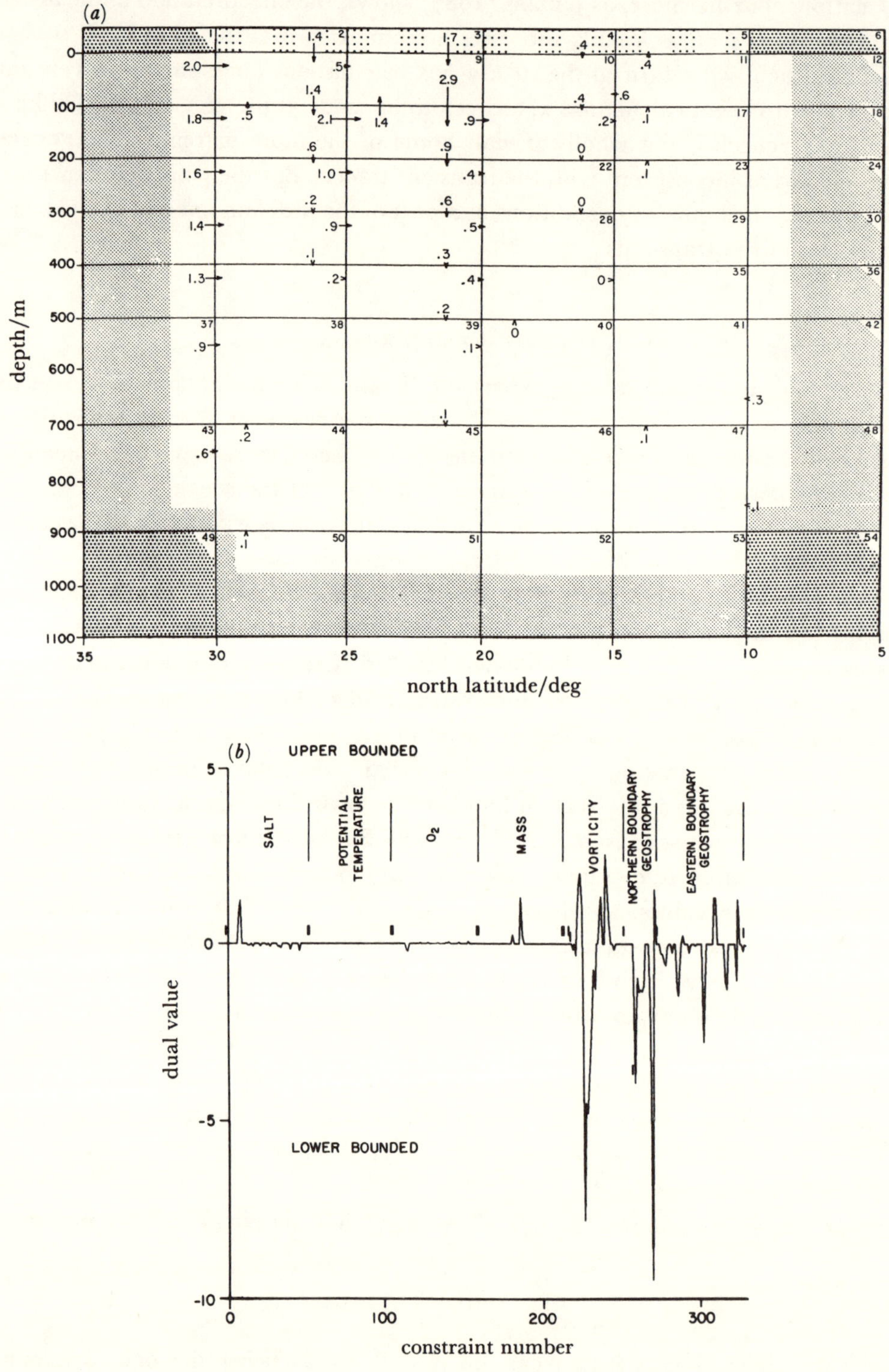

FIGURE 16. (*a*) A reference solution, with minimum $\sum_{i,j} J_{ij}$, for which the tracer distributions in figures 14 and 15 were computed. (*b*) Value of the dual to the solution of (*a*). Positive values denote constraints at their upper limits, negative at the lower limits.

interior boxes to the northern boundary is required. Injection of fluid from the Ekman boxes is also unidirectional and downward, apart from the flow into box 5 demanded by the Ekman suction there. There is little or no flow across the property front, with much of the flow having been ejected to the west before reaching the strong-gradient region. A quite strong flow along the front (6 Sverdrups) is demanded in box 11 with a strong zonal divergence. The high OUR in the two-dimensional flow is thus reduced through the physical divergence now permitted.

The dual solution found by the linear-programming procedure (Simplex method) provides at least a partial answer to the question of which of the constraints were important in determining the value of the objective function. The dual for this particular solution is shown in figure 16*b*. Each equation in the system has a non-zero dual if the constraint is 'binding', i.e. if the particular balance has been forced to either its upper or lower limit. In figure 16*b*, positive values correspond to box balances at their upper limits, negative values to those at their lower limits. The value of the dual is the sensitivity of the objective function to perturbation in the bound, and can be interpreted as a measure of the relative importance of the particular balance. Non-binding, 'basic', constraints have zero dual value. The dual can be interpreted as the Lagrange multipliers of the system.

For the solution in figure 16*a*, the most important factors in setting the value of the minimum of H are, first, mass balance, led by that for box 21, followed by the geostrophic velocities at the northern boundary and the vorticity balances. Increases or decreases in these binding equalities would make a much larger change in the particular solution than would for example, a modification of the salt balances.

Although each of the solutions we will display has a different dual, the general characterization given by figure 16*b* seems universal; the quantities that really determine the fluxes in the system are geostrophy, mass balance, and the linear-vorticity balances.

5.3. *Ventilation extremes*

We can now explore the limits, within this model, of movement of water in and out. The first solution to be compared to figure 16 is the one shown in figure 17 in which we sought to maximize and in figure 18, to minimize the downward mass flux from the surface layer into the layer below (that is from boxes 8–11 into boxes 14–17, remembering that the top tier of boxes is an artefact). One might crudely view these limits as those reflecting buoyancy injection of fluid into the interior. Whether the Ekman flow, imposed at the tops of these boxes penetrates through them vertically, or becomes deflected to the adjacent box on the north or south, is determined by the model itself. This approach was taken because the correct driven depth of Ekman flux is unknown. The flow in figure 17 was just feasible when tested against the $[^3H]/[ST]$ terminal distribution. A more extreme solution, when the vertical J_{ij} were permitted to grow to 10 Sverdrups (*ca.* 10^{-3} cm s^{-1}) was seriously inconsistent with the ^{3}He data. In that solution (which is not displayed), so much low ^{3}He water was carried into the interior that the system was unable to supply enough high ^{3}He water from elsewhere without violating the ^{3}H distributions. The transient tracers thus permit us to rule out solutions with extremely large vertical transfers (it appeared unlikely, from the form of the infeasibility, that the introduction of additional degrees of freedom in the boundary conditions would have made the solution acceptable).

From (9), we can relate this result to conventional vertical eddy coefficients. We take the flows at the base of box 9 in figure 17 as an example. Substituting into (9), we obtain

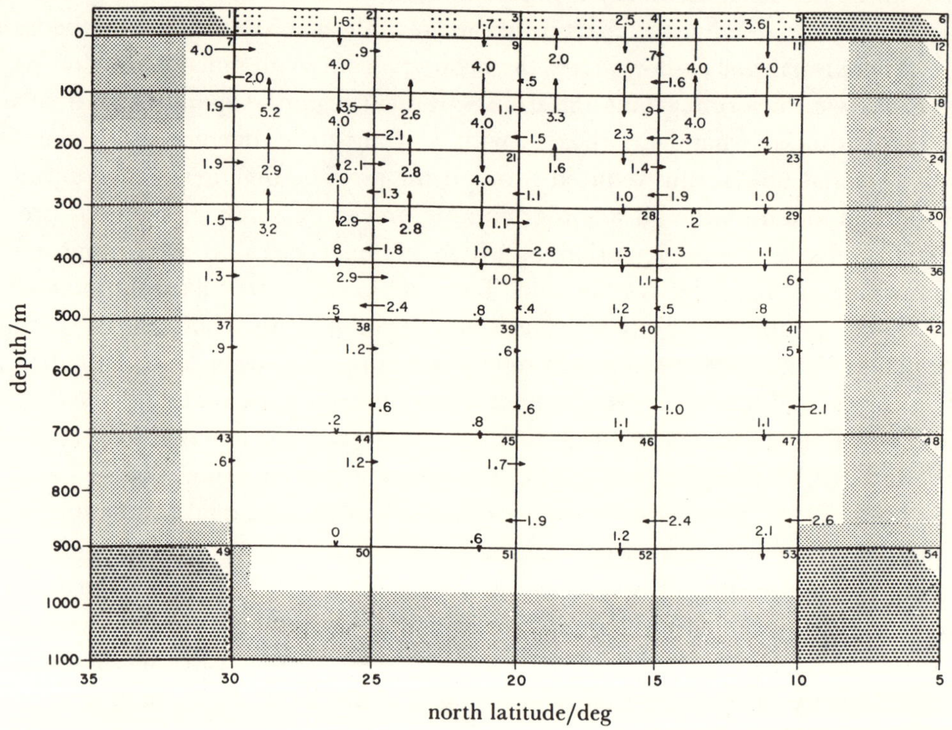

FIGURE 17. Solution in which the injection from the top layer was maximized (i.e. from boxes 8–11 into the layer beneath).

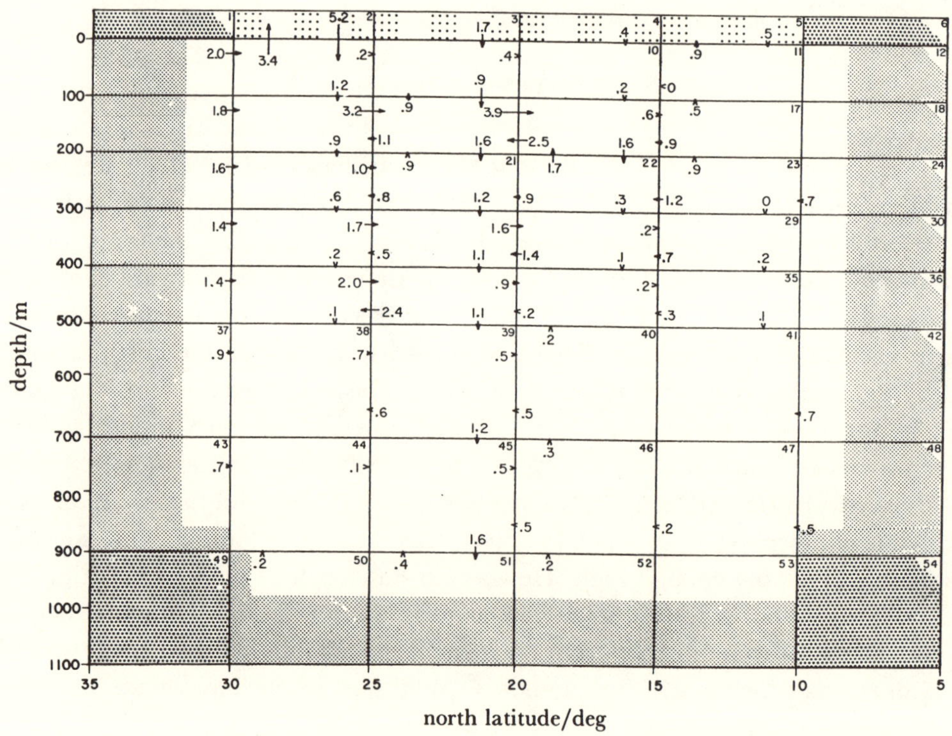

FIGURE 18. Same as figure 17, but a minimum flux.

$K \approx 3\ \text{cm}^2\ \text{s}^{-1}$. Because figure 17 is close to an upper bound for vertical transfer rates, the $^3H/^3He$ data suggestion that the upper-ocean apparent vertical eddy coefficient on this spatial scale must be less than about $3\ \text{cm}^2\ \text{s}^{-1}$.

Figures 19 and 20 display the maximization and minimization of ventilation from the north.

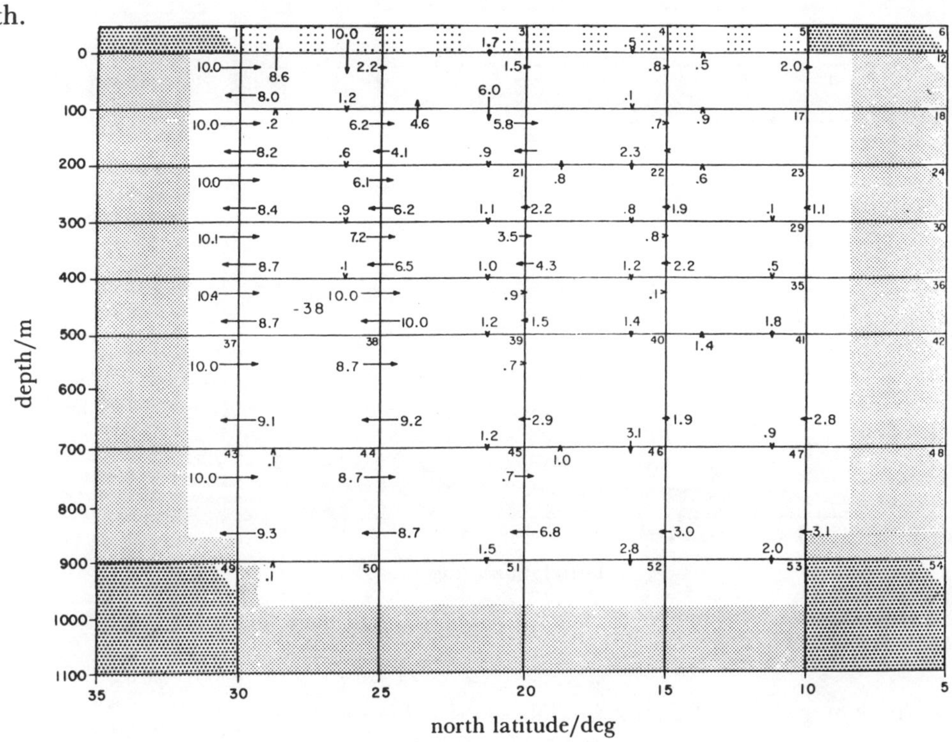

FIGURE 19. Solution showing maximum input from the northern boundary.

The difference between the minimizing and maximizing solutions is largely the difference between the flows that are unidirectional from the surface and northern boundary boxes and those in which a substantial amount of 'backflow' occurs. For example, the difference between figures 17 and 18 is essentially that in the former, the 'Ekman layer' receives substantial amounts of fluid from the layer below, whereas in the latter, the amount of export of fluid into the surface layer is much less. In both cases, the net transfer is set by the Ekman-flux values.

Comparing figures 16*a* and 20, we see that the solution of minimum overall circulation has a northern injection which is the same as the minimum northern injection, but the interior solutions are (slightly) different because of the different objective functions.

5.4. *Sarmiento's estimates*

Sarmiento (1983), using a box model of the tritium distribution in the sub-tropical North Atlantic, concluded that exchange between the surface waters and the upper thermocline had to be much greater (by a factor of 5) than that provided by Ekman pumping alone. He suggested that buoyancy forcing provided much of the excess injection. Because his model includes, for example, the region of 18 °C water formation and the Gulf Stream, the inference is reasonable.

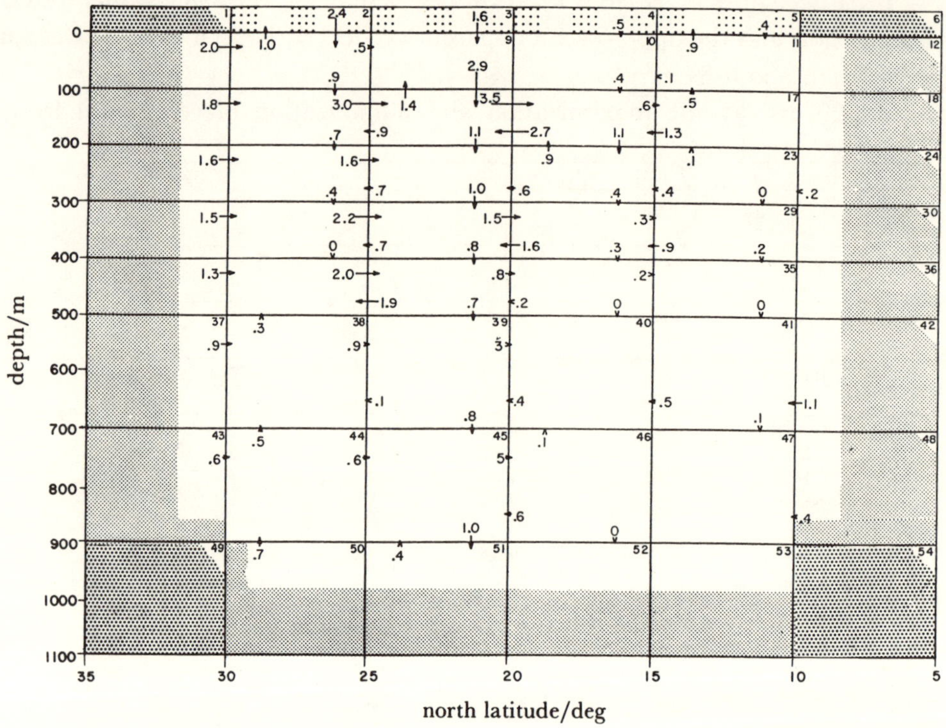

FIGURE 20. Solution of minimum input from the northern boundary.

But in the much more localized box model used in this present paper, there is no need to provide extra injection over and above that given by the estimates of Ekman pumping. As is clear from figure 14, most of the ^3H–^3He enters from the north (and east) rather than from above. This is not to say that excess invasion from above (as in the solution in figure 17) is not possible, merely that the ^3H–^3He distribution does not require it. Given the geometry of Sarmiento's (1983) model (his figure 2), it is possible that, even in the much larger domain over which he calculates the inventories, the missing time history at the northern isopycnal base of his outcropping layers could greatly influence the estimates of the injection necessary from above.

6. Discussion

The specific result of this paper is an estimate of the possible range of renewal rates of the water in the Eastern Atlantic thermocline. Although these may prove useful, the most interesting part of the work has been the formulation of a system in which transient tracers could be used to constrain and test the fluid flow. A formal identification can be made with the problem of distributed-system boundary control in which the 1979–1980 ^3H–^3He survey provides a terminal constraint on the system evolution. The open-ocean boundary conditions, and the uncertainties in the surface-boundary conditions correspond with the need to find a control function to drive the system to the observed terminal state within observational errors and subject to various auxiliary constraints (e.g. positivity). In the tracer problem we are not seeking to actually control the system, but asking for realizable boundary conditions that could have driven it to the observed terminal state.

The analogue with control problems makes it possible to adapt many of the methods and deductions from that subject. In the present particular case, we drew the conclusion that the transient tracers would serve mainly as a not very stringent consistency check on flows more tightly constrained by dynamical measurements (geostrophy) and steady tracer balances. They do serve to cap the maximum near-surface vertical exchange.

For the present region, with the particular model, and the data as they exist, it is difficult to escape the conclusion that the transient-tracer observations are less informative than the dynamical and steady balances. It is premature, however, on the basis of this one example, to conclude that this relative importance is universal. Almost surely there are situations in which the consistency check will fail badly, thus constraining the circulation directly. We can, even in the present context, examine ways in which the existing drastic underdeterminancy owing to the control variables could be reduced.

It is clear that as the number of boxes used increases, that the ratio of boundary/control unknowns to interior flow parameters decreases rapidly. The distance (in non-dimensional terms) of a given interior box from the boundary boxes would also increase. It may well be that a point would be reached in which the interior time histories would remain unaffected by the boundary histories for sufficiently long times that severe constraints would be placed on the interior fields. But one cannot arbitrarily decrease the box size to bring about this reduction in underdeterminancy. In the present case, the $[^3H]/[^3He]$ eddy noise visible in Jenkins's (1987) maps means that the average concentration variances would increase as the box size decreased, and the inequalities would widen, with the result that the underdeterminancy may well remain unchanged or even worsen.

It should be pointed out, moreover, that there has been an implicit, enormous reduction in system degrees of freedom through the expedient assumption of a steady flow field. Relaxation of this assumption opens up vast new vistas of difficulty, the conquest of which is better postponed.

In the formal inverse problem, observations of the transient tracer C are used to make demands upon the flow field $\boldsymbol{J}$. That practical transient-tracer data are less demanding of $\boldsymbol{J}$ than are steady tracers is manifested in two ways: the difficulty of defining the control variables, and the necessity for having an estimate of the $\partial C/\partial t$ term, neither of which is an issue for the steady tracers. Constraints on $\boldsymbol{J}$ from the steady nutrients or oxygen are of course, gravely weakened by uncertainty over the biological source/sink terms that appear in the equations. The inert transient tracers are not subject to this latter uncertainty. Overall, the impression that one gains from the present exercise is that physical constraints involving neither biology nor chemistry (geostrophy, vorticity, mass and salt conservation) are much more powerful general tools for estimating $\boldsymbol{J}$ than are any of the biologically active, or transient, tracers.

The determination of a strategy for transient-tracer observations thus remains an important challenge. Perhaps a few tentative, nearly speculative, conclusions are permissible. The greatest difficulties emerged when the boundary-value histories had to be computed as part of the set of unknowns. This problem would be much ameliorated by mapping distributions over regions bounded by physical régimes (e.g. the continents) where the time histories can either be measured easily, or inferred with little uncertainty. It may be sensible to focus the observational efforts on developing time histories critical to large interior volumes. For example, the ^3H–^3He the history of the narrow Iceland–Scotland overflows would constrain any model of the flow in the interior to the south. Such histories would also be demanding of any model that claimed to compute the wintertime injection of fluid from the surface poleward of this region.

Finally, the unique capability of transient tracers is their ability to provide estimates of the very-large-scale, global, transfers from atmosphere to ocean, as a function of time. For climate purposes, and the CO_2 problem, a strategy of sampling directed at accurate inventory estimates, as opposed to interior flow estimates, seems plausible. To the extent that global inventories can be estimated over decadal timescales, their changes would appear to powerfully constrain air–sea transfers of properties.

Charmaine King provided all the computational help. The paper would not have been possible without extensive discussions with Dr William Jenkins. The work was supported in part by grant OCE8521685 of the U.S. National Science Foundation, and N00014-85-G-0241 of the U.S. Office of Naval Research.

Appendix

Table 1 listed the constraints and the formal error used in most of the steady inverse calculations described in this paper. Minor deviations permitted in some cases are described in the text. Unless otherwise specified, the row constraints apply to the interior boxes of figures 2–6.

Table A1 describes the constraints used in the transient tracer control problems. Again, minor deviations are described in the text.

Table A1

row constraints

tritium
- all interior boxes within 0.5 TU[1] of observed means in 1979
- surface box concentration within 0.5 TU of observed history (figure 8)
- open boundary boxes with 0.5 TU of 1979 observations at end of integration

stable tritium
- all interior boxes within 0.5 TU of observed means in 1979
- surface box concentration with 0.5 of observed ^3H concentration
- open boundary boxes with 0.5 TU of 1979 values
- $[ST] \geqslant [^3H]$ on all boundaries
- $[ST] = [^3H]$ on surface boundary

column constraints

$0 \leqslant \gamma_{1,2} \leqslant 10$ $\quad\quad$ $-0.1 \leqslant \gamma_3 \leqslant 10$

(1) TU represents tritium units.

References

Armi, L. & Stommel, H. 1983 Four views of a portion of the North Atlantic subtropical gyre. *J. phys. Oceanogr.* **13**, 828–857.

Brewer, P. G., Sarmiento, J. L. & Smethie, W. M. Jr 1985 The transient tracers in the ocean (TTO) program: The North Atlantic study. *J. geophys. Res.* **90**, 6903–6905.

Broecker, W. S. & Östlund, H. G. 1979 Property distributions along the sigma-theta = 26.8 isopycnal in the Atlantic Ocean. *J. geophys. Res.* **84**, 1145–1154.

Broecker, W. S. & Peng, T. H. 1982 *Tracers in the sea.* (690 pages.) New York: Eldigio Press.

Cox, M. D. & Bryan, K. 1984 A numerical model of the ventilated thermocline. *J. phys. Oceanogr.* **14**, 674–687.

Dietrich, G., Kalle, K., Krauss, W. & Siedler, G. 1980 *General oceanography, an introduction*, 2nd edn. New York: Wiley.

Dreisigacker, E. & Roether, W. 1978 Tritium and 90-Sr in North Atlantic surface water. *Earth planet. Sci. Lett.* **38**, 301–312.

Jenkins, W. J. 1987 3-H and 3-He in the beta triangle: Observations of gyre ventilation and oxygen utilization rates. *J. phys. Oceanogr.* **17**, 763–783.

Jenkins, W. J., Lott, D. E., Davis, M. W. & Boudreau, R. D. 1985 *W.H.O.I. helium isotope data report no. 2, tritium and 3-He data from the beta triangle* (AII-107, October 1979 and OC-78, March 1980). Technical Report 85-40 Woods Hole Oceanographic Institution. (32 pages.)

Kawase, M. & Sarmiento, J. L. 1985 Nutrients in the Atlantic thermocline. *J. geophys. Res.* **90**, 8961–8979.
Keeling, C. D. & Bolin, B. 1967 The simultaneous use of chemical tracers in oceanic studies. I. General theory of reservoir models. *Tellus* **19**, 566–581.
Leetmaa, A. & Bunker, A. F. 1978 Updated charts of the mean annual wind stress, convergences in the Ekman layers and Sverdrup transports in the North Atlantic. *J. mar. Res.* **36**, 311–322.
Luenberger, D. G. 1979 *Introduction to dynamic systems. Theory, models and applications.* (466 pages.) New York: John Wiley.
Luyten, J. R., Pedlosky, J. & Stommel, H. 1983 The ventilated thermocline. *J. phys. Oceanogr.* **13**, 292–309.
Olbers, D. J., Wenzel, M. & Willebrand, J. 1985 The inference of North Atlantic circulation patterns from climatological hydrographic data. *Rev. Geophys.* **23**, 313–356.
Östlund, H. G. 1984 *NAGS tritium. North Atlantic gyre studies and associated projects.* Tritium Laboratory Data Report no. 13. (324 pages.) University of Miami, Rosenstiel School of Marine and Atmospheric Science, Miami.
Roache, P. J. 1976 *Computational fluid dynamics.* (446 pages.) Albuquerque: Hermosa.
Roemmich, D. & Wunsch, C. 1985 Two transatlantic sections: Meridional circulation and heat flux in the subtropical North Atlantic Ocean. *Deep Sea Res.* **32**, 619–664.
Salmon, R. 1986 A simplified linear ocean circulation theory. *J. mar. Res.* **44**, 695–711.
Sarmiento, J. L. 1983 A tritium box model of the North Atlantic thermocline. *J. phys. Oceanogr.* **13**, 1269–1274.
Sarmiento, J. L., Thiele, G., Toggweiler, J. R., Key, R. M. & Moore, W. S. 1988 Thermocline ventilation and oxygen utilization rates obtained from multiple tracers. (In preparation.)
Stengel, R. F. 1986 *Stochastic Optimal Control.* (638 pages.) New York: Wiley-Interscience.
Tikhonov, A. N. & Arsenin, V. Y. 1977 *Solutions of ill-posed problems.* (258 pages.) Washington, D.C.: V. H. Winston.
Weiss, W., Roether, W. & Dreisigacker, E. 1979 Tritium in the North Atlantic Ocean. In *Behavior of Tritium in the Environment*, pp. 315–336. Vienna: IAEA.
Wen, C. Y. & Fan, L. T. 1975 *Models for flow systems and chemical reactors.* (570 pages.) New York: Marcel Dekker.
Wunsch, C. 1984*a* An eclectic Atlantic Ocean circulation model. Part 1: The meridional flux of heat. *J. phys. Oceanogr.* **14**, 1712–1733.
Wunsch, C. 1984*b* An estimate of the upwelling rate in the equatorial Atlantic based on the distribution of bomb radiocarbon and quasi-geostrophic dynamics. *J. geophys. Res.* **89**, 7971–7978.
Wunsch, C. 1988*a* Using transient tracers: the regularization problem. *Tellus.* (In the press.)
Wunsch, C. 1988*b* Transient tracers as a problem in control theory. (Submitted.)
Wunsch, C. & Grant, B. 1982 Towards the general circulation of the North Atlantic Ocean. *Prog. Oceanogr.* **11**, 1–59.
Wunsch, C. & Roemmich, D. 1985 Is the North Atlantic in Sverdrup balance? *J. phys. Oceanogr.* **15**, 1876–1880.

Discussion

M. Whitfield (*Marine Biological Association, Plymouth, U.K.*). Would not Professor Wunsch's model be more closely constrained if OUR values had been obtained? Does this not suggest that a closer coupling between tracer movements and process-orientated studies would be valuable?

C. Wunsch. I would say 'yes' to both questions. If OUR could be reliably specified it would be very useful in constraining the physical flow field. However, my impression is that most OUR estimates have been obtained from physical flow models not unlike my own, and are therefore not independent pieces of information. I am unable to say whether there is any real prospect of purely biochemical estimates in which one would have quantitatively useful confidence.

D. J. Webb (*Institute of Oceanographic Sciences, Wormley, Surrey, U.K.*). Is it not slightly unfair to concentrate on a section across the main current flow, where the information on the velocities and diffusion rates is negligible compared with the geostrophic velocity-information gained from the density field? With a section taken along the core of a major current, where errors due to tracer advection out of the plane of the section are less, the tracer data would be much more useful in determining diffusion rates and velocities within the plane of the section.

C. Wunsch. Exactly to whom or what I am being unfair is not clear to me. I took the data that existed and tried to make inferences from them. The challenge to modellers is to use real data distributions, not those that exist only in some theoretical world.

It is certainly true that if one knew where the axis of a flow was, and if one could then apply the tracer distributions wholly to determining (say) the mixing, that the inferences about mixing rates would be stronger. But the assumptions that the axis of flow is known *a priori* to the observer who made the measurements, and that two dimensional tracer distributions necessarily imply steady two-dimensional flow fields, are very stringent ones. In practice we prefer to deduce the flow axis from the tracers, rather than to assume it.

W. J. Jenkins (*Woods Hole Oceanographic Institution, Massachusetts, U.S.A.*). I understand the main message of Professor Wunsch's presentation here is the limitations imposed by the lack of tracer data, and the strength of better-known physical constraints such as quasi-geostrophic balance and conservation of potential vorticity. As he correctly pointed out, we are hampered by incomplete knowledge of time variations in the transient-tracer distributions. All these point to a pressing need to increase the database in this regard.

My major reservation regarding Professor Wunsch's calculation stems from the physical construction of his model. To generalize, I feel that the model as posed is not sufficiently 'expert' to answer the questions he asks of it. His calculations represent an important first step toward quantitative evaluation of the utility (or lack thereof) of tracers in studying ocean circulation and mixing, but I would think it fair to stress that they are clearly preliminary and should not be regarded as a final judgement.

I think that the primary structure of his model should be oriented along isopycnals. Most of us believe that flow, to a first approximation, occurs along isopycnals. Diapycnic processes can be introduced as a perturbation. Second, I would be inclined to ignore the section data and concentrate on the β-triangle itself, where the data are more fully three dimensional. The section, aside from being two dimensional, crosses a property front in the ocean which is known from the days of Wüst and Defant to be controlled by the flow of waters from regions far outside the purview or this limited data set. I would expect any tracer analysis in such a situation to fail.

The choice of a box-model formulation, although mathematically 'reasonable', is not optimal for the task at hand. Professor Wunsch's procedure or averaging over box domains and ascribing an uncertainty associated with the internal variance is destroying important information. The primary information in the ^{3}He distribution lies in its gradients. There are gradients within the box domains that are not only lost by the averaging process, but contribute to his estimated 'noise', thus weakening confidence in the larger scale gradient estimates. I would go further to state that the ^{3}H/^{3}He age, although complicated by mixing does offer a unique advantage of having a built-in normalization regarding the time history. Put another way, as evidenced by Professor Roether's presentation, the ^{3}H–^{3}He age is relatively steady on the shallow isopycnals and hence may be easier to use in the inverse calculations.

The presence of eddy noise in the data makes it difficult to estimate gradients on small scales in a reliable way. None the less, the mean gradients across the triangle are determinable to order 10% in ^{3}He. Perhaps a more sensible approach would be to reduce the degrees of freedom with a least-squares or objective mapping procedure.

Finally, my original purpose in sampling the β-triangle area was to use it as a test bed for the ^3H–^3He dating technique because the area was so well characterized hydrographically. As I have tried to demonstrate in my recent study, the age data are approximately consistent with the climatologically averaged geostrophic velocities. The small inconsistencies are in themselves interesting, but I should also point out that if one were to reduce the problem to one of cost-benefit analysis, one must weigh on the one side the cost of acquiring and analysing the tracer data, and on the other side the cost of the many hydrographic cruises that were necessary to achieve the climatological mean estimates. More importantly, the tracers yield information not simply related to the velocity field *per se*, but also to the complex amalgam of processes involved in ventilation. The data set that Professor Wunsch has used does not contain sufficient information to solve all the problems, but this does not mean that one has placed firm limits on what tracers in general can do.

C. Wunsch. I am not aware of any problem in which the answer is determined by the coordinate system chosen. Proper choice of coordinate system can make the solution much easier to get, but the actual answer should be independent of it. In the present case, I wanted to use the vorticity constraints, which are much simpler to apply in geographical coordinates. If one wishes to add constraints that cross-isopycnal mixing is very weak, those statements are additional linear constraints upon the J_{ij}, involving local projections onto the isopycnals in each box. One can be confident that addition of such constraints would tighten the solution bounds. They would be easier to write down in isopycnal coordinates, but then the vorticity constraints would have to be written through projections onto the geographical coordinate system. A decision as to whether one set of constraints or the other is more fundamental is somewhat arbitrary. My own view is that a conclusion that cross-isopycnal mixing is small is something that should be deduced from the observations, not imposed, whereas the vorticity constraints are inferred systematically from Newton's laws of motion and are more difficult to escape.

Certainly one could reduce the box size somewhat. But the model is a climatological one referring to multi-year integrated behaviour. I chose the box size used on the basis of a subjective judgement as to the smallest region over which I was comfortable that I was not just looking at eddy noise. Given the importance of the heat, salt and vorticity constraints, and that we know so much more about spatial scales of eddy variability for these physical fields, I would be reluctant to reduce the box size by very much. A different set of boxes could be used for ^3He than for other tracers, within the same model, but in the last analysis, there will be no substitute for a good time series of ^3He, at mesoscale time–space resolution, to demonstrate that reduction in scale does not increase the noise faster than information is being gained.

I have tried to be very careful not to overemphasize the generality of these results. In that sense, I regard this model as a test bed too. In my mind, the major results are two. (1) In the spirit of what I have called 'eclectic modelling', one can use quantitatively all sorts of different data with irregular distributions in space and time. I recognize the realities of oceanic observations, and the whole eclectic idea was to learn how to employ observations as we actually get them, not as a mathematician classically assumes them to be. (2) That surveys of transient tracers provide a 'terminal constraint' seems to be a useful mathematical concept now permitting us to use the control mathematics in other situations to better understand the relative information content of different types of observation. The mathematical issue would be no different had the calculation been restricted to the triangle area itself, although the result might have been more constraining. Reduction in degrees of freedom would also obviously lead to tighter bounds from the tracers, but which degrees of freedom depicted in the figures in the paper should be rejected as physically unreasonable?

Nothing would have pleased me more than to find in the present situation that the existing ^{3}H–^{3}He data strongly constrained the flow, mixing, OUR, etc. There may be some much cleverer way to use these data than I have thought of (but just making many more restrictive assumptions about how the ocean behaves is not what I mean), and different regions may yield very different results. The challenge now is to demonstrate that.